SRA

Connecting Math Concepts

Level E Teacher's Guide

COMPREHENSIVE EDITION

A DIRECT INSTRUCTION PROGRAM

McGraw Hill Education

Bothell, WA • Chicago, IL • Columbus, OH • New York, NY

Acknowledgments

The authors are grateful to the following people for their input in the field-testing and preparation of *SRA Connecting Math Concepts: Comprehensive Edition Level E:*

Daniel Alig
Jon Bellino
Amilcar Cifuentes
Ashly Cupit
Andy Frankel
Crystal Hall
Evan Haney
Joanna Jachowicz
Margie Mayo
James Ouano
Oliva Velasquez
Vanessa Williams
Jason Yanok

MHEonline.com

Send all inquiries to:
McGraw-Hill Education
8787 Orion Place
Columbus, OH 43240

ISBN: 978-0-02-103620-2
MHID: 0-02-103620-9

Printed in the United States of America.

5 6 7 8 9 10 QVS 20 19 18 17 16

The McGraw·Hill Companies

Contents

Program Overview

The *Connecting Math Concepts: Comprehensive Edition* is a six-level series that will accelerate the math learning performance of students in grades K through 5. Levels A through F are suitable for regular-education students in Kindergarten through fifth grade. The series is also highly effective with at-risk students.

Connecting Math Concepts: Comprehensive Edition is based on the fact that understanding mathematics requires making connections

- among related topics in mathematics, and
- between procedures and knowledge.

Connecting Math Concepts does more than expose students to connections. It stresses understanding and introduces concepts carefully, then weaves them together throughout the program. Once something is introduced, it never goes away. It often becomes a component part of an operation that has several steps.

The organization of *Connecting Math Concepts: Comprehensive Edition* is powerful because lessons have been designed to

- Teach explicit strategies that all students can learn and apply.
- Introduce concepts at a reasonable rate, so all students make steady progress.
- Help students make connections between important concepts.
- Provide the practice needed to achieve mastery and understanding.
- Meet the math standards specified in the Common Core State Standards for Mathematics.

The program's Direct Instruction design permits significant acceleration of student performance. The instructional sequences are the same for all students, but the rate at which students proceed through each level should be adjusted according to student performance. Higher performers proceed through the levels faster. Lower performers receive more practice. Benchmark in-program Mastery Tests provide information about how well students are mastering what has been taught most recently. Students' daily performance and test performance disclose whether they need more practice or whether they are mastering the material on the current schedule of lesson introduction.

The program enables the teacher to teach students at a faster rate and with greater understanding than they probably ever achieved before. The scripted lessons have been shaped through extensive field-testing and classroom observation. The teacher individualizes instruction to accommodate different groups that make different mistakes and require different amounts of practice to learn the material.

Introduction to 2013 CMC Level E

- *CMC Level E* is designed for students who have successfully completed *CMC Level D* or who pass the Placement Test for *Level E* (see page 195).
- *CMC Level E* instruction meets all requirements of the Common Core State Standards for Mathematics for fourth grade.

Program Information

The following summary table lists facts about 2013 *CMC Level E.*

Students who are appropriately placed in *CMC Level E*	Pass Placement Test (p. 195)
How students are grouped	Instructional groups should be as homogeneous as possible.
Number of lessons	• 130 regular lessons • 13 Mastery Tests • 2 Cumulative Tests
Schedule	• 50 minutes for structured work • Additional 20 minutes for students' Independent Work • 5 periods per week
Teacher Material	• Teacher's Guide • Presentation Book 1: Lessons 1–70, Tests 1–7 • Cumulative Test 1 (Lessons 1–70) • Presentation Book 2: Lessons 71–130, Tests 8–13 • Cumulative Test 2 (Lessons 1–130) • Answer Key • Board Displays CD
Student Material	• Workbook: Lessons 1–130 • Textbook Lessons 3–130 • Student Assessment Book: Mastery Tests 1–13; Remedies worksheets for Mastery Tests 1–13, Cumulative Tests 1 and 2
In-Program Tests	• 13 ten-lesson Mastery Tests • Administration and Remedies are specified in the Teacher Presentation Books. • Tests and Remedies worksheets are in the Student Assessment Book.
Optional Cumulative Tests	• 2 Cumulative Tests • Administration is specified in Teacher Presentation Books 1 and 2. • Tests are in the Student Assessment Book.
Additional Teacher/ Student Material	• Student Practice Software (accessed online via ConnectED) • Math Fact Worksheets (Online Blackline Masters via ConnectED) • Access to *CMC* content online via ConnectED • *SRA 2Inform* available on ConnectED for online progress monitoring

TEACHER MATERIAL

The teacher material consists of:

The Teacher's Guide: This guide explains the program and how to teach it properly. The Scope and Sequence chart on pages 12–13 shows the various tracks (topics or strands) that are taught; indicates the starting lesson for each track/strand; and shows the lesson range. This guide calls attention to potential problems and provides information about how to present exercises and how to correct specific mistakes the students may make. The guide is designed to be used to help you teach more effectively.

Two Teacher Presentation Books: These books specify each exercise in the lessons and tests to be presented to the students. The exercises provide scripts that indicate what you are to say, what you are to do, the responses students are to make, and correction procedures for common errors. (See Teaching Effectively, **Using the Teacher Presentation Scripts,** for details about using the scripts.)

Answer Key: The answers to all of the problems, activities, and tests appear in the Answer Key to assist you in checking the students' classwork and Independent Work and for marking tests. The Answer Key also specifies the remedy exercises for each test and provides a group summary of test performance.

Board Displays CD: The teacher materials include the Board Displays CD, which shows all the displays you present during the lessons. This component is flexible and can be utilized in different ways to support the instruction—via a computer hooked up to a projector, to a television, or to any interactive white board. The electronic Board Displays are also available online on ConnectED. You can navigate through the displays with a touch of the finger if you have an interactive white board or with the click of a button from a mouse (wired or wireless) or remote control.

Practice Software: The *CMC Level E* Practice Software provides students additional practice with the skills and concepts taught in *CMC Level E.* It is a core component for meeting several Common Core State Standards for Mathematics. Students apply their skills to tasks presented onscreen. The tasks are governed by an algorithm that adjusts the amount of practice students receive according to how well they perform. Games and reward screens provide students with reinforcement for meeting performance goals. The software is organized into blocks, each presenting activities for a 30-lesson segment of the program as students proceed through the lessons.

The Math Facts strand of the software is organized into sets of facts that follow the instructional sequence in the lessons. It is designed to facilitate continuous review and reinforcement of the math facts as they are introduced and practiced. It is available via ConnectED with 10 student seat licenses per every Teacher Materials Kit purchase.

ConnectED: On McGraw-Hill/SEG's ConnectED platform you can plan and review *CMC* lessons and see correlations to Common Core State Standards for Mathematics. Access the following *CMC* materials from anywhere you have an Internet connection: eBooks of the Presentation Books and Teacher's Guides, PDFs of the Answer Keys, an online planner, online printable versions of the Board Displays CD, student Practice Software, student eTextbook, Math Fact Worksheets, and correlations. *CMC* on ConnectED also features a progress monitoring application called *SRA 2Inform* that stores student data and provides useful reports and graphs about student progress. Refer to the card you received with your Teacher Materials Kit for more information about redeeming your access code, good for one six-year teacher subscription and 10 student seat licenses, which provides access to the Practice Software and eTextbook.

STUDENT MATERIAL

The student materials include a Workbook for each student, a Textbook, and a Student Assessment Book. The Textbook and Workbooks contain writing activities, which the students do as part of the structured presentation of a lesson and as independent seatwork. The Student Assessment Book contains material for the Mastery Tests as well as test Remedies worksheet pages and optional Cumulative Test pages.

Textbook: Lessons 1–130

Workbook: Lessons 1–130

Student Assessment Book: Mastery Tests 1–13, Cumulative Tests 1 and 2, and test Remedies worksheets for Mastery Tests

WHAT'S NEW IN 2013 *CMC LEVEL E*

Most instructional strategies are the same as those of the earlier *CMC Level E;* however, the procedures for teaching these strategies have been greatly modified to address problems teachers had teaching the content of the previous editions to at-risk students. The 2013 edition of *CMC Level E* has also been revised on the basis of field-testing.

- The 2013 edition provides far more oral work than earlier editions. This work is presented as "hot series" of tasks. The series are designed so that students respond to ten or more related questions or directions per minute; therefore, these series present a great deal of information about an operation or discrimination in a short period of time.
- The content is revised so that students learn not only the basics but also the higher-order concepts. The result is that fourth-grade students who complete *CMC Level E* are able to work the full range of problems and applications that define understanding of fourth-grade math.
- The hallmark of Direct Instruction mathematics programs is that they teach all the component skills and operations required to provide a solid foundation in topics involving addition, subtraction, multiplication, and division; fractions; ratios; word problems for all of the above; inverse operations; complementary, supplementary, and vertical angles; functions and coordinate system. The *CMC Level E* program addresses all standards specified in the Common Core State Standards for Mathematics for fourth-grade math. (pages 14–15 and 169–194).
- *CMC Level E* has support/enhancements, including technology components, for teachers and students. These enhancements include displays in the Teacher Presentation Books, a Board Displays CD (also available online), Workbook and Answer Key pages reduced in the Teacher Presentation Books, a Student Assessment Book with all program assessments in one location, student Practice Software, and the ability to plan and review lessons online via ConnectED.

The Structure of Connecting Math Concepts Level E

Connecting Math Concepts Level E is appropriate for students who complete *Level D* or who pass the *Level E* Placement Test.

Level E has two starting points: Lesson 1 and Lesson 31. Lessons 1–30 are designed to acquaint new students with the conventions and information that continuing students learned in *Level D*. Continuing students start at Lesson 31. Lessons are designed to review facts and other information presented in *Level D* and to introduce new material at a rate that would be appropriate for both the continuing students and the students who started at Lesson 1.

If you have a group that has new and continuing students, the simplest solution is to start all students at Lesson 1 and proceed as quickly as the lower performers, who are appropriately placed in the program, are able to proceed.

Reproducible copies of the Placement Test appear on page 195 of this guide.

SCHEDULING

The program contains 130 lessons and 13 in-program Mastery Tests. The ideal goal is to teach one lesson each period. If students are not firm on content that is being introduced, you will need to repeat parts of lessons or entire lessons. Particularly early in the program, you may need to repeat entire lessons because students will perform much better on subsequent lessons if all lessons are taught to mastery.

Also, some lessons are longer and may require more than a period to complete. Following long lessons, try to get back on a schedule of teaching a lesson a day. This pattern assures that students receive daily practice in skills or operations that have been recently introduced.

The program is to be taught daily. Periods for structured work are 50 minutes. Students need an additional 20 minutes or more to complete their Independent Work. If the Independent Work cannot be completed in school, it may be assigned as homework, but this is not an attractive alternative, particularly on Fridays. It's important for students to bring back their work on the following school day. If you assign Textbook Independent Work as homework, the Textbooks should remain in the classroom.

A final note: *CMC Level E* is not designed as a supplemental program and should not be used as one.

How the Program Is Different

Connecting Math Concepts Level E differs from traditional approaches in the following ways.

Field Tested

CMC Level E has been shaped through field testing and revision based on difficulties students and teachers encountered. The field-test philosophy is that if teachers or students have trouble with material presented, the program is at fault. Revisions are made to alleviate observed problems.

The field-test results of *Connecting Math Concepts Level E* disclose that if the teacher implements the program according to the presentation detail provided for each exercise, students will learn the content and become proficient in the content so they can advance to the next levels of math instruction.

Organization and Instructional Design

CMC Level E represents a sharp departure from the idiom of how to teach math through "discovery" or even through programs that have a progression of "units" and some form of cumulative review. These programs severely underestimate the amount of practice students need to attain fluency with problem-solving steps, such as translating word problems into equations, solving them, and answering the question the problem asks. The programs also don't have a good scheme for teaching the essential component skills of complex operations before these operations are introduced.

All levels of *CMC* strictly follow the practice of first introducing all the component discriminations and skills students need and then combining them into an operation. All that remains to be taught is the sequence of steps and the difference between the newly taught operation and similar operations students have been taught.

The design of lessons is based largely on the following considerations:

- During a period, it is not productive to work only on a single topic. If a lot of new information is being presented, students may become overwhelmed. A better procedure—one that has been demonstrated to be superior in studies of learning and memory—is to distribute the practice; so instead of working for 50 minutes on a single topic, students work each day for possibly 10 minutes on each of four or five topics.
- When full-period topics are presented, it becomes very difficult for a teacher to provide sufficient practice and review on the latest skills that have been taught. A more productive organization works on a particular problem type for a small part of many consecutive lessons. This organization presents each topic as a **track.**
- The lessons that result from the track organization rather than the single-topic organization are designed so that only about 10–15% of the lesson material is new—introduced for the first time. The rest of the material is either work on problem types that have been introduced in the preceding lessons, slight expansions, or new applications that build on what was taught earlier. For instance, Lesson 42 (see page 209 of this guide) presents seven exercises, from six different tracks: Facts, Fractions, Column Multiplication, Division, Word Problems, and Decimals. The only tracks that have anything new are Fractions, Multiplication, and Word Problems.

The new material in two tracks provides extension of what students already know. For instance, for the first fraction exercise students have learned to write fractions for whole numbers, and they've added and subtracted fractions with like denominators. On Lesson 42, students apply both skills to problems that add whole numbers and fractions. They change the whole number into a fraction with the same denominator and then combine the fractions. The **Tracks** section of this guide (page 46) presents an overview of the design strategies used in each track to minimize the amount of new material that is taught.

Uniform Strategies

The Fraction and Word Problem tracks illustrate the instructional efficiency that is achieved by teaching students uniform strategies. The procedures students learn allow them to process a full range of problems of a given type.

Students are taught to use a strategy for finding equivalent fractions to solve the following types of fraction problems, percent problems, and ratio word problems.

a. What fraction equals 80/100 with a numerator of 4?

b. What percent does 4/5 equal?

c. What fraction with a denominator of 5 equals 80%?

d. In a garden, the ratio of roses to irises is 4 to 5. There were 80 roses. How many irises were in the garden?

e. On a ranch, there were 4 cows to every 5 sheep. There were 100 sheep.
   1. How many cows were on the ranch?
   2. How many cows and sheep were there altogether?
   3. How many fewer cows were there than sheep?

f. 4/5 of the flowers in a garden were roses. There were 80 roses.
   1. How many flowers were in the garden?
   2. How many other flowers were in the garden?

Students in fourth grade often don't have a strategy for figuring out the answers to these problems. Students who attempt to solve problems without a solid strategy usually start by guessing which of the smaller numbers in each problem divides the biggest number. They then multiply that quotient by the other smaller number. This process yields answers that are randomly incorrect.

CMC presents a uniform strategy—the strategy for finding equivalent fractions—as the foundational strategy for working these problem types.

Students figure out equivalent fractions by multiplying a fraction they start with by a fraction that equals 1. The fraction in the answer is equal to the fraction the equation starts with, because the starting value wasn't changed. It was multiplied by 1.

For the following equation students learn that 3/7 and 18/35 are not equal because 3/7 is not multiplied by a fraction that equals 1.

$$\frac{3}{7} \times \frac{6}{5} = \frac{18}{35}$$

For this equation students learn that 3/7 and 18/42 are equal because the value of 3/7 wasn't changed to get 18/42. 3/7 was multiplied by 1.

$$\frac{3}{7} \times \frac{6}{6} = \frac{18}{42}$$

Using this relationship, students learn to construct fractions that are equivalent. For this problem students figure out the fraction with a denominator of 100 that is equivalent to 4/5.

$$\frac{4}{5} \times \frac{\quad}{\quad} = \frac{\quad}{100}$$

$$\frac{4}{5} = \frac{\quad}{100}$$

Students can work the problem in the denominators. 5 times what number equals 100? The answer is 20. So the fraction that equals 1 is 20 twentieths.

$$\frac{4}{5} \times \frac{20}{20} = \frac{\quad}{100}$$

$$\frac{4}{5} = \frac{\quad}{100}$$

The problem for the numerators is 4 times 20, which equals 80.

$$\frac{4}{5} \times \frac{20}{20} = \frac{80}{100}$$

$$\frac{4}{5} = \frac{80}{100}$$

The fraction that is equivalent to 4/5 is 80/100.

a. What fraction equals 80/100 with a numerator of 4?

Students learn to work problem A using the same basic strategy: What fraction equals 80/100 with a numerator of 4?

Students set up the fraction multiplication equation based on the numbers the problem gives.

$$\frac{4}{\ } \times \frac{\ }{\ } = \frac{80}{100}$$

Students can work the problem in the numerators: 4 times what number equals 80? The answer is 20. So the fraction that equals 1 is 20 twentieths.

$$\frac{4}{\ } \times \frac{20}{20} = \frac{80}{100}$$

The problem for the denominators is what number times 20 equals 100? The answer is 5.

$$\frac{4}{5} \times \frac{20}{20} = \frac{80}{100}$$

The fraction that is equivalent to 80/100 with a numerator of 4 is 4/5.

b. What percent does 4/5 equal?

Students learn that a percent converts to a fraction with a denominator of 100. 9 hundredths equals 9 percent and 45 percent equals 45 hundredths. For this problem, students write a denominator of 100 for the equivalent fraction they're solving for. Then, they use the strategy for equivalent fractions to complete the equation.

$$\frac{4}{5} \times \frac{20}{20} = \frac{80}{100}$$

$$\frac{4}{5} = \quad \%$$

To complete the simple equation, students convert the equivalent fraction back to percent.

$$\frac{4}{5} \times \frac{20}{20} = \frac{80}{100}$$

$$\frac{4}{5} = 80\%$$

4/5 equals 80%.

c. What fraction with a denominator of 5 equals 80%?

Students set up the equivalent fraction problem.

$$\frac{\ }{5} \times \frac{\ }{\ } = \frac{80}{100}$$

$$\frac{\ }{5} = 80\%$$

Students complete the equations.

$$\frac{4}{5} \times \frac{20}{20} = \frac{80}{100}$$

$$\frac{4}{5} = 80\%$$

d. In a garden, the ratio of roses to irises is 4 to 5. There were 80 roses. How many irises were in the garden?

To work this problem, students learn to write a ratio equation based on the equivalent fraction equation.

The names are *roses* and *irises*. The numbers for the first fraction are 4 and 5. The number after the equal sign is 80. 80 tells about roses, so it goes in the numerator. Here's the ratio set-up:

$$\begin{matrix}\text{Roses} \\ \text{Irises}\end{matrix} \frac{4}{5} \times \frac{\ }{\ } = \frac{80}{\ }$$

Students solve the equivalent fraction equation and box the number that tells about the answer.

$$\begin{matrix}\text{Roses} \\ \text{Irises}\end{matrix} \frac{4}{5} \times \frac{20}{20} = \frac{80}{\boxed{100}}$$

The answer is 100. 100 tells about irises because it goes with the name that's on the bottom.

e. On a ranch, there were 4 cows to every 5 sheep. There were 100 sheep.

1. **How many cows were on the ranch?**
2. **How many cows and sheep were there altogether?**
3. **How many fewer cows were there than sheep?**

To work problem E, students follow the same steps as problem D for the first part of the problem.

Here's the ratio equation for problem E:

$$\frac{\text{Cows}\ 4}{\text{Sheep}\ 5} \times \frac{20}{20} = \frac{80}{100}$$

Students use the information contained in the ratio equation to answer questions 1–3.

1. How many cows were on the ranch?

The answer is the number for cows in the ratio equation after the equal sign—80.

2. How many cows and sheep were there altogether?

To answer this question, students add the number of cows and the number of sheep after the equals sign. 80 + 100 = 180. There were 180 cows and sheep altogether.

3. How many fewer cows were there than sheep?

To answer this question, students subtract the number of cows from the number of sheep after the equal sign. 100 – 80 = 20. There were 20 fewer cows than sheep.

f. 4/5 of the flowers in a garden were roses. There were 80 roses.

1. How many flowers were in the garden?

2. How many other flowers were in the garden?

For the ratio word problems introduced before the fraction wording in problem F, the names and the numbers for the first fraction are always parallel. For the type, "There are 4 roses for every 5 flowers," the number for roses and the name roses come first. The number and name for flowers comes next. For the type, "The ratio of roses to flowers is 4 to 5," the name *roses* comes first and the name *flowers* comes next. Then the number for roses is followed by the number for flowers.

To work problem F, students learn to write the ratio equation for statements that tell about a fraction.

The fraction wording, "4/5 of the flowers were roses," the order in which the names appear, does not correspond to the first fraction. The first name is flowers. The next name is roses. Writing the names in order and the fraction, yields an incorrect ratio:

$$\frac{\text{Flowers}\ 4}{\text{Roses}\ 5}$$

There are more flowers than roses, so the name flowers must go with the bigger number. Roses goes with the smaller number. Here is the correct ratio for problem F.

$$\frac{\text{Roses}\ 4}{\text{Flowers}\ 5}$$

Once the ratio equation is set up, students work the rest of the problems just like they worked problem E. They write the number after the equal sign and complete the equivalent fraction equation.

$$\frac{\text{Roses}\ 4}{\text{Flowers}\ 5} \times \frac{20}{20} = \frac{80}{100}$$

Question 1 asks, "How many flowers were in the garden?" The answer is the number in the ratio equation after the equal sign: 100.

Question 2 asks, "How many other flowers were in the garden?" To answer this question, students subtract the number of roses from the number of flowers after the equal sign: 100 – 80 = 20. There were 20 other flowers in the garden.

(For a more thorough discussion of equivalent fractions and how they are used, see Equivalent Fractions on page 95; see Percent on page 110; see Ratio Word Problems on page161.)

The point of explaining how the equivalent fraction strategy is used to solve a variety of problem types is to illustrate a core feature of the *CMC* programs. Students are taught a strategy that permits them to handle an assortment of problems that would otherwise be difficult to solve for fourth graders. The program does not teach this strategy all at once, but starts with the component skills, and carefully adds new discriminations after students have mastered earlier ones. After applying the strategy in a variety of contexts, students develop a conceptual model for how fractions, percents, and ratios are related. This conceptual model not only enables students to succeed in *CMC Level E,* but serves as a solid reference point for all future work in math.

Scripted Presentations

All exercises in each lesson are scripted. The script indicates the wording you use in presenting the material and correcting student errors. Once you are familiar with the program, you may deviate some from the exact wording; however, until you know why things are phrased as they are, you should follow the exact wording. The most common mistakes teachers make in presenting the material is to rephrase some instructions. Later, when the original instructions become components of more complicated operations, the students are not prepared to respond to steps that have variant wording.

In *Connecting Math Concepts Level E,* you first present material in a structured sequence that requires students to respond verbally. This technique permits you to present tasks at a high rate so it is very efficient for teaching. Also, it provides you with information about which students are responding correctly and which need more repetition.

Typically, after students respond to a series of verbal tasks, you present written work.

Here's part of an exercise with a verbal series in which students discriminate between problems that multiply fractions or that add or subtract fractions. Then students determine if they can work the problem that's written.

a. (Display:) [62:7A]

$\frac{5}{4} \times \frac{3}{8}$ $\frac{5}{4} - \frac{3}{4}$ $\frac{7}{7} + \frac{7}{7}$

$\frac{5}{4} - \frac{3}{8}$ $\frac{6}{9} \times \frac{5}{8}$ $\frac{2}{7} + \frac{5}{7}$

Some of these problems add, some subtract, and some multiply. You can't work some of the addition and subtraction problems because the denominators are not the same. Remember, if you add or subtract, the denominators must be the same.

b. (Point to $\frac{5}{4} \times \frac{3}{8}$.) Read the problem. (Signal.) *5/4 x 3/8.*
- Does the problem add or subtract? (Signal.) *No.*
- So do the denominators have to be the same? (Signal.) *No.*
- Can you work the problem? (Signal.) *Yes.*

c. (Point to $\frac{5}{4} - \frac{3}{8}$.) Read the problem. (Signal.) *5/4 – 3/8.*
- Does the problem add or subtract? (Signal.) *Yes.*
- So do the denominators have to be the same? (Signal.) *Yes.*
- Can you work the problem? (Signal.) *No.*

d. (Point to $\frac{5}{4} - \frac{3}{4}$.) Read the problem. (Signal.) *5/4 – 3/4.*
- Does the problem add or subtract? (Signal.) *Yes.*
- So do the denominators have to be the same? (Signal.) *Yes.*
- Can you work the problem? (Signal.) *Yes.*

A great advantage of sequences like these is that if students later make mistakes involving the discriminations the series addressed, you have information about exactly how to correct the mistakes. For instance, if students make mistakes writing the wrong denominator, your correction would be to ask:

- Does the problem add or subtract?
- So do the denominators have to be the same?

If the answer to the above questions is "Yes," then you ask, "Can you work the problem?"

If the answer to the above questions is "No," then you say, "Say the problem for the denominators."

The scripted presentation is designed to help you present the key discriminations quickly and with consistent language, which helps maximize the efficiency of your teaching.

Language

Connecting Math Concepts Level E does not always initially use the traditional mathematical vocabulary associated with some content. The reason is that what is being taught occurs in stages over many lessons—not all at once in a single lesson or several lessons. The language students need to solve traditional problems will ultimately be taught. The general format of introduction, however, calls for a minimum of vocabulary and a strong emphasis on demonstrating how the operation works, what the discriminations are, and which steps are needed to solve problems. Vocabulary that is not essential to solving a problem type will probably be introduced after students have applied the strategies to many problem sets.

Scope and Sequence for Connecting Math Concepts Level E

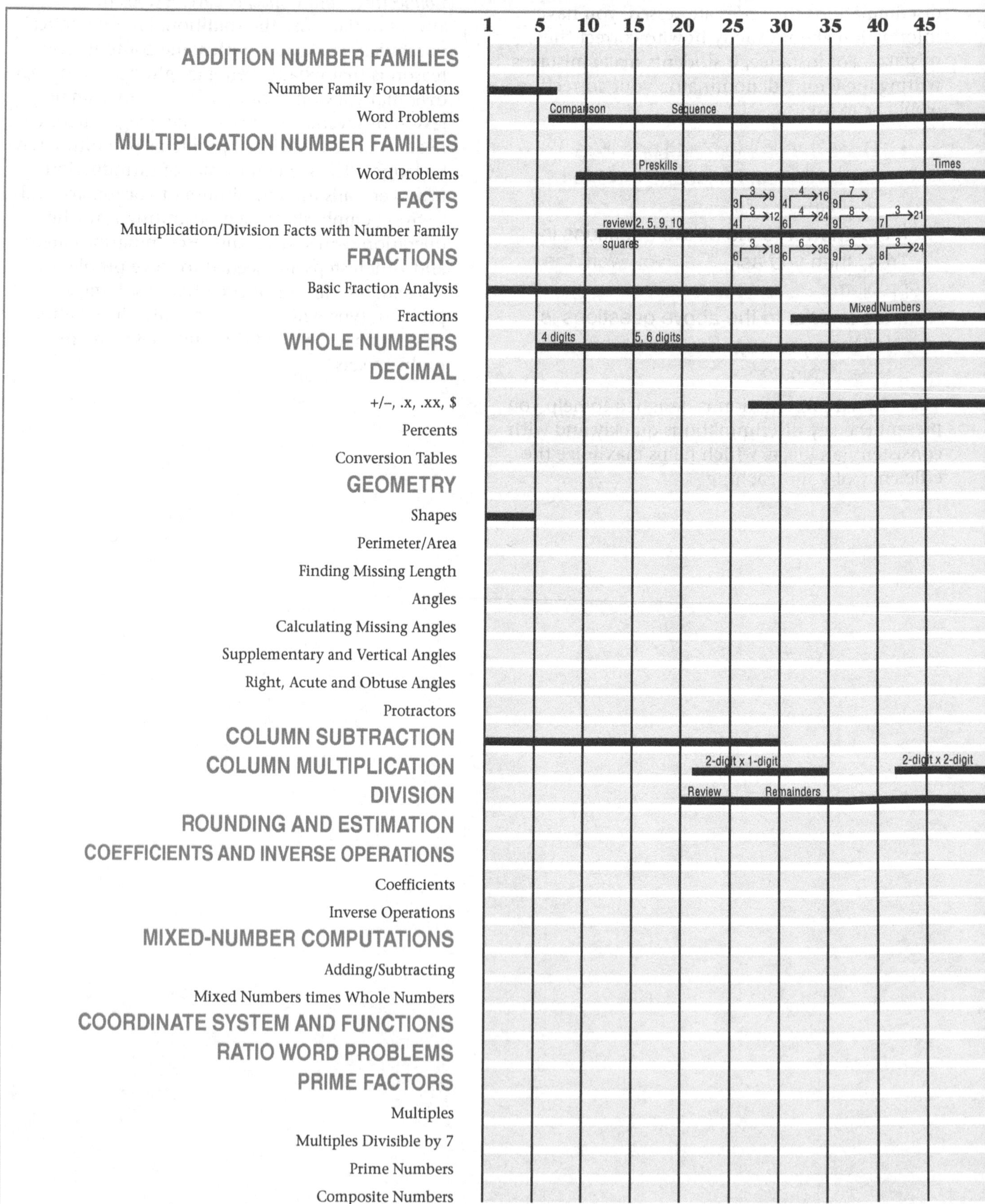

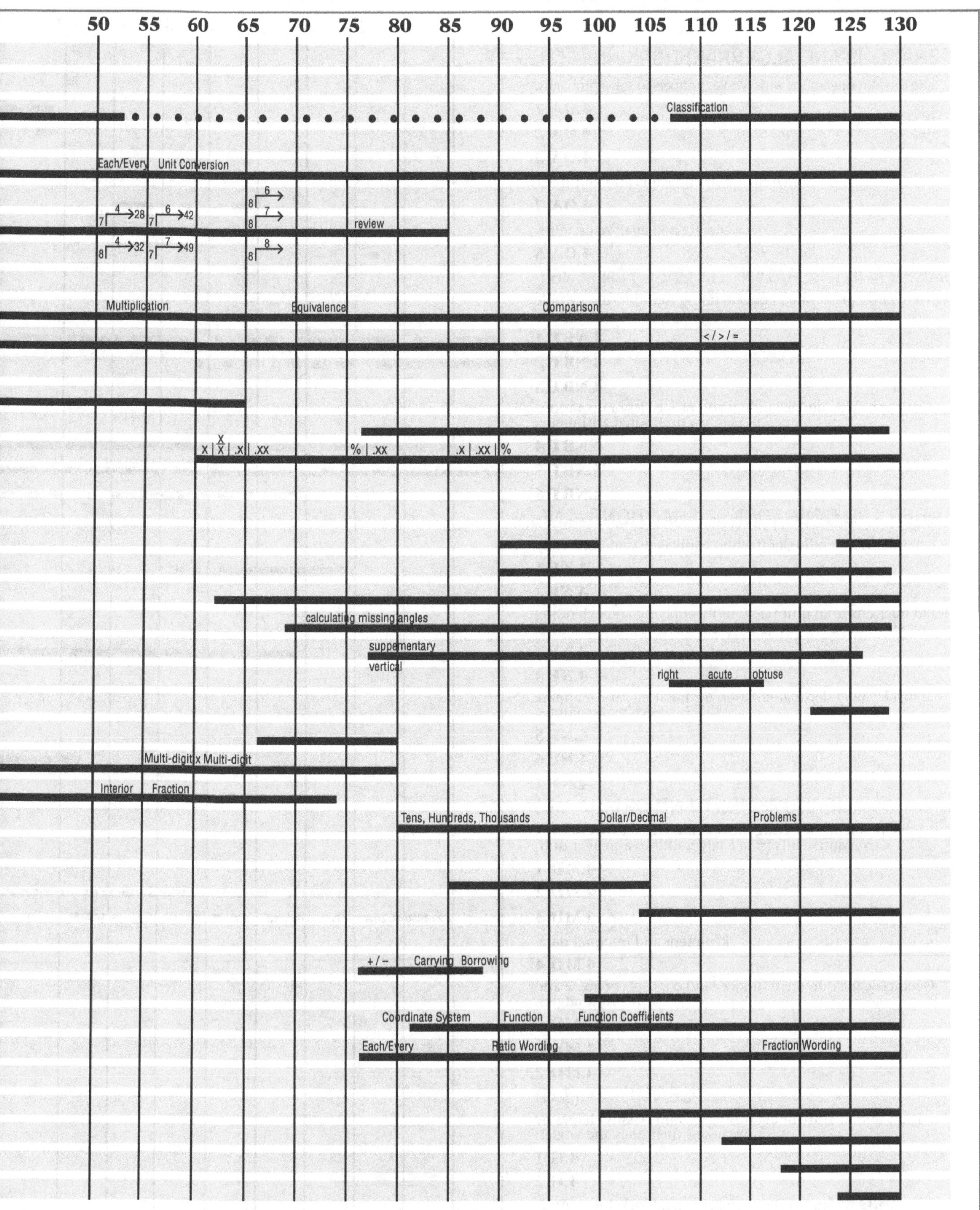
50 55 60 65 70 75 80 85 90 95 100 105 110 115 120 125 130
Classification
Each/Every
Unit Conversion
review
Multiplication
Equivalence
Comparison
</>/=
x | x/x | .x | .xx
% | .xx
.x | .xx | %
calculating missing angles
suppementary
vertical
right
acute
obtuse
Multi-digit x Multi-digit
Interior
Fraction
Tens, Hundreds, Thousands
Dollar/Decimal
Problems
+ / −
Carrying
Borrowing
Coordinate System
Function
Function Coefficients
Each/Every
Ratio Wording
Fraction Wording

Common Core State Standards Chart and CMC Level E

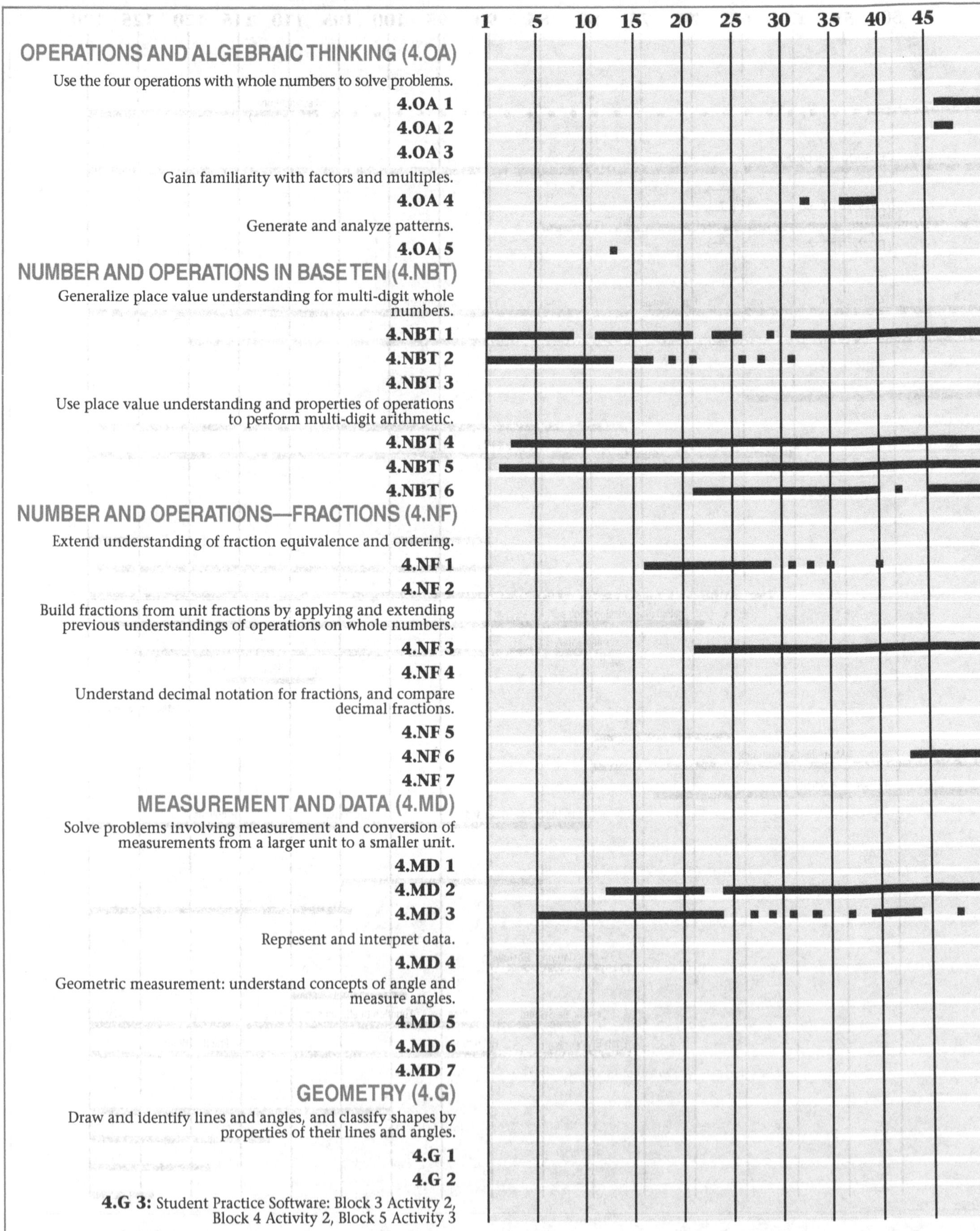

1 5 10 15 20 25 30 35 40 45
OPERATIONS AND ALGEBRAIC THINKING (4.OA)
Use the four operations with whole numbers to solve problems.
4.OA 1
4.OA 2
4.OA 3
Gain familiarity with factors and multiples.
4.OA 4
Generate and analyze patterns.
4.OA 5
NUMBER AND OPERATIONS IN BASE TEN (4.NBT)
Generalize place value understanding for multi-digit whole numbers.
4.NBT 1
4.NBT 2
4.NBT 3
Use place value understanding and properties of operations to perform multi-digit arithmetic.
4.NBT 4
4.NBT 5
4.NBT 6
NUMBER AND OPERATIONS—FRACTIONS (4.NF)
Extend understanding of fraction equivalence and ordering.
4.NF 1
4.NF 2
Build fractions from unit fractions by applying and extending previous understandings of operations on whole numbers.
4.NF 3
4.NF 4
Understand decimal notation for fractions, and compare decimal fractions.
4.NF 5
4.NF 6
4.NF 7
MEASUREMENT AND DATA (4.MD)
Solve problems involving measurement and conversion of measurements from a larger unit to a smaller unit.
4.MD 1
4.MD 2
4.MD 3
Represent and interpret data.
4.MD 4
Geometric measurement: understand concepts of angle and measure angles.
4.MD 5
4.MD 6
4.MD 7
GEOMETRY (4.G)
Draw and identify lines and angles, and classify shapes by properties of their lines and angles.
4.G 1
4.G 2
4.G 3: Student Practice Software: Block 3 Activity 2, Block 4 Activity 2, Block 5 Activity 3

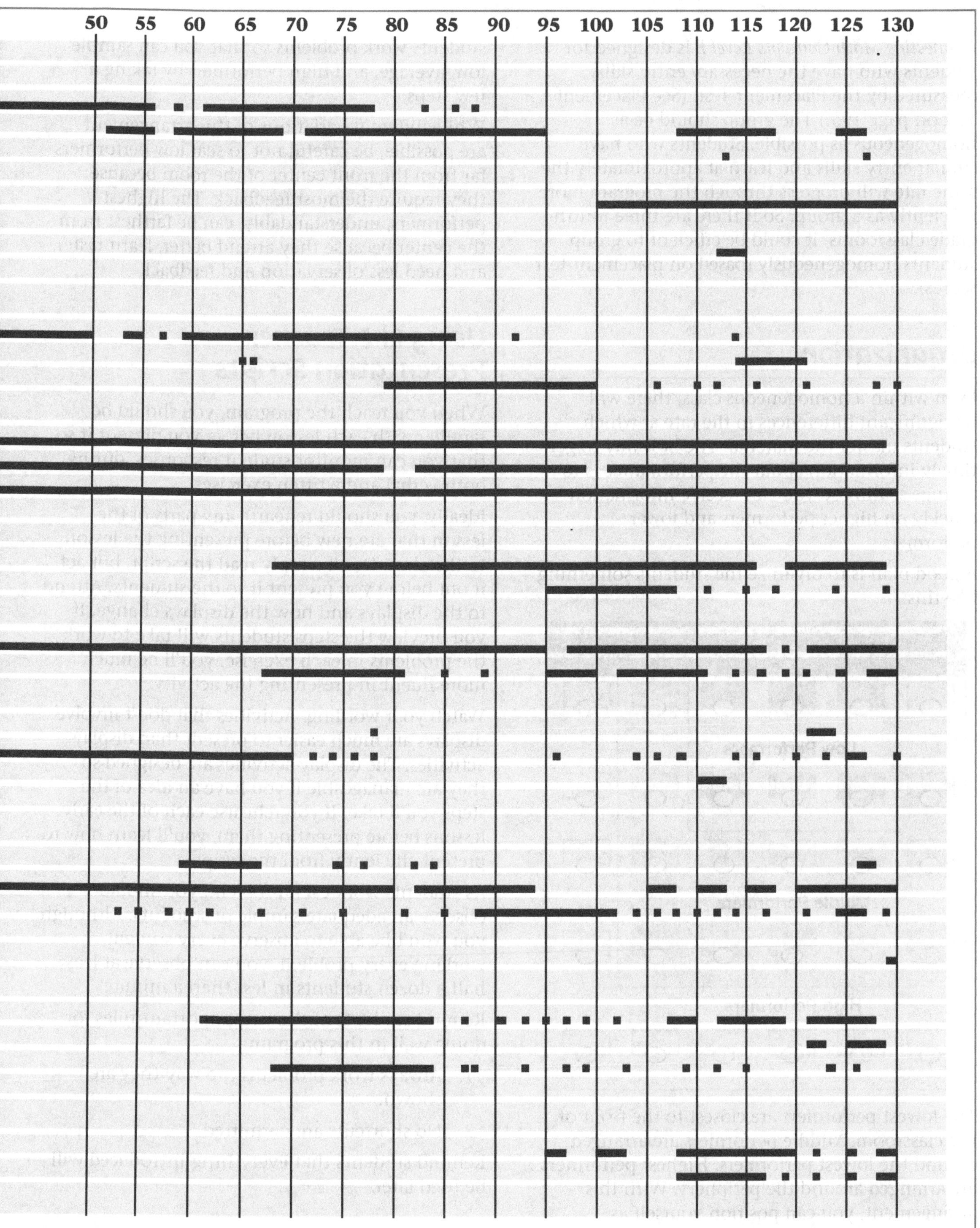
50
55
60
65
70
75
80
85
90
95
100
105
110
115
120
125
130

Teaching Effectively

Connecting Math Concepts Level E is designed for students who have the necessary entry skills measured by the Placement Test. (See Placement Test on page 195.) The group should be as homogeneous as possible. Students who have similar entry skills and learn at approximately the same rate will progress through the program more efficiently as a group. So if there are three fourth-grade classrooms, it could be efficient to group students homogeneously (based on placement-test scores).

Organization

Even within a homogeneous class, there will be significant differences in the rate at which students master the material. The best way to get timely information about the performance is to arrange seating so you can receive information quickly on higher performers and lower performers.

A good plan is to organize the students something like this:

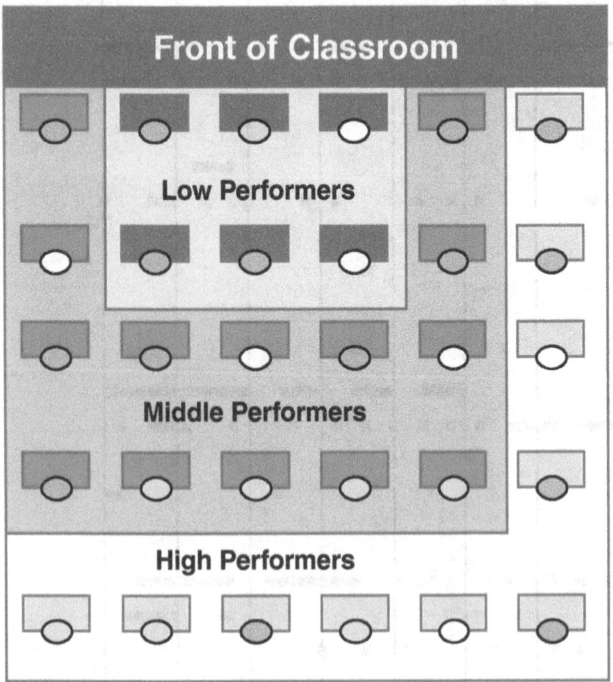

The lowest performers are closest to the front of the classroom. Middle performers are arranged around the lowest performers. Highest performers are arranged around the periphery. With this arrangement, you can position yourself as students work problems so that you can sample low, average, and high performers by taking a few steps.

While different variations of this arrangement are possible, be careful not to seat low performers far from the front center of the room because they require the most feedback. The highest performers, understandably, can be farthest from the center because they attend better, learn faster, and need less observation and feedback.

Using the Teacher Presentation Scripts

When you teach the program, you should be familiar with each lesson before you present it so that you can monitor student responses, during both verbal and written exercises.

Ideally, you should rehearse any parts of the lesson that are new before presenting the lesson to the class. Don't simply read the script, but act it out before you present it to the students. Attend to the displays and how the displays change. If you preview the steps students will take to work the problems in each exercise, you'll be much more fluent in presenting the activity.

Watch your wording. Activities that don't involve displays are much easier to present than display activities. The display activities are designed so they are manageable if you have an idea of the steps you'll take. If you rehearse each of the early lessons before presenting them, you'll learn how to present efficiently from the script.

As students work each problem, you should observe an adequate sample of students. Although you won't be able to observe every student working every problem, you can observe at least half a dozen students in less than a minute.

Remind students of the two important rules for doing well in this program:

1. Always work problems the way they are shown.
2. No shortcuts are permitted.

Remind students that everything introduced will be used later.

Reinforce students who apply what they learn.

Always require students to rework incorrect problems.

The script for each lesson indicates precisely how to present each structured activity. The script shows what you say, what you do, and what the students' responses should be.

What you say appears in blue type:

You say this.

What you do appears in parentheses:

(You do this.)

The responses of the students are in italics. Things students may say that are acceptable but not required are in parenthesis.

Students say this (but may or may not say this).

Although you may feel uncomfortable "reading" a script (and you may feel that the students will not pay attention), try to present the exercises as if you're saying something important to the students. If you do, you'll find that working from a script is not difficult and that students respond well to what you say.

This Teacher's Guide contains two sample lessons on pages 209–233: Lesson 42 and Lesson 103. We'll use parts of exercises from Lesson 42 to illustrate the important script conventions for *CMC Level E.* We'll also use part of an exercise from Lesson 97. Here's part of Exercise 2 from Lesson 42 that illustrates some of the important signaling and display conventions:

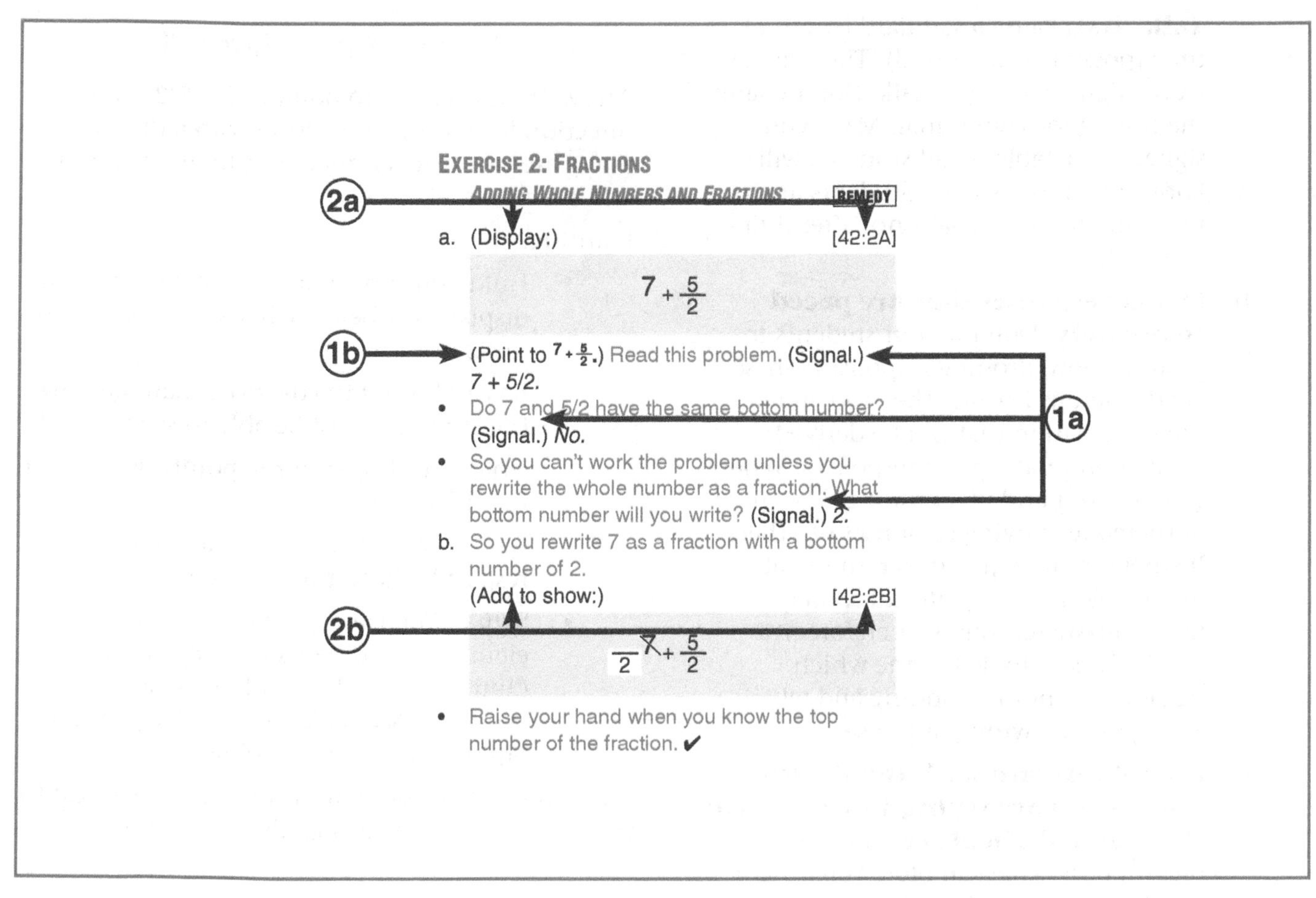

ARROW 1: SIGNALS

There are two types of signals teachers use to present tasks in *CMC Level E.* As indicated above, all signals have the same purpose: to trigger a simultaneous response from the group. All signals have the same rationale: If you can get the group to respond simultaneously (with no student leading the others), you will get information about the performance of all the students—not just those who happen to respond first.

The simplest way to signal students to respond together is to adopt a timing practice—just like the timing in a musical piece.

General Rules for Signals

Students will not be able to initiate responses together at the appropriate rate unless you follow these rules:

a. **Talk first.** Pause a standard length of time (possibly one second). Then signal. Never signal when you talk. Don't change the timing for your signal. Make your signal predictable, so all students will know when to respond. Students are to respond on your signal—not after it or before it.

b. **Model responses that are paced reasonably.** Don't permit students to produce slow, droning responses. These are dangerous because they rob you of the information that can be derived from appropriate group responses. When students respond in a droning way, many of them are copying responses of others. If students are required to respond at a reasonable speaking rate, all students must initiate responses; therefore, it's relatively easy to determine which students are not responding and which are saying the wrong response.

c. **Do not respond with the students (unless you are trying to work with them on a difficult response).** You present only what's in blue. You do not say the answers with the students unless the script specifically tells you to. You should not move your lips or give other spurious clues about what the answer is.

Think of signals this way: If you use them correctly, they provide you with much diagnostic information. A weak response suggests that you should repeat a task and provides information about which students may need more help. Signals are, therefore, important early in the program. After students have learned the routine, the students will be able to respond on cue with no signal. That will happen, however, only if you always give your signals at the end of a constant time interval after you complete what you say.

Basic Signal: Arrow 1a

A basic signal follows a question, a direction, or the words, "Get ready." Just before each response for the three bullets in step A, the script indicates to signal. You can signal for arrow 1a by tapping what you're pointing to, clapping one time, snapping your fingers, or tapping your foot. After initially establishing the timing for signals, you can signal through voice inflection only.

Point-and-Touch Signal: Arrow 1b

Arrow 1b directs you to point at 7 + 5/2. This direction is used in connection with a display. There are two teacher directions for every point-and-touch signal.

Pointing:

- Hold your finger about 6 inches from the display, just below what students should focus on.
- Be careful not to cover the material—all the students must be able to see it.
- Hold your finger in the pointing position for at least one second.
- Say the verbal cue for students to respond, "Read this problem."
- Signal. You can signal by touching elements as students respond to them. (Touch 7, +, 5/2 as students read symbols.) *Seven plus 5 halfs.* Or you can tap once below the problem.

The timing for these signals should be identical to the timing for all other signals to the group.

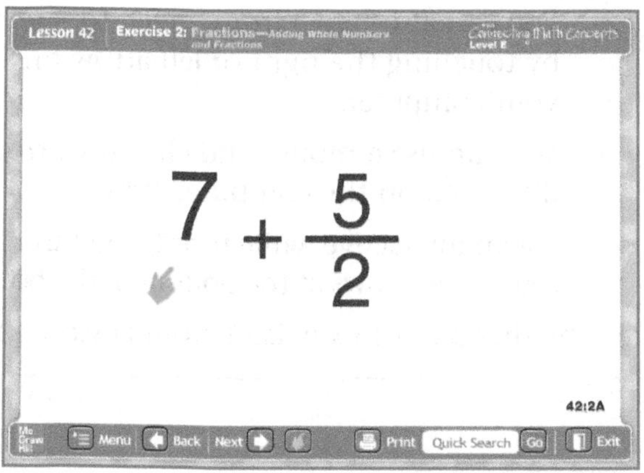

The Board Display CD (as well as the identical Board Displays on ConnectED) has a special feature that can be used for signaling. The cursor shown on the screen can be replaced by an orange hand icon that, in turn, can be used to signal. You can replace the cursor on the screen by clicking on the icon of a hand at the bottom of the screen. The cursor will turn into a picture of a hand. You can move the hand cursor to the part of the display students should focus on. Click to signal and the hand moves to mimic a touch. You can use the hand cursor to mimic the timing, pointing, and touching of your point and touch signal. Practice using the optional hand cursor before using it during a lesson.

Arrow 2: Boardwork/Board Displays CD

In many exercises, you will display material on the board or on a screen.

The word **(Display:)** appears in the script when a display is to be shown to students (arrow 2a). The words **(Add to Show:)** appear in the script when something is to be added to an existing display (arrow 2b).

The program has been designed so that you can

a) show all displays and additions or changes to displays by using the Board Displays CD that comes with the Teacher Presentation Book, or

b) write most displays on the board and project selected displays with an overhead projector or a document camera. Some displays will be difficult to write on a board.

Using the Board Display CD to Show All Displays

The Board Displays CD and the online versions contain all the displays for every lesson. The displays are labeled consecutively for each lesson. Note that the display code (arrow 2a) shown for each display begins with a number that indicates the lesson. The next number indicates the exercise on that lesson. The letters at the end of the code indicate the order of the displays. The codes for the displays shown in Exercise 2 are **42:2A** and **42:2B.** The 2 indicates that it is Exercise 2. The **A** indicates that it is the first display in that exercise. The **B** identifies the next display in the exercise. The letters **A** and **B** in the display codes do not correspond to the step of the exercise. It is coincidental that in step A display **A** appears and in step B display **B** appears. ***Note:*** The identification code appearing on each CD display corresponds to the code shown in the *Level E* Teacher Presentation Book.

The best way to use the Board Display CD is to stand where the images are projected and use a remote device to direct the presentation. (If you are using an interactive white board, you can simply touch the screen.) Being close to the image allows you to point to details of the display as you signal.

Follow these procedures when presenting a new lesson:

- After inserting the disc, click on the icon for the CD. The computer will display the main menu. Select the desired lesson from the main menu. If you select Lesson 42, the computer screen will show the exercises for Lesson 42, *starting with the first exercise that has a board display.*
- When you click on an exercise, the display codes will be listed for that exercise.
- When you reach the first display in the Teacher Presentation Book, click twice on the appropriate display code to present that display.

If you aren't going to use a computer for displays, preview lessons to determine which displays you'll write and which will be shown on a document projector.

You can make printed copies of all the displays by accessing McGraw-Hill Education's ConnectED platform.

Whatever system you use, your goal should be to keep the presentation moving without serious interruptions. Make sure you have displays you'll project ready before presenting the lesson so you can maintain a good pace.

Using a Computer and Projector

You can move from display to display in several ways.

- You can use a remote control. Pressing the forward arrow on the remote calls up the next display. Pressing the back arrow returns to the previous display.
- You can also move from screen to screen by touching the right or left arrow on your computer.
- You can use a mouse and click the arrows displayed on the computer screen.
- For an interactive white board, you can touch the arrows at the bottom of the board.

Here's the first part of Exercise 1 from Lesson 42:

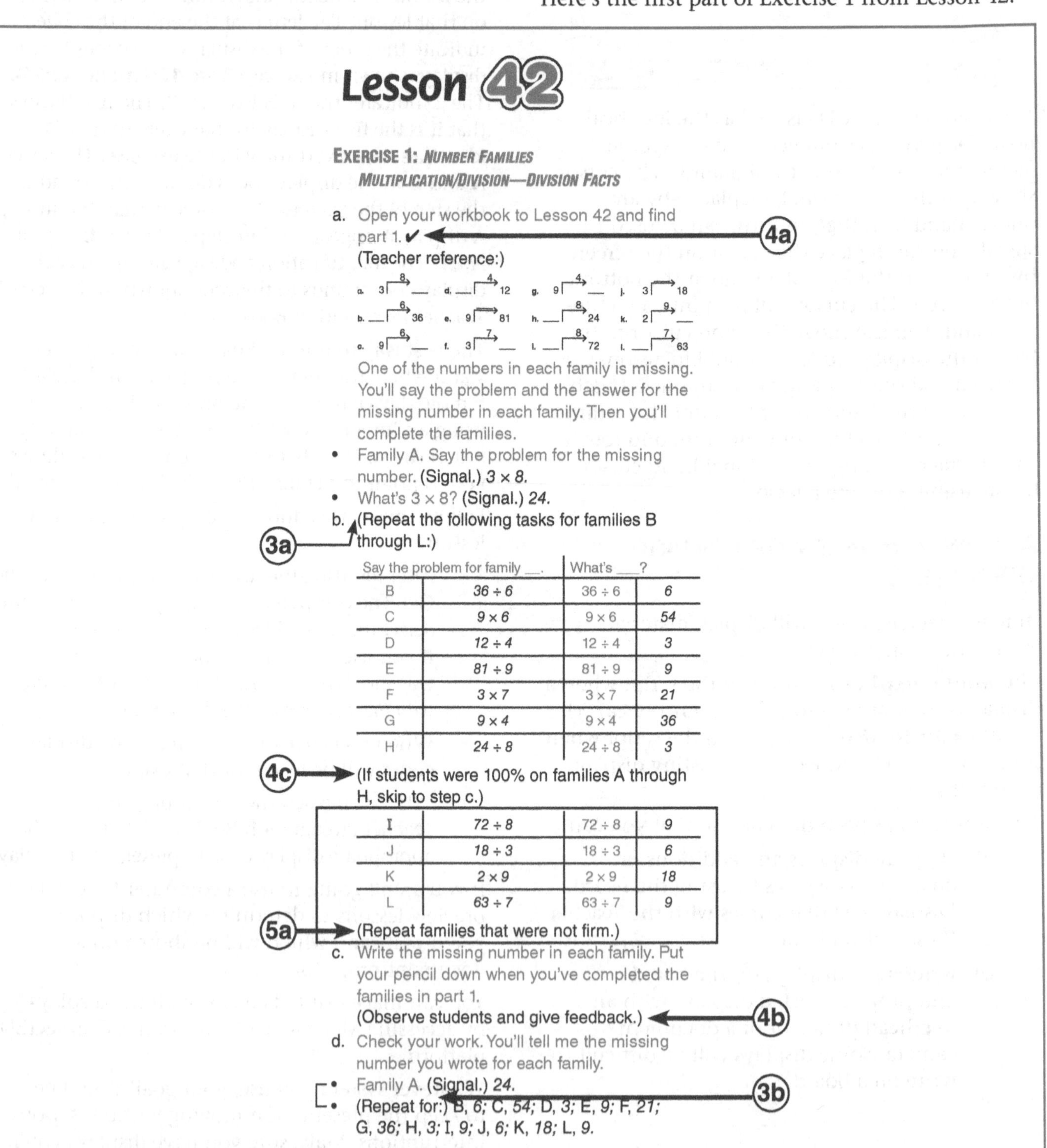

Lesson 42

EXERCISE 1: *NUMBER FAMILIES*
MULTIPLICATION/DIVISION—DIVISION FACTS

a. Open your workbook to Lesson 42 and find part 1. ✔ ← (4a)
(Teacher reference:)

a. 3 —8→ ___	d. ___ —4→ 12	g. 9 —4→ ___	j. 3 —→ 18
b. ___ —6→ 36	e. 9 —→ 81	h. ___ —8→ 24	k. 2 —9→ ___
c. 9 —6→ ___	f. 3 —7→ ___	i. ___ —8→ 72	l. ___ —7→ 63

One of the numbers in each family is missing. You'll say the problem and the answer for the missing number in each family. Then you'll complete the families.

- Family A. Say the problem for the missing number. **(Signal.)** *3 × 8.*
- What's 3 × 8? **(Signal.)** *24.*

b. **(Repeat the following tasks for families B through L:)** ← (3a)

Say the problem for family ___.		What's ___?	
B	*36 ÷ 6*	36 ÷ 6	*6*
C	*9 × 6*	9 × 6	*54*
D	*12 ÷ 4*	12 ÷ 4	*3*
E	*81 ÷ 9*	81 ÷ 9	*9*
F	*3 × 7*	3 × 7	*21*
G	*9 × 4*	9 × 4	*36*
H	*24 ÷ 8*	24 ÷ 8	*3*

(4c) → **(If students were 100% on families A through H, skip to step c.)**

I	*72 ÷ 8*	72 ÷ 8	*9*
J	*18 ÷ 3*	18 ÷ 3	*6*
K	*2 × 9*	2 × 9	*18*
L	*63 ÷ 7*	63 ÷ 7	*9*

(5a) → **(Repeat families that were not firm.)**

c. Write the missing number in each family. Put your pencil down when you've completed the families in part 1.
(Observe students and give feedback.) ← (4b)

d. Check your work. You'll tell me the missing number you wrote for each family.

- Family A. **(Signal.)** *24.*
- **(Repeat for:)** *B, 6; C, 54; D, 3; E, 9; F, 21; G, 36; H, 3; I, 9; J, 6; K, 18; L, 9.* ← (3b)

Arrow 3: Abbreviated Scripts:

Arrow 3a shows a table that is used to abbreviate the teacher presentation scripts in *CMC Level E.* In step A, the script specifies the following tasks for problem A in bullets 1 and 2.

- Family A. Say the problem for the missing number. (Signal.) *3 × 8.*
- What's 3 × 8? (Signal.) *24.*

The table is a shorthand way of specifying the same tasks for the remaining families. The top row shows the generic wording with a blank for each task. Below the first row, the first column shows what you say for the blank when you present the first task for each family. The next column shows what students should respond to the first tasks. The next column shows what you say for the blank in the second task for each family. The next column shows what students should respond to the second tasks.

The presentation specifies the following presentation for family B:

- Family B. Say the problem for the missing number. (Signal.) *36 ÷ 6.*
- What's 36 ÷ 6? (Signal.) *6.*

The presentation specifies the following presentation for family H:

- Family H. Say the problem for the missing number. (Signal.) *24 ÷ 8.*
- What's 24 ÷ 8? (Signal.) *3.*

Arrow 3b shows a repeat-for abbreviation used in the teacher presentation scripts. In step D, bullet 1, the script directs students to say the missing number they wrote for each family. Then the script specifies the following wording for family A:

- Family A. (Signal.) *24.*

The repeat-for direction that follows tells you to repeat the task for the remaining families.

Here is the presentation spelled out for families B through E:

- Family B. (Signal.) *6.*
- Family C. (Signal.) *54.*
- Family D. (Signal.) *3.*
- Family E. (Signal.) *9.*

Arrow 4: Pacing Your Presentation and Interacting with Students as They Work

You should pace your verbal presentation at a normal speaking rate—as if you were telling somebody something important.

Arrows 4a, 4b, and 4c designate three ways to pace your presentation. One is marked with a ✔. Another is a note to **(Observe students and give feedback.)**. Both of these designations indicate that you will monitor students. The third designation is a note, **"(If students were (100%) on (problems a through e), skip to step (c).)"**, followed by tasks that are boxed. This note directs you to use the information about student performance to determine if the students can be accelerated and skip the instruction in the box. More information about the skip boxes appears later on.

Arrow 4a shows a ✔. A ✔ is a note to check students' performance on a task that requires only a second or two. If you are positioned close to several lower-performing students, quickly check whether two or three of them are responding appropriately. If they are, proceed with the presentation.

Arrow 4b shows an observe and give feedback direction. The **(Observe students and give feedback.)** direction requires more careful observation. You sample more students and you give feedback—not only to individual students but to the group. Here are the basic rules for what to do and what not to do when you observe and give feedback.

- If the task is one that takes no more than 30 seconds, observe and give feedback to several students. Focus on the lower performers.
- If the task requires considerably more time, move from the front of the room to a place where you can quickly sample the performance of low, middle, and high performers.
- As you observe, make comments to the whole class. Focus these comments on students who are following directions, working quickly, and working accurately: "Wow, a couple of students are almost finished. I haven't seen one mistake so far."

- Students put their pencils down to indicate that they are finished. Acknowledge students who are finished. They are not to work ahead.
- If you observe mistakes, do **not** provide a great deal of individual help. Point out any mistake, but do not work the problems for the students. For instance, if a student gets one of the problems wrong, point to it and say, "You made a mistake. Fix it." Observe some other students. Then return to make sure the student fixed the mistake. Make sure that you check the lower performers and give them feedback. When you show them what they did wrong, keep your explanation simple. The more involved your explanations, the more likely they are to get confused.
- If you observe a problem that several students are having, tell the class, "Stop. We have a problem." Point out the mistake. Repeat the part of the exercise that gives them information about what they are to do. Do not provide new teaching or new problems. Simply repeat the part of the exercise that gives students the information they need and reassign the work. Then say, "Now let's work it the right way."
- Allow students a reasonable amount of time. Do not wait for the slowest students to complete the problems before presenting the workcheck during which students correct their work and fix any mistakes. You can usually use the middle performers as a gauge for what is reasonable. As you observe that they are completing their work, announce, "Okay, you have about 20 seconds to finish up." At the end of that time, begin the workcheck.

If you follow the procedures for observing students and giving feedback, your students will work faster and more accurately. They will also become facile at following your directions.

- If you wait a long time period before presenting the workcheck, you punish those who worked quickly and accurately. Soon, they will learn that there is no payoff for doing well—no praise, no recognition—but instead a long wait while you give attention to those who are slow.
- If you don't make announcements about students who are doing well and working quickly, the class will not understand what's expected. Students will probably not improve as much.
- If you provide extensive individual help on written tasks, you will actually reinforce students for not listening to your directions and for being dependent on your help. Furthermore, this dependency becomes contagious. It doesn't take other students in the class long to discover that they don't have to listen to your directions, that they can raise their hand and receive help that shows them how to do the assigned work.

These expectations are the opposite of the ones you want to induce. You want students to be self-reliant and to have reasons for learning and remembering what you say when you instruct them. The simplest reasons are that they will use what they have just been shown and that they will receive reinforcement for performing well.

If you follow the management rules outlined above, all students who are properly placed in the program should be able to complete assigned work within a reasonable period of time and have reasons to feel good about their ability to do math. That's what you want to happen. As students improve, you should tell them about it. "What's this? Everybody's finished with that problem already? That's impressive."

Arrow 4c shows a criterion followed by a skip box. All skip boxes work the same way. The teacher direction specifies a criterion for student performance. If students have met that criterion **and** you are confident that students do not need additional practice, then you skip the part of the presentation that's boxed and start at the step that's specified in the direction. There are three basic criteria for the teacher directions: 1) perfect performance for that exercise to that point; 2) perfect performance for all of the exercises of that type in the last two lessons, including that exercise to that point; 3) confidence that the students are firm and have mastered it.

The criterion specified in Exercise 1 is 100% for problems A through E. If students make a mistake on problem D, you would present the tasks for all of the families.

If students performed perfectly the first time through on problems A through E, you would skip to step C. However, if you knew that students had trouble with problems that multiply by 6, you might not skip to step C. A good idea would be to present problems I and L to give them practice on the problems that are difficult for students.

Usually, parts that are boxed contain verbal firming of material. If students are at mastery on performing these tasks, skipping the parts does not affect students' performance.

Sometimes problems in the Workbook and Textbook are boxed. This box indicates problems that may be skipped. If students have met the criterion for skipping a Textbook problem that they've already copied on their lined paper, reinforce them for their perfect performance. "I had you copy that problem, but you've been doing so well on these types of problems, that we can skip it."

Some teachers like to assign these skipped problems as extra-credit independent work. "You can earn an extra point for correctly working each of the problems we were able to skip today."

Be conscientious about not skipping steps when the class doesn't meet the criteria. The amount of practice specified in the teacher presentation is enough to enable virtually all appropriately placed students to master all of the material. Average classes usually have several students who benefit greatly from this practice on most exercises. Some classes won't meet the criterion for any skip boxes. Some classes will meet the criterion for 10 to 15% of the skip boxes. Only very high performing classes will meet the criterion for more than 30% of the skip boxes.

Do not use the skip boxes to make up for lesson progress. Skipping boxed parts of exercises when students do not meet the criterion will impede the class's progress in the long run. If a class skips instruction that is boxed, but does not meet the criterion, they will miss critical practice and application of the knowledge, strategies, and discriminations to perform later tasks fluently. As a result of missing this practice, students' performance will be slower on many later lessons. The lack of fluency ends up taking a lot more time than skipping the parts saved.

Skip boxes are a great way to accelerate the class and reinforce them for performing perfectly. When students have met the criterion for a skip box, and you are confident that they've mastered the tasks presented in the exercise, let them know that they don't need instruction that a lot of other classes need. "Nice work! You're 100% on those problems, so we get to skip going over the rest of them."

Reinforcing students with specific praise communicates the performance you expect from them and gives them a payoff for meeting or exceeding your expectations—skipping parts.

Arrow 5: Firming Repeats

A special correction is needed when correcting mistakes on tasks that teach a sequence or a relationship. This type of correction is marked with one of three notes.

- (Repeat steps that were not firm.) Illustrated by arrow 5a on Lesson 42, Exercise 1.
- (Repeat until firm.) Illustrated by arrow 5b on Lesson 42, Exercise 3.
- (Repeat step _ until firm.) Illustrated by arrow 5c on Lesson 42 Exercise 6.

Here are parts of Exercises 3 and 6 from Lesson 42:

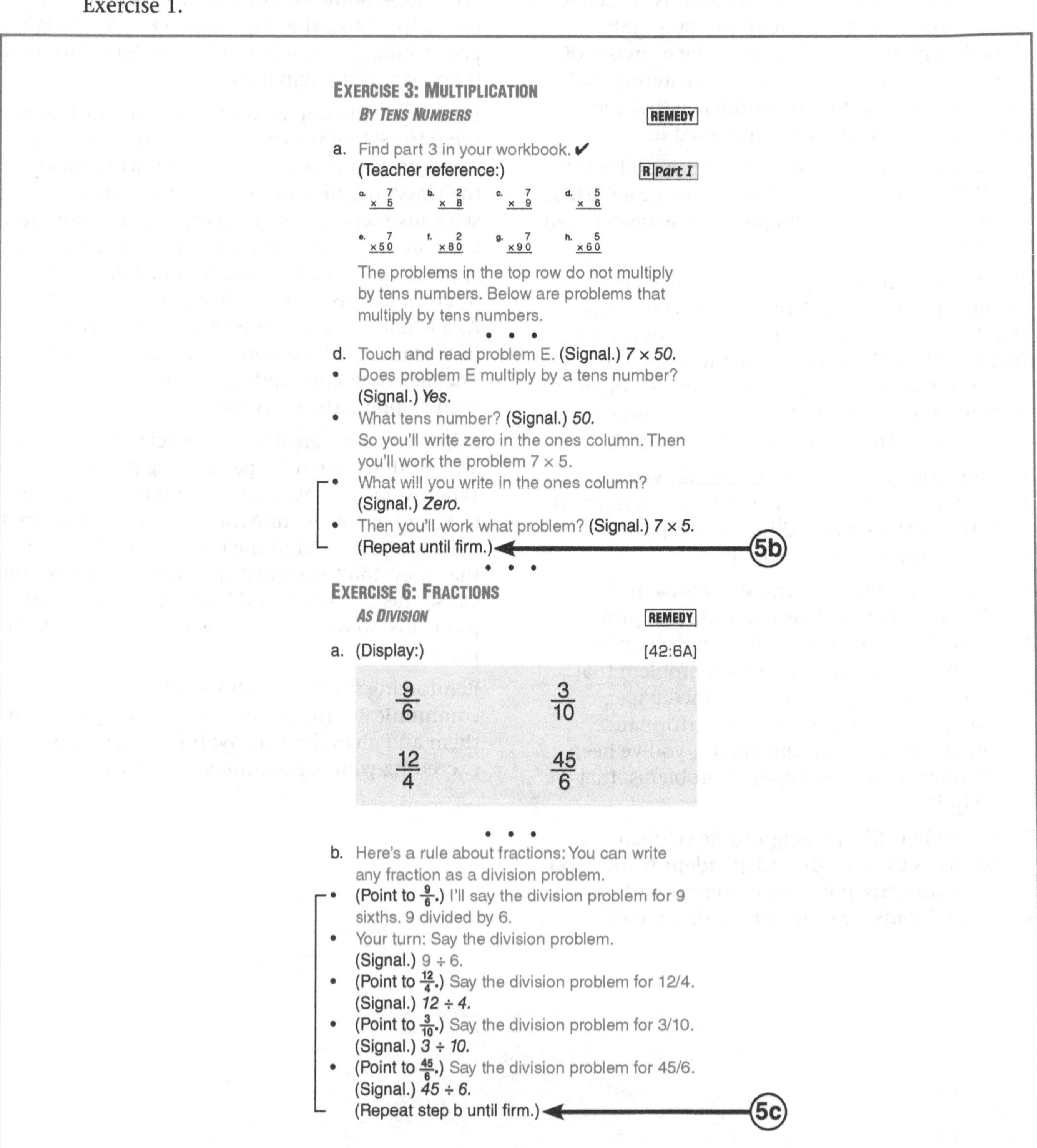

Exercise 3: Multiplication
By Tens Numbers REMEDY

a. Find part 3 in your workbook. ✔
(Teacher reference:) R Part I

a. 7×5	b. 2×8	c. 7×9	d. 5×6
e. 7×50	f. 2×80	g. 7×90	h. 5×60

The problems in the top row do not multiply by tens numbers. Below are problems that multiply by tens numbers.

• • •

d. Touch and read problem E. (Signal.) *7 × 50.*
- Does problem E multiply by a tens number? (Signal.) *Yes.*
- What tens number? (Signal.) *50.*
So you'll write zero in the ones column. Then you'll work the problem 7 × 5.
- What will you write in the ones column? (Signal.) *Zero.*
- Then you'll work what problem? (Signal.) *7 × 5.*
(Repeat until firm.) ← 5b

• • •

Exercise 6: Fractions
As Division REMEDY

a. (Display:) [42:6A]

$\frac{9}{6}$ $\frac{3}{10}$

$\frac{12}{4}$ $\frac{45}{6}$

• • •

b. Here's a rule about fractions: You can write any fraction as a division problem.
- (Point to $\frac{9}{6}$.) I'll say the division problem for 9 sixths. 9 divided by 6.
- Your turn: Say the division problem. (Signal.) *9 ÷ 6.*
- (Point to $\frac{12}{4}$.) Say the division problem for 12/4. (Signal.) *12 ÷ 4.*
- (Point to $\frac{3}{10}$.) Say the division problem for 3/10. (Signal.) *3 ÷ 10.*
- (Point to $\frac{45}{6}$.) Say the division problem for 45/6. (Signal.) *45 ÷ 6.*

(Repeat step b until firm.) ← 5c

All repeat directions appear at the end of the tasks that are to be repeated. For arrows 5b and 5c, Repeat-until-firm and Repeat-step-__-until-firm directions, the tasks and the direction are bracketed on the left to clarify the tasks that need to be repeated. The tasks must be mastered before material that follows is presented. Repeat-until-firm and Repeat-step-__-until-firm directions are used: after a sequence of tasks teaching a relationship that students may not understand if each of the tasks aren't responded to correctly and fluently; when students must produce a series of responses in a consistent sequence (as in counting).

For (Repeat until Firm.) and (Repeat step __ until firm.) follow these steps:

1. Correct the mistake. (Tell the answer and repeat the task that was missed.)
2. Return to the beginning of the specified step and present the entire step.

In Exercise 6, step B, you present four different fractions. For each fraction, you direct students to say the division problem for it.

If students are confused, if some don't answer a question, or if the answer to any of the questions is wrong, you repeat the bracketed part of step B after you have performed the normal correction procedure for each mistake.

When you hear a mistake, you say the correct answer and repeat the task. However, make sure that students are firm in all of the problems you present in step B. You cannot be sure that students are firm unless you repeat the step.

Here's a summary of the steps you follow when repeating a part of the exercise until firm:

Correct the mistake.

(Tell the correct answer.) The division problem for 3 tenths is 3 divided by 10.

Repeat the task.

What's the division problem for 3 tenths? (Signal.)

Repeat the step.

Let's do those again.

(Start at the beginning of the bracket in step B and present the entire task.)

Repeat the step again to make sure students are firm.

You did it without any errors. Let's do it again perfectly.

(Start at the beginning of the bracket in step B and present the entire task.)

Repeat until firm is based on the information you need about the students. When the students made the mistake, you told the answer. Did they remember the answer? Would they now be able to perform the procedure or sequence of responses correctly? The repeat-until-firm procedure provides you with answers to these questions. Students show you through their responses whether or not the correction worked, whether or not they are firm. Arrow 5c directs you to repeat Exercise 6, step B until students can say fractions as division problems without making a mistake. This repeat direction firms a series of responses that must be produced in a consistent sequence.

A (Repeat _ that were not firm) direction occurs when students are expected to apply procedures to a set of examples. Arrow 5a on Exercise 1 (page 20) directs you to repeat problems that required you to make a correction. For each mistake students make, you would perform the normal correction: tell the answer; repeat the task; start again at the top of the column and repeat the column. After completing through problem L, you would go back and present the problems that required corrections. You would continue repeating these problems until students respond perfectly to all problems on which errors were made.

ARROW 6: CALL ON INDIVIDUAL STUDENTS

In *Levels A, B,* the scripts specified many individual turns which usually appeared at the end of exercises. In *Levels C, D, E,* and *F,* three types of individual turns that require verbal responses.

Here is part of Lesson 42 Exercise 7 and part of Lesson 97 Exercise 1:

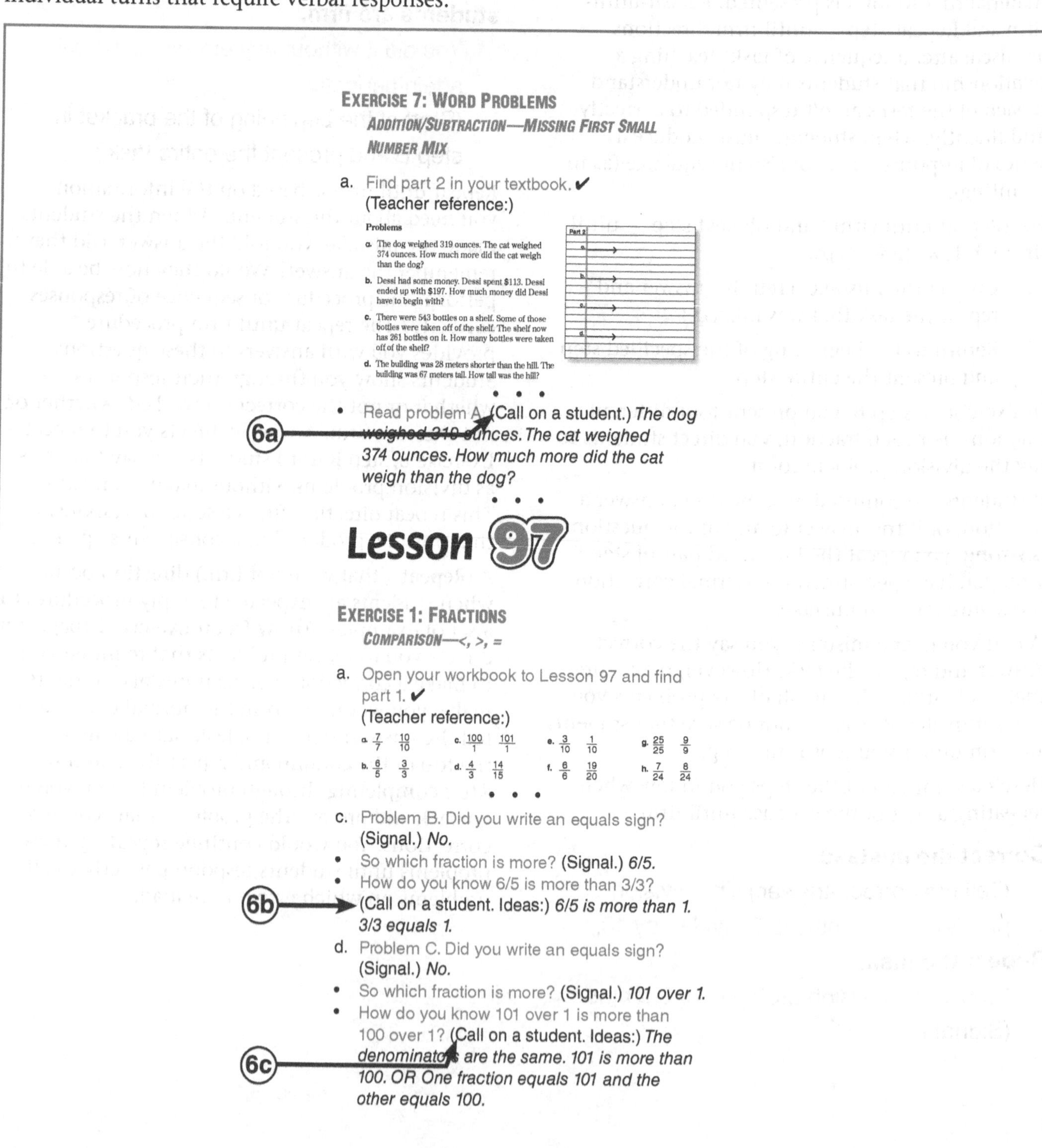

EXERCISE 7: WORD PROBLEMS

ADDITION/SUBTRACTION—MISSING FIRST SMALL NUMBER MIX

a. Find part 2 in your textbook. ✔
(Teacher reference:)

Problems

a. The dog weighed 319 ounces. The cat weighed 374 ounces. How much more did the cat weigh than the dog?

b. Dessi had some money. Dessi spent $113. Dessi ended up with $197. How much money did Dessi have to begin with?

c. There were 543 bottles on a shelf. Some of those bottles were taken off of the shelf. The shelf now has 261 bottles on it. How many bottles were taken off of the shelf?

d. The building was 28 meters shorter than the hill. The building was 67 meters tall. How tall was the hill?

• • •

- Read problem A. (Call on a student.) *The dog weighed 319 ounces. The cat weighed 374 ounces. How much more did the cat weigh than the dog?*

• • •

Lesson 97

EXERCISE 1: FRACTIONS

COMPARISON—<, >, =

a. Open your workbook to Lesson 97 and find part 1. ✔
(Teacher reference:)

a. $\frac{7}{7}$ $\frac{10}{10}$ c. $\frac{100}{1}$ $\frac{101}{1}$ e. $\frac{3}{10}$ $\frac{1}{10}$ g. $\frac{25}{25}$ $\frac{9}{9}$

b. $\frac{6}{5}$ $\frac{3}{3}$ d. $\frac{4}{3}$ $\frac{14}{15}$ f. $\frac{6}{6}$ $\frac{19}{20}$ h. $\frac{7}{24}$ $\frac{8}{24}$

• • •

c. Problem B. Did you write an equals sign? (Signal.) *No.*
- So which fraction is more? (Signal.) *6/5.*
- How do you know 6/5 is more than 3/3? (Call on a student. Ideas:) *6/5 is more than 1. 3/3 equals 1.*

d. Problem C. Did you write an equals sign? (Signal.) *No.*
- So which fraction is more? (Signal.) *101 over 1.*
- How do you know 101 over 1 is more than 100 over 1? (Call on a student. Ideas:) *The denominators are the same. 101 is more than 100. OR One fraction equals 101 and the other equals 100.*

Arrow 6a shows the call-on individual turn that requires an exact answer. This call-on direction specifies that a student read problem A. For call-on individual turns that direct a student to read a word problem or directions, unlike other individual turns, call on students who read well and audibly. Do not follow the normal correction procedure if a student misreads a direction. Read that part correctly, and then direct the student to continue. For all other call-on individual turns that call for an exact response, correct the mistake immediately and include the entire class when you repeat the task.

Arrow 6b shows a form of the call-on individual turn that provides an idea of what students should respond. This call-on direction provides two thoughts that must go together to adequately answer the question, "How do you know 6/5 is more than 3/3?" The thoughts are: 6/5 is more than one; 3/3 is equal to 1. As long as students communicate both those thoughts, regardless in which order, the answer is acceptable. If a student responds, " 3/3 is 1, and 1 is less than 6/5," the answer is acceptable. The student communicated the thought that 6/5 is more than 1. The student also communicated that 3/3 is equal to 1. Acceptable answers should not be corrected and firmed.

Arrow 6c shows a call-on individual turn that contains an **OR.** For this call-on direction, there are two possible acceptable ideas provided. An acceptable response communicates both or one of the ideas. If a student responds, "one of the fractions equals 101 and the denominators are the same," the answer is not correct. This response did not communicate both or one of the ideas. However, if the student responds, "The parts are the same size and 102 is bigger than 101," the response is acceptable. The parts in each unit refer to the denominator, and the student indicated that they were the same size. The student also identified the larger numerator. Other call-on directions that have ORs require exact answers.

Workchecks (Check Your Work.)

The purpose of the workcheck is to give students timely feedback on their work. You will observe and give feedback to many students as they work. Mark correct problems with a C and incorrect problems with an X. It is important for all of the students to correct mistakes on written work, and to do it in a way that allows you to see what mistakes they made.

Students should write their work in pencil so they can erase and make any corrections that are necessary as they work; however, they are not to change their work once it is finished.

Students are to use a colored pen or pencil for checking their work. When you indicate that it is time to "Check your work," they put down their pencils and pick up their marking pens.
If their work is correct, they mark a **C** (or a ✔) for the item. If their work is wrong, they write an **X,** make a line through incorrect work and write the correct answer above, below or next to the crossed out answer using their marking pens. Students hand in their marked-up lined-paper work and Workbooks daily. The errors that students make provide information on what needs to be firmed or repeated.

During workchecks that involve several problems, circulate among students and check their work. Praise students who are fixing mistakes. Allow a reasonable amount of time for them to check each problem.

Do not wait for the slowest students to finish their check. Keep the workcheck moving as quickly as possible.

Independent Work

The goal of the Independent Work is to provide review of previously taught content. The time required for Independent Work ranges from 15–30 minutes, with the time requirements increasing later in the program. Ideally, all Independent Work is completed in class, but not necessarily during the period in which the structured part of each lesson is presented. If it is not practical for students to complete the work at school, it may be assigned as homework. The cautions are that students are not to take their Textbook home. The eTextbook is available on ConnectED for this purpose. Students can access the eTextbook from ConnectED at home after the student seat licenses have been activated. Students are not to tear out Workbook pages because they are not perforated as they were in the lower levels of *CMC*. They need to understand that if they take their Workbook home, they must bring it back the following school day.

Each newly introduced problem type becomes part of the Independent Work after it has appeared several times in structured teacher presentations. Everything that is taught in the program becomes part of the Independent Work. As a general rule, all major problem types that are taught in the program appear at least 10 times in the Independent Work. Some appear as many as 30 or more times. Early material is included in later lessons so that the Independent Work becomes relatively easy for students and provides them with evidence that they are successful.

Unacceptable Error Rates

Students' Independent Work should be monitored, and remedies should be provided for error rates that are too high. As a rule, if more than 30 percent of the students miss more than one or two items in any part of the Independent Work, provide a remedy for that part. The first lesson in which a recently taught skill is independent may have error rates of more than 30 percent of the students. Don't provide a remedy for these situations, but point out to the students that they had trouble with this part and possibly go over the most frequently missed problem. If an excessive error rate continues, provide a systematic correction.

High error rates on independent practice may be the result of the following:

a. The students may not be placed appropriately in the program.

b. The initial presentation may not have been adequately firmed. (The students made mistakes that were not corrected. The parts of the teacher presentation in which errors occurred were not repeated until firm. Skip-box parts were not presented when criteria was not met.)

c. Students may have received inappropriate help. (When they worked structured problems earlier, they received too much help and became dependent on the help.)

d. Students may not have been required to follow directions carefully.

The simplest remedy for unacceptably high error rates on Independent Work is to repeat the structured exercises that occurred immediately before the material became independent. For example, if students have an unacceptable error rate on a particular kind of word problem, go to the last one or two exercises that presented the problem type as a teacher-directed activity. Repeat those exercises until students achieve a high level of mastery. Follow the script closely. Make sure you are not providing a great deal of additional prompting. Then assign the Independent Work for which their error rate was too high. Check to make sure students do not make too many errors.

Grading Papers and Feedback

The teacher material includes a separate Answer Key. The key shows the work for all problems presented during the lesson and as Independent Work. When students are taught a particular method for working problems, they should follow the steps specified in the key. You should make sure students know that the work for a problem is wrong if the procedure is not followed.

After completing each lesson and before presenting the next lesson, follow these steps:

1. Check for excessive error rates for any parts of the written work from the structured part of the lesson. Note parts that have excessive error rates for more than 30% of the students. For instance, if a particular skill has had high error rates for more than two consecutive days, provide a remedy. Reteach as described above.

2. Conduct a workcheck for the Independent Work. One procedure is to provide a structured workcheck of Independent Work at the beginning of the period. Do not attempt to provide students with complete information about each problem. Read the answers. Students are to mark each item as correct or incorrect. The workcheck should not take more than five minutes. Students are to correct errors at a later time and hand in the corrected work. Keep records that show for each lesson whether students handed in corrected work. Attend to these aspects of the student's work:
   a. Were all the mistakes corrected?
   b. Is the appropriate work shown for each correction (not just the right answer)?
   c. Did the student perform acceptably on tasks that tended to be missed by other students? The answer to this question provides you with information on the student's performance on difficult tasks.
3. Award points for Independent Work performance. A good plan is to award one point for completing the Independent Work, one point for correcting all mistakes, and three points for making no more than four errors on the Independent Work. Students who do well can earn five points for each lesson. These points can be used as part of the basis for assigning grades. The Independent Work should be approximately one-third of the grade. The rest of the grade would be based on the Mastery Tests. (The Independent Work would provide students with up to 50 points for a ten-lesson period; each Mastery Test provides another possible 100 points—100% for a perfect test score.) An online progress-monitoring application, *SRA 2Inform,* is available on ConnectED.

Inducing Appropriate Learning Behaviors

Lower Performers

Here are other guidelines for reinforcing appropriate learning behaviors for lower performers who start at Lesson 1.

- During the first 10 lessons, hold students to a high criterion of performance. Remind them that they are to follow the procedures you show them.
- If they do poorly on the first Mastery Test, which follows Lesson 10, provide the specified Remedies (see the following section, In-Program Mastery Tests), then repeat Lessons 9 and 10. Tell students, "We're going to do these lessons again. This time, we'll do them perfectly."

Be positive. Reinforce students for following directions and not making the kinds of mistakes they had been making. Understand, however, that for some students, the relearning required to perform well is substantial, so be patient, but persistent. After students have completed lessons with a high level of success, they will understand your criteria for what they should do to perform acceptably. Retest them and point out that their performance shows that they are learning.

Following Directions

Students who have not gone through earlier levels of *CMC* may have strategies for approaching math that are not appropriate. Most notably, they may be very poor at following directions, even if they understand them. For instance, if you instruct the students to work problem C and then stop, a high percentage of them will not follow this direction.

Throughout *CMC Level E,* you will give students precise directions that may be quite different from those they have encountered earlier in their school experience. The most common problems occur when you direct students to work part of a problem and then stop, work one problem and then stop, or set up a series of problems without solving them.

The simplest remedy is to tell students early in the program that they are to listen to directions and follow them carefully. A good plan is to award points to the group (which means all members of the group earn the same number of points) for following directions. Praise students for attending to directions. If you teach students to follow directions quickly and accurately early in the school year, students will progress much faster later in the program, and you will not have to nag them as much about following directions.

Exercises that are the most difficult for these students, and the most difficult for the teacher to present effectively, are those that have long explanations. An effective procedure is to move fast enough to keep the students attentive.

Go fast on parts that do not require responses from the students; go more slowly and be more emphatic when presenting directions for what the students are to do. If you follow this general guideline, students will attend better to what you say.

Following Lined-Paper Icons

Students are to write answers to Textbook items on their lined paper. The general rules students follow are:

- They never write anything in their Textbook.
- They write their name at the top of the lined paper.
- They follow the layout icons shown in the Textbook.

Here are Textbook icons from different Textbook parts of Lesson 42. These icons show students exactly what to write on their lined paper.

Expect students to require considerable practice to become proficient in reproducing the layouts for the different lined-paper icons. In addition to not lining up columns, some students won't skip lines according to the icon layout. Students often crowd items on a line close together; some students will write illegibly.

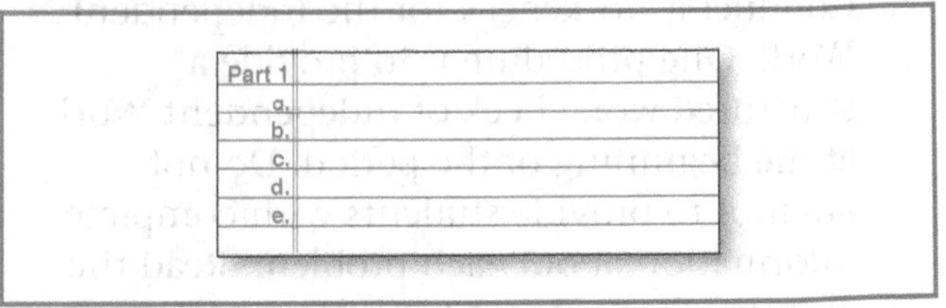

For Part 1, students write Part 1 on a line, and then write the letters **a** through **e** on the lines below.

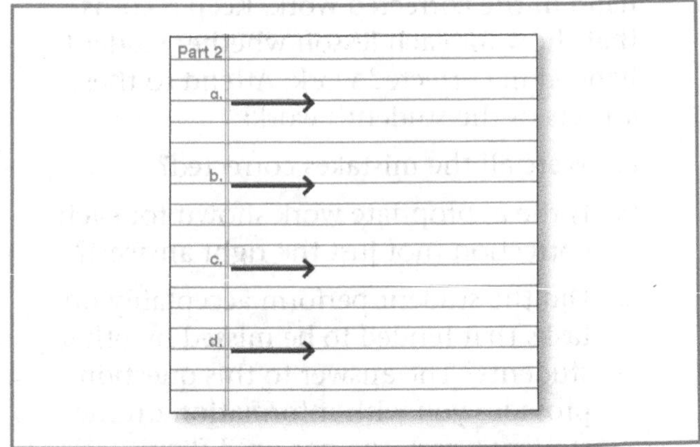

For Part 2, students write the letter **a** and an addition number family arrow two lines below the line on which they write Part 2. They go down four more lines and write the letter **b** with an addition number family arrow.

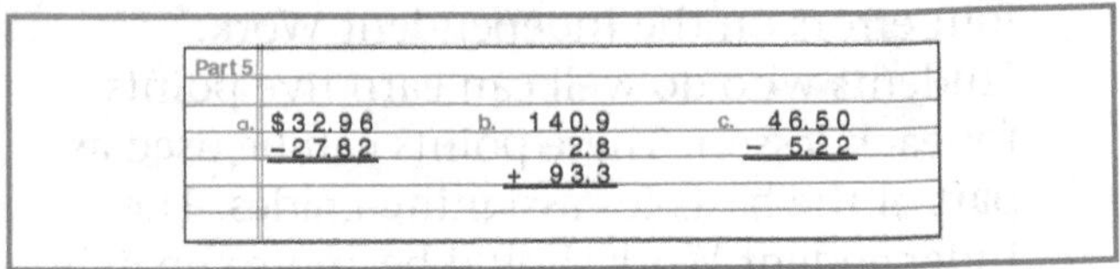

For Part 5, students write Part 5, skip a line, write the letter **a** and copy the problem $32.96 minus $27.82. Next to problem **a,** students write problem **b** and then problem **c.**

Students will tend to make many errors if they don't line up the columns in column problems accurately. Starting with the very first lessons, prompt students to line up their columns. Reinforce students who line them up, and direct students who do not line them up accurately to rewrite the problem with the columns lined up. Part 5 is a good example of a Textbook part students should copy so the columns are lined up.

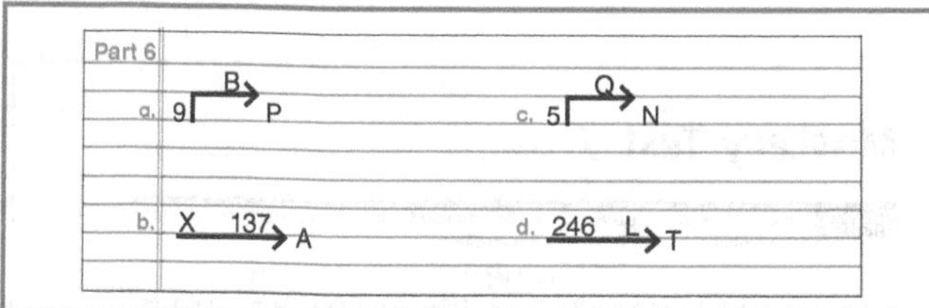

For Part 6, students write Part 6, write the letter **a** two lines below and make the multiplication number family for **a.** Students go down four more lines and make the addition number family for **b.** Students make the multiplication number family for **c** and the addition number family for **d** next to the families for **a** and **b** with sufficient space for students to work a column problem between each pair of families.

Following Solution Steps

An important behavior to address early is that students are to follow the solution steps that are taught. For much of their work, students make a number family, substitute numbers for one or more letters, write the number problem, and write the answer with a unit name. Students who follow these steps are able to solve problems far more reliably than students who do various calculations without properly representing the problem as a number family.

Do not permit shortcuts or working the problem with steps missing. Mark problems with the missing steps as incorrect even if students arrive at the correct answer. At first, the solution conventions may strike some students as being laborious. Tell them early in the sequence that if they learn these steps, they will later avoid many difficulties students have when trying to work problems that involve a lot of steps. Point out to students that they will learn "real math strategies" that will permit them to have far less difficulty with higher math than they would if they were not well versed in solving problems by following the procedures you teach.

Long Lessons

Expect some lessons to run long. Do not hurry to complete a long lesson in one period if it would realistically take more than the period to present and check the material.

Complete the long lesson during the next period, and then start the next lesson during that period and go as far as you can during the allotted time. If you are running long on most of the lessons, the group may not be at mastery, or too much time is being spent on each problem set.

The simplest solution is to select one lesson and repeat it as many as two more times. If the students do not perform rapidly and accurately on all the exercises and work, they are misplaced in the program (which should have been apparent from Mastery Test performance). The best solution is to move the group back to an earlier lesson. Administer the previous two Mastery Tests to determine where students should be placed. Then re-teach the program starting with the lesson for the new placement. Skip exercises that students have mastered.

In-Program Mastery Tests

Connecting Math Concepts Level E provides 13 in-program tests that permit you to assess how well each student is mastering the program content. The tests are packaged in the separate Student Assessment Book. Tests are scheduled to follow every tenth lesson, starting with Lesson 10 and continuing through the end of the program. The primary purpose of the tests is to provide you with information about how well each student is performing on the most recent things that have been taught in the program.

If the information shows that the group did not pass parts of the test, the program provides a specific remedy for each part. The Answer Key shows a passing criterion for each part of the test. A Remedy Table indicates the exercises that are to be repeated before the students are retested. Before presenting the next lesson, provide Remedies for parts students fail.

A good method for minimizing the possibility of students copying from each other is to **maximize** the distance between them when they take the test. Discrepancies in the test performance and daily performance of some students pinpoint which students may be copying.

Below is Test 7, which is scheduled after Lesson 70.

Mastery Test 7 Name ______________

Part 1

a. 9 × 3 =	h. 4 × 5 =	o. 7 × 6 =	w. 7 × 5 =
b. 3 × 8 =	i. 4 × 2 =	p. 7 × 4 =	x. 6 × 6 =
c. 6 × 3 =	j. 1 × 10 =	q. 3 × 4 =	y. 4 × 9 =
d. 7 × 8 =	k. 8 × 4 =	r. 8 × 3 =	z. 2 × 8 =
e. 8 × 6 =	l. 2 × 9 =	s. 7 × 7 =	A. 3 × 7 =
f. 7 × 9 =	m. 8 × 9 =	t. 5 × 3 =	B. 4 × 4 =
g. 6 × 4 =	n. 8 × 8 =	u. 8 × 7 =	C. 9 × 0 =
		v. 6 × 9 =	D. 6 × 2 =

Part 2

a. 8)64	g. 4)0	m. 9)36	s. 6)30	y. 2)6
b. 4)28	h. 7)56	n. 6)24	t. 6)18	z. 6)42
c. 7)63	i. 5)25	o. 6)48	u. 4)32	A. 5)5
d. 10)100	j. 2)18	p. 3)24	v. 6)54	B. 8)24
e. 3)12	k. 5)45	q. 7)42	w. 4)16	C. 8)72
f. 2)14	l. 7)49	r. 3)21	x. 8)56	D. 9)81

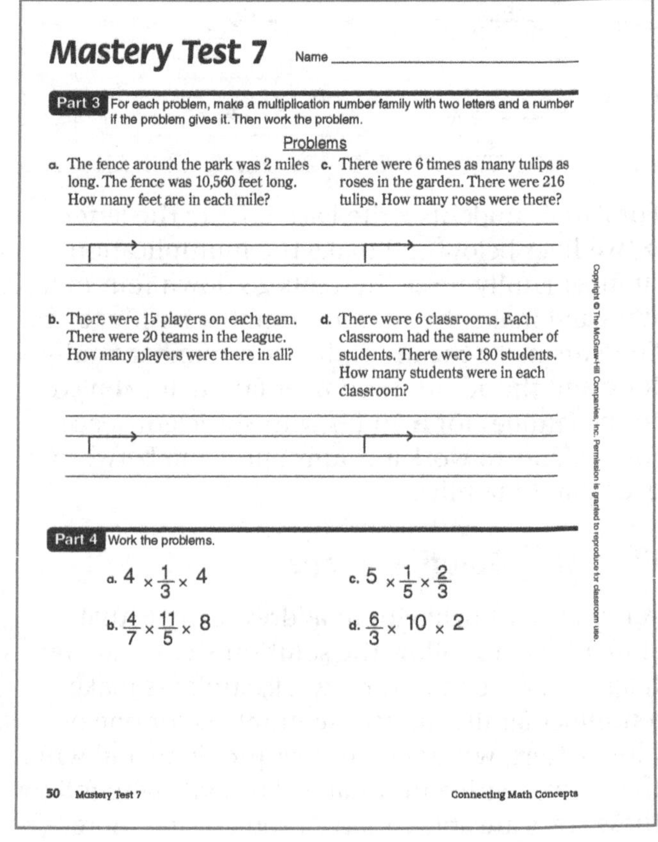

Mastery Test 7 Name ______________

Part 3

For each problem, make a multiplication number family with two letters and a number if the problem gives it. Then work the problem.

Problems

a. The fence around the park was 2 miles long. The fence was 10,560 feet long. How many feet are in each mile?

c. There were 6 times as many tulips as roses in the garden. There were 216 tulips. How many roses were there?

b. There were 15 players on each team. There were 20 teams in the league. How many players were there in all?

d. There were 6 classrooms. Each classroom had the same number of students. There were 180 students. How many students were in each classroom?

Part 4

Work the problems.

a. $4 \times \frac{1}{3} \times 4$

c. $5 \times \frac{1}{5} \times \frac{2}{3}$

b. $\frac{4}{7} \times \frac{11}{5} \times 8$

d. $\frac{6}{3} \times 10 \times 2$

Mastery Test 7 Name ______________

Part 5

The list of angles shows the degrees for the pictures. Write the degrees for each picture.

List
50°
90°
160°
180°
230°
270°

a. ___ b. ___ c. ___ d. ___ e. ___ f. ___

Part 6

Complete the table so each row shows the fraction equation, the division problem and the answer, and the mixed or whole number.

	Fraction	Division	Mixed or Whole Number	
a.	$\frac{240}{8}$			
b.			$10\frac{3}{4}$	
c.		3)172		
d.			$5\frac{8}{10}$	
e.	$\frac{129}{4}$			

Mastery Test 7 Name ______________

Part 7

Work the problems.

a. 306 − 49

b. 104 − 9

c. 501 − 314

d. 105 − 99

Part 8

Write the fraction for each problem.

Descriptions

a. The numerator of fraction A is 30. The denominator is 2.

b. The numbers for fraction B are 8 and 15. The numerator is smaller than the denominator.

c. The numerator of a fraction is 6 times the denominator. The denominator is 4.

d. There are 8 parts in each unit. 12 parts are shaded.

e. The numbers for a fraction are 17 and 18. The denominator is larger than the numerator.

f. A fraction is less than 1. The numbers for the fraction are 4 and 9.

Scoring the Test and Providing Remedies

The Answer Key for each Mastery Test provides the correct answers and shows the work for each item.

Tables that accompany each Mastery Test show the passing score for each part and indicate the percentages for different total test scores.

Here are the tables for Mastery Test 7.

Passing Criteria Table — Mastery Test 7			
Part	Score	Possible Score	Passing Score
1	1 for every 5 problems	6	5
2	1 for every 5 problems	6	5
3	1 for each family 1 for each problem and answer	8	6
4	1 for each problem	4	3
5	1 for each angle	6	6
6	1 for every row	5	4
7	2 for each problem	8	6
8	1 for each fraction	6	6
	Total	49	

Test 7 Percent Summary					
Score	%	Score	%	Score	%
49	100	43	88	37	76
48	98	42	86	36	73
47	96	41	84	35	71
46	94	40	82		
45	92	39	80		
44	90	38	78		

The Passing Criteria Table gives the possible points for each item, the possible points for the part, and the passing criterion.

Students fail a part of the test if they score fewer than the specified number of passing points. For example, Part 4 has 4 possible points. A passing score is 3. If a student scores 3 or 4, the student passes the part. A student with a score of less than 3 fails the part.

Note that points are sometimes awarded for working different parts of the problem. For example, in Part 3 students can earn 2 points for each item—1 point for the family and 1 point for the problem and answer.

The scoring sheet indicates for Parts 1 and 2 students earn 1 point for every 5 problems. On other Mastery Test parts, the scoring sheets could indicate that students earn 1 point for every 4 problems, 1 point for every 3 problems, and 1 point for every 2 problems. The scoring for these parts works the same way. The number of points awarded equals the whole number part of:

$$\frac{\text{(the number of correct answers)}}{\text{(the problems per points)}}$$

If students answer 22 of the facts correctly, students earn 4 points.

$$\frac{22}{5} = 4\frac{2}{5}$$

22 divided by 5 equals 4 and 2/5. The whole number part of the answer is 4, so that's how many points the students earn. The passing criterion for the part is 5 points. So students earn 5 points and pass Parts 1 and 2 if they answer 25 or more of the 30 facts in each part correctly. If students answer fewer than 25 of the facts correctly in one of the parts, they do not pass.

Provide remedies for each of the parts students fail. If less than 25% of your students fail a part, it may be easier to present the remedy to just the students who need the remedy while other students are doing Independent Work. If 25% or more of your students miss a part, present the remedy to the entire class.

After students have performed acceptably on the remedies for the parts they failed, re-test students. Students only need to be tested on parts they failed.

The Percent Summary table shows the percentage grade you'd award students who have a perfect score of 50, a score of 49, and so forth.

Record each student's performance on the Group Summary of Mastery Test Performance (provided on pages 203–208 of this Teacher's Guide.) The Group Summary accommodates up to 30 students. The sample below shows only 6 students.

Here's how the results could be summarized following Mastery Test 7:

Remedy Summary—Group Summary of Test Performance

Note: Test remedies are specified in the Answer Key. Percent Summary is also specified in the Answer Key.

Name	Test 7								
	Check parts not passed								Total %
	1	2	3	4	5	6	7	8	
1. Amanda			✓		✓		✓		
2. Karen			✓						
3. Adam	✓		✓			✓	✓	✓	
4. Chan									
5. Felipe	✓								
6. Jack			✓				✓		
Number of students Not Passed = NP	2	0	4	0	1	1	3	1	
Total number of students = T	6	6	6	6	6	6	6	6	
Remedy needed if NP/T = 25% or more	Y	N	Y	N	N	N	Y	N	

The summary sheet provides you with a cumulative record of each student's performance on the in-program Mastery Tests.

Summarize each student's performance by making a check mark for each part failed.

At the bottom of each column, write the total number of failures for that part and the total number of students in the class. Then divide the number of failures by the number of students to determine the failure rate for each part.

Provide a group remedy for each part that has a failure rate of more than 25% (.25).

Test Remedies

The Answer Key specifies Remedies for each test. Any necessary Remedies should be presented before the next lesson (Lesson 71.)

Here are the remedies for Mastery Test 7:

Remedy Table — Mastery Test 7

Part	Test Items	Remedy		Remedies Worksheet	Textbook
		Lesson	Exercise		
1	Multiplication Facts	66	1 (through WB step e)	Part A	—
		66	3	Part B	—
		67	1 (through WB step e)	Part C	—
		67	3	Part D	—
2	Division Facts	66	1	Parts A and E	—
		67	1	Parts C and F	—
		68	1	Parts G and H	—
3	Word Problems	66	5	—	Part 2
		67	4	—	Part 1
		68	6	—	Part 2
		69	5	—	Part 2
4	Fractions—Multiplication by Whole Numbers	64	7	—	Part 3
		65	7	—	Part 3
		68	8	—	Part 4
		69	7	—	Part 4
5	Angles	63	6	—	Part 2
		63	9	—	Part 6
		65	6	—	Part 2
		68	2	Part I	—
6	Fractions—Division, Mixed Numbers	61	2	Part J	—
		63	3	Part K	—
		64	2	Part L	—
7	Column Subtraction —Borrowing from Zero	66	4	—	Part 1
		68	7	—	Part 3
		69	8	—	Part 5
8	Fractions—Numerators	67	5	—	—
		68	5	—	Part 1
		69	4	—	Part 1

Retest

Retest individual students on any part failed.

If the same students frequently fail parts of the Mastery Tests, those students must master the skills that are taught or they will have trouble later in the program when those skills become a component in larger operations or more complex applications.

If students consistently fail tests, they are probably not placed appropriately in the program.

On the completed Group Summary of Test Performance for Mastery Test 7 on page 34, more than one quarter of the students failed Parts 1, 3, and 7. You provide group Remedies by re-teaching the exercises specified as the remedy for those parts. You may also need to provide individual Remedies for Amanda and Adam because they failed additional parts.

For Part 1, you'd re-teach to the group Lesson 66, Exercise 1 through step E, Lesson 66, Exercise 3, Lesson 67, Exercise 1 through step E and Lesson 67, Exercise 3. For Part 3, you'd re-teach Lesson 66, Exercise 5, Lesson 67, Exercise 4, Lesson 68, Exercise 6 and Lesson 69, Exercise 5. For Part 7, you'd re-teach Lesson 66, Exercise 4, Lesson 68, Exercise 7 and Lesson 69, Exercise 8.

A good plan for presenting the Remedies to the class is to present them in the order that they appear in the program: Lesson 66, Exercise 1; Lesson 66, Exercise 3; Lesson 66, Exercise 4; Lesson 66, Exercise 5; Lesson 67, Exercise 1; Lesson 67, Exercise 3; Lesson 67, Exercise 4; Lesson 68, Exercise 6; Lesson 68, Exercise 7; Lesson 69, Exercise 5; Lesson 66, Exercise 8. With this plan, students receive practice on the parts that is sequential and somewhat distributed between the parts: Part 1; Part 1; Part 7; Part 3; Part 1; Part 1; Part 3; Part 3; Part 7; Part 3; Part 7.

It may not be possible or practical to follow all the steps indicated for correcting individual students who make too many mistakes. For these cases, provide test Remedies to the *entire class* and move on in the program, attending to those students who make chronic mistakes without significantly slowing the group's progress.

Remedy Worksheets

All the Workbook parts needed for remedies appear in the Student Assessment Booklet immediately after each test. Below are remedy parts for Test 7.

Remedies

Name ____________________

Part A

a. 6 —3→ ___	d. 7 ⟹ 56	g. 6 ⟹ 24	j. 7 ⟹ 49
b. ___ —8→ 64	e. 6 —8→ ___	h. 7 —8→ ___	k. ___ —5→ 20
c. 3 —8→ ___	f. ___ —8→ 56	i. 8 ⟹ 32	l. 6 —7→ ___

Part B

a. 9 × 7 =	g. 6 × 7 =	m. 3 × 8 =	s. 9 × 9 =	y. 8 × 1 =
b. 3 × 4 =	h. 8 × 8 =	n. 4 × 6 =	t. 4 × 7 =	z. 5 × 6 =
c. 2 × 2 =	i. 10 × 10 =	o. 6 × 8 =	u. 8 × 3 =	A. 7 × 3 =
d. 7 × 8 =	j. 3 × 0 =	p. 3 × 7 =	v. 7 × 7 =	B. 6 × 6 =
e. 9 × 6 =	k. 8 × 4 =	q. 8 × 7 =	w. 6 × 3 =	C. 9 × 3 =
f. 8 × 6 =	l. 4 × 9 =	r. 9 × 8 =	x. 10 × 9 =	D. 4 × 8 =

Part C

a. ___ —8→ 48	d. ___ —8→ 24	g. 6 ⟹ 48	j. ___ —7→ 21
b. ___ —6→ 24	e. 3 —9→ ___	h. 8 ⟹ 64	k. 8 —4→ ___
c. 8 —7→ ___	f. ___ —7→ 42	i. 3 —6→ ___	l. ___ —9→ 81

Part D

a. 1 × 10 =	g. 8 × 8 =	m. 7 × 7 =	s. 8 × 9 =	y. 8 × 7 =
b. 3 × 8 =	h. 8 × 6 =	n. 6 × 4 =	t. 3 × 7 =	z. 2 × 8 =
c. 2 × 9 =	i. 3 × 4 =	o. 8 × 4 =	u. 4 × 9 =	A. 5 × 3 =
d. 7 × 8 =	j. 9 × 3 =	p. 6 × 9 =	v. 7 × 4 =	B. 4 × 4 =
e. 4 × 5 =	k. 7 × 6 =	q. 4 × 2 =	w. 6 × 2 =	C. 9 × 0 =
f. 7 × 9 =	l. 6 × 3 =	r. 8 × 3 =	x. 6 × 6 =	D. 7 × 5 =

Remedies

Remedies CONTINUED

Name ____________________

Part E

a. 8⟌56 g. 6⟌48 m. 3⟌18 s. 4⟌12 y. 9⟌45
b. 4⟌28 h. 12⟌12 n. 7⟌49 t. 7⟌56 z. 8⟌8
c. 6⟌30 i. 6⟌54 o. 8⟌72 u. 4⟌24 A. 3⟌21
d. 7⟌42 j. 9⟌36 p. 10⟌0 v. 8⟌16 B. 5⟌15
e. 5⟌35 k. 3⟌24 q. 7⟌21 w. 8⟌72 C. 7⟌70
f. 1⟌9 l. 8⟌64 r. 3⟌27 x. 8⟌32 D. 7⟌56

Part F

a. 8⟌64 g. 6⟌24 m. 6⟌54 s. 6⟌48 y. 6⟌18
b. 2⟌18 h. 7⟌63 n. 4⟌0 t. 2⟌6 z. 6⟌42
c. 7⟌56 i. 7⟌42 o. 6⟌30 u. 4⟌32 A. 4⟌16
d. 5⟌45 j. 4⟌28 p. 3⟌24 v. 9⟌36 B. 8⟌24
e. 3⟌12 k. 10⟌100 q. 5⟌25 w. 5⟌5 C. 8⟌56
f. 7⟌49 l. 2⟌14 r. 3⟌21 x. 8⟌72 D. 9⟌81

Part G

a. 7 ⟶ 49 d. 6 ⟶ 42 g. ___ 8 ⟶ 56 j. ___ 6 ⟶ 36
b. 4 8 ⟶ ___ e. 8 8 ⟶ ___ h. 3 ⟶ 27 k. ___ 7 ⟶ 21
c. 2 ⟶ 18 f. 6 ⟶ 48 i. 7 9 ⟶ ___ l. 3 6 ⟶ ___

Remedies

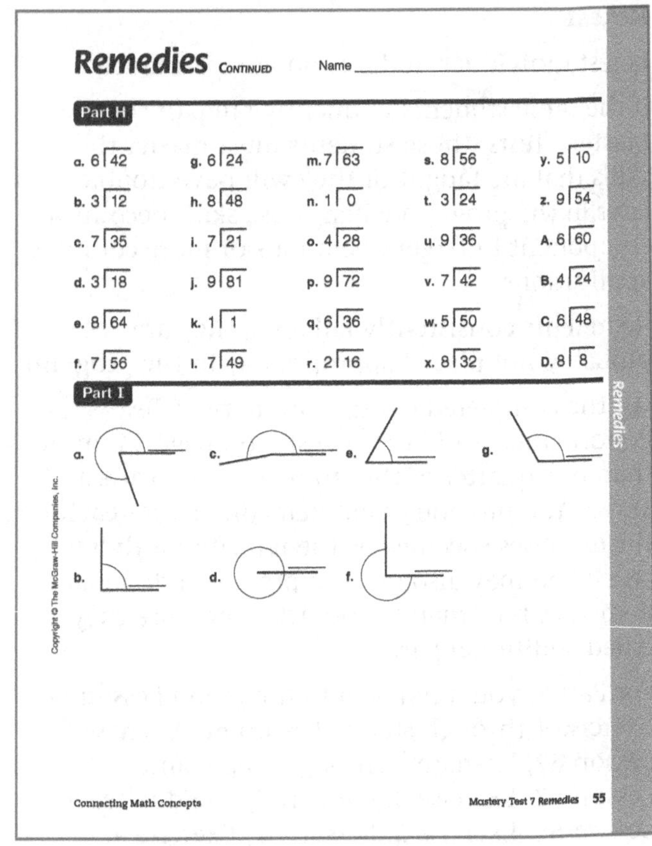

Remedies CONTINUED

Name ____________________

Part H

a. 6⟌42 g. 6⟌24 m. 7⟌63 s. 8⟌56 y. 5⟌10
b. 3⟌12 h. 8⟌48 n. 1⟌0 t. 3⟌24 z. 9⟌54
c. 7⟌35 i. 7⟌21 o. 4⟌28 u. 9⟌36 A. 6⟌60
d. 3⟌18 j. 9⟌81 p. 9⟌72 v. 7⟌42 B. 4⟌24
e. 8⟌64 k. 1⟌1 q. 6⟌36 w. 5⟌50 C. 6⟌48
f. 7⟌56 l. 7⟌49 r. 2⟌16 x. 8⟌32 D. 8⟌8

Part I

Remedies

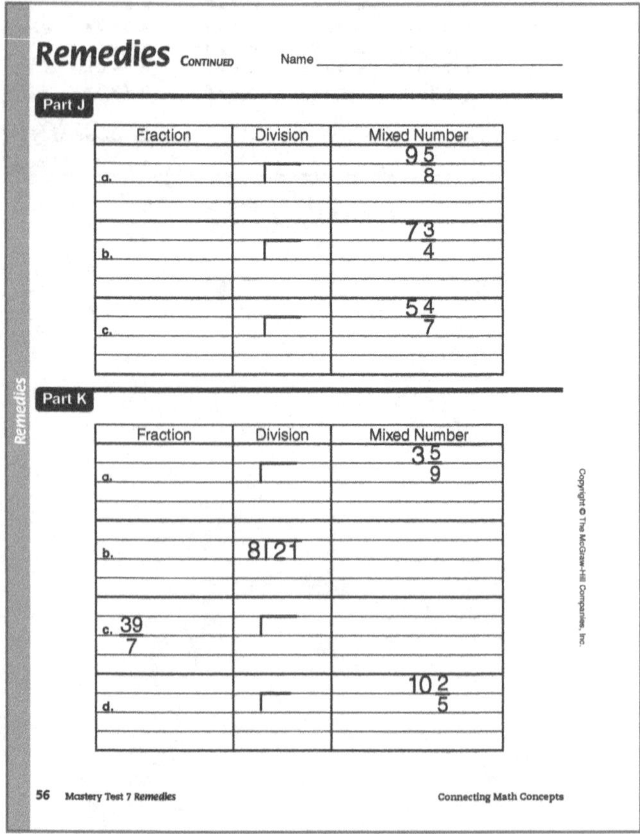

Remedies CONTINUED

Name ____________________

Part J

	Fraction	Division	Mixed Number
a.		⟌	$9\frac{5}{8}$
b.		⟌	$7\frac{3}{4}$
c.		⟌	$5\frac{4}{7}$

Part K

	Fraction	Division	Mixed Number
a.		⟌	$3\frac{5}{9}$
b.		8⟌21	
c.	$\frac{39}{7}$	⟌	
d.		⟌	$10\frac{2}{5}$

Remedies

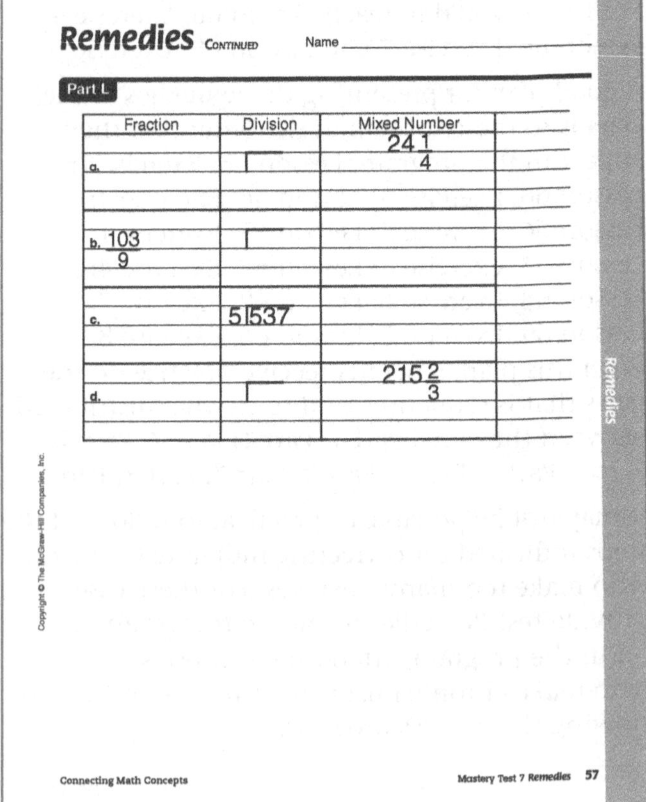

Remedies CONTINUED

Name ____________________

Part L

	Fraction	Division	Mixed Number
a.		⟌	$24\frac{1}{4}$
b.	$\frac{103}{9}$	⟌	
c.		5⟌537	
d.		⟌	$215\frac{2}{3}$

Remedies

For Textbook remedies, students use their Textbook, not the Student Assessment Book.

Exercises that are specified as remedies for a Mastery Test have a **REMEDY** icon next to the exercise heading.

Remedy worksheet parts are also identified with an icon, e.g. **R Part B**.

Cumulative Tests

Level E has two Cumulative Tests. Cumulative Test 1 appears after Mastery Test 7 in the Student Assessment Book. Cumulative Test 2 appears after Mastery Test 13, at the end of the program.

Note that you don't have to present the Cumulative Test immediately after students complete Mastery Test 7. You may present this test any time before students reach Lesson 75 in the program.

These tests sample critical skills and concepts that have been taught earlier. Cumulative Test 1 assesses content from the beginning of the program through Lesson 70.

Cumulative Test 2 samples critical items from the beginning of the level, but it is weighted for the lesson range from 70 to the end of the program. The presentation scripts for the Cumulative Tests appear at the end of each Presentation Book.

These Cumulative Tests may require more than one period to administer, but try to minimize, as much as possible, interrupting the schedule for teaching lessons.

Test remedies are not specified; however, the test scoring chart shows corresponding Mastery Test parts that may be used to identify possible remedies.

The Answer Keys for Cumulative Tests 1 and 2 follow.

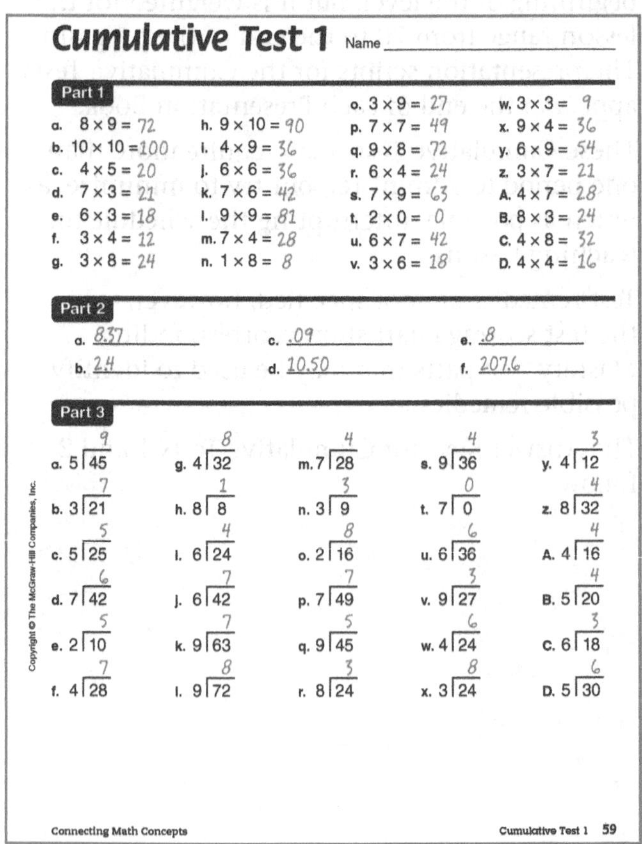

Cumulative Test 1 Name ________

Part 1

a. 8 × 9 = 72
b. 10 × 10 = 100
c. 4 × 5 = 20
d. 7 × 3 = 21
e. 6 × 3 = 18
f. 3 × 4 = 12
g. 3 × 8 = 24
h. 9 × 10 = 90
i. 4 × 9 = 36
j. 6 × 6 = 36
k. 7 × 6 = 42
l. 9 × 9 = 81
m. 7 × 4 = 28
n. 1 × 8 = 8
o. 3 × 9 = 27
p. 7 × 7 = 49
q. 9 × 8 = 72
r. 6 × 4 = 24
s. 7 × 9 = 63
t. 2 × 0 = 0
u. 6 × 7 = 42
v. 3 × 6 = 18
w. 3 × 3 = 9
x. 9 × 4 = 36
y. 6 × 9 = 54
z. 3 × 7 = 21
A. 4 × 7 = 28
B. 8 × 3 = 24
C. 4 × 8 = 32
D. 4 × 4 = 16

Part 2

a. 8.37 b. 24 c. .09 d. 10.50 e. .8 f. 207.6

Part 3

a. 5⟌45 = 9
b. 3⟌21 = 7
c. 5⟌25 = 5
d. 7⟌42 = 6
e. 2⟌10 = 5
f. 4⟌28 = 7
g. 4⟌32 = 8
h. 8⟌8 = 1
i. 6⟌24 = 4
j. 6⟌42 = 7
k. 9⟌63 = 7
l. 9⟌72 = 8
m. 7⟌28 = 4
n. 3⟌9 = 3
o. 2⟌16 = 8
p. 7⟌49 = 7
q. 9⟌45 = 5
r. 8⟌24 = 3
s. 9⟌36 = 4
t. 7⟌0 = 0
u. 6⟌36 = 6
v. 9⟌27 = 3
w. 4⟌24 = 6
x. 3⟌24 = 8
y. 4⟌12 = 3
z. 8⟌32 = 4
A. 4⟌16 = 4
B. 5⟌20 = 4
C. 6⟌18 = 3
D. 5⟌30 = 6

Copyright © The McGraw-Hill Companies, Inc.

Connecting Math Concepts Cumulative Test 1 59

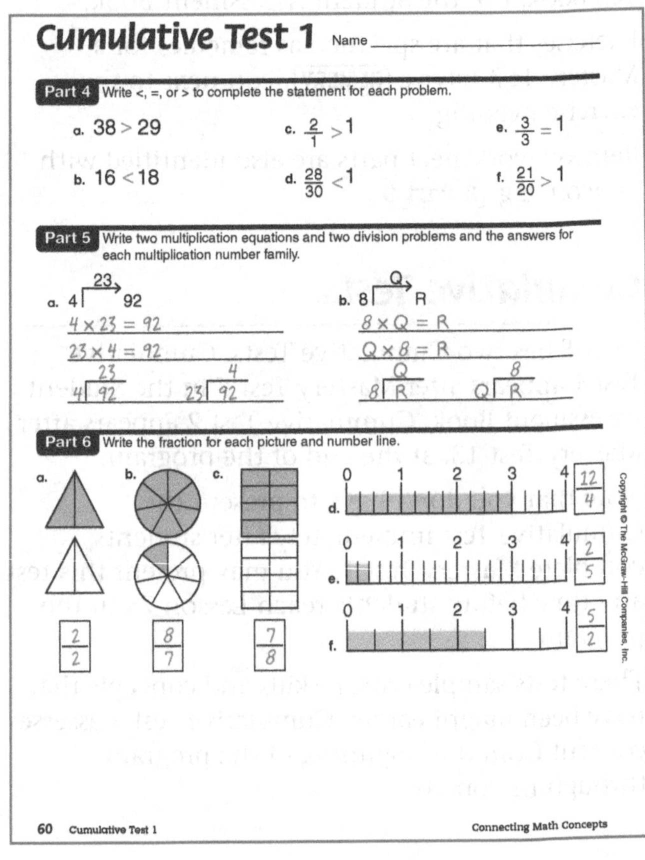

Cumulative Test 1 Name ________

Part 4 Write <, =, or > to complete the statement for each problem.

a. 38 > 29
b. 16 < 18
c. $\frac{2}{1} > 1$
d. $\frac{28}{30} < 1$
e. $\frac{3}{3} = 1$
f. $\frac{21}{20} > 1$

Part 5 Write two multiplication equations and two division problems and the answers for each multiplication number family.

a. 4 ⟶ 23, 92
4 × 23 = 92
23 × 4 = 92
4⟌92 = 23
23⟌92 = 4

b. 8 ⟶ Q, R
8 × Q = R
Q × 8 = R
8⟌R = Q
Q⟌R = 8

Part 6 Write the fraction for each picture and number line.

a. $\frac{2}{2}$
b. $\frac{8}{7}$
c. $\frac{7}{8}$
d. $\frac{12}{4}$
e. $\frac{2}{5}$
f. $\frac{5}{2}$

Copyright © The McGraw-Hill Companies, Inc.

60 Cumulative Test 1 Connecting Math Concepts

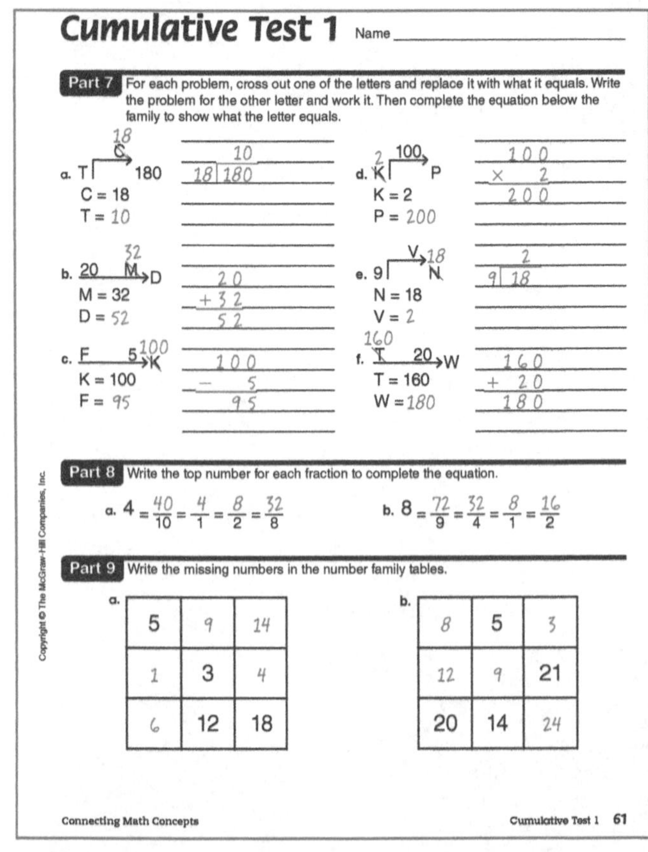

Cumulative Test 1 Name ________

Part 7 For each problem, cross out one of the letters and replace it with what it equals. Write the problem for the other letter and work it. Then complete the equation below the family to show what the letter equals.

a. T ⟶ C, 180
C = 18
T = 10
18⟌180 = 10

b. 20 ⟶ M, D
M = 32
D = 52
20 + 32 = 52

c. F ⟶ 5, K
K = 100
F = 95
100 − 5 = 95

d. K ⟶ 100, P
K = 2
P = 200
100 × 2 = 200

e. 9 ⟶ V, N
N = 18
V = 2
9⟌18 = 2

f. T ⟶ 20, W
T = 160
W = 180
160 + 20 = 180

Part 8 Write the top number for each fraction to complete the equation.

a. $4 = \frac{40}{10} = \frac{4}{1} = \frac{8}{2} = \frac{32}{8}$

b. $8 = \frac{72}{9} = \frac{32}{4} = \frac{8}{1} = \frac{16}{2}$

Part 9 Write the missing numbers in the number family tables.

a.

5	9	14
1	3	4
6	12	18

b.

8	5	3
12	9	21
20	14	24

Copyright © The McGraw-Hill Companies, Inc.

Connecting Math Concepts Cumulative Test 1 61

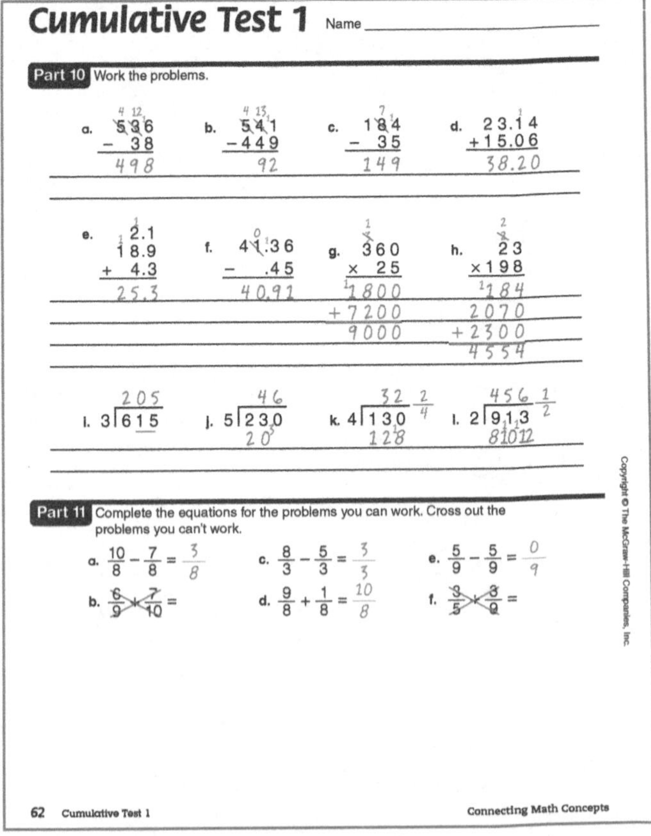

Cumulative Test 1 Name ________

Part 10 Work the problems.

a. 536 − 38 = 498
b. 541 − 449 = 92
c. 184 − 35 = 149
d. 23.14 + 15.06 = 38.20
e. 2.1 + 18.9 + 4.3 = 25.3
f. 41.36 − .45 = 40.91
g. 360 × 25 = 1800 + 7200 = 9000
h. 23 × 198 = 184 + 2070 + 2300 = 4554
i. 3⟌615 = 205
j. 5⟌230 = 46
k. 4⟌130 = 32 $\frac{2}{4}$
l. 2⟌913 = 456 $\frac{1}{2}$

Part 11 Complete the equations for the problems you can work. Cross out the problems you can't work.

a. $\frac{10}{8} - \frac{7}{8} = \frac{3}{8}$
b. $\frac{6}{9} \times \frac{7}{10} =$ (crossed out)
c. $\frac{8}{3} - \frac{5}{3} = \frac{3}{3}$
d. $\frac{9}{8} + \frac{1}{8} = \frac{10}{8}$
e. $\frac{5}{9} - \frac{5}{9} = \frac{0}{9}$
f. $\frac{3}{5} \times \frac{3}{9} =$ (crossed out)

Copyright © The McGraw-Hill Companies, Inc.

62 Cumulative Test 1 Connecting Math Concepts

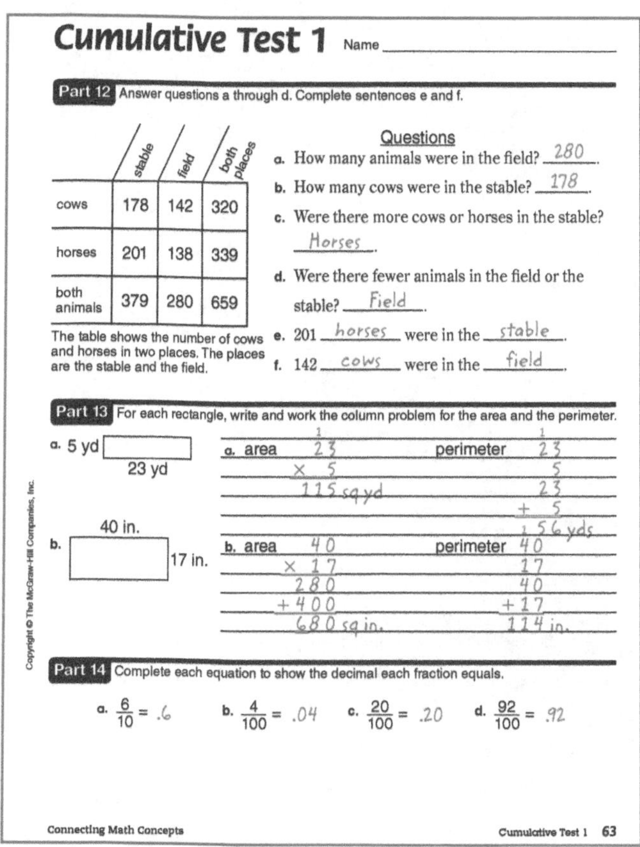

Cumulative Test 1

Name ____________

Part 12 Answer questions a through d. Complete sentences e and f.

	stable	field	both places
cows	178	142	320
horses	201	138	339
both animals	379	280	659

The table shows the number of cows and horses in two places. The places are the stable and the field.

Questions

a. How many animals were in the field? 280.

b. How many cows were in the stable? 178.

c. Were there more cows or horses in the stable? Horses.

d. Were there fewer animals in the field or the stable? Field.

e. 201 horses were in the stable.

f. 142 cows were in the field.

Part 13 For each rectangle, write and work the column problem for the area and the perimeter.

a. 5 yd, 23 yd

a. area: 23 × 5 = 115 sq yd

perimeter: 23 + 5 + 23 + 5 = 56 yds

b. 40 in., 17 in.

b. area: 40 × 17; 280 + 400 = 680 sq in.

perimeter: 40 + 17 + 40 + 17 = 114 in.

Part 14 Complete each equation to show the decimal each fraction equals.

a. $\frac{6}{10} = .6$ b. $\frac{4}{100} = .04$ c. $\frac{20}{100} = .20$ d. $\frac{92}{100} = .92$

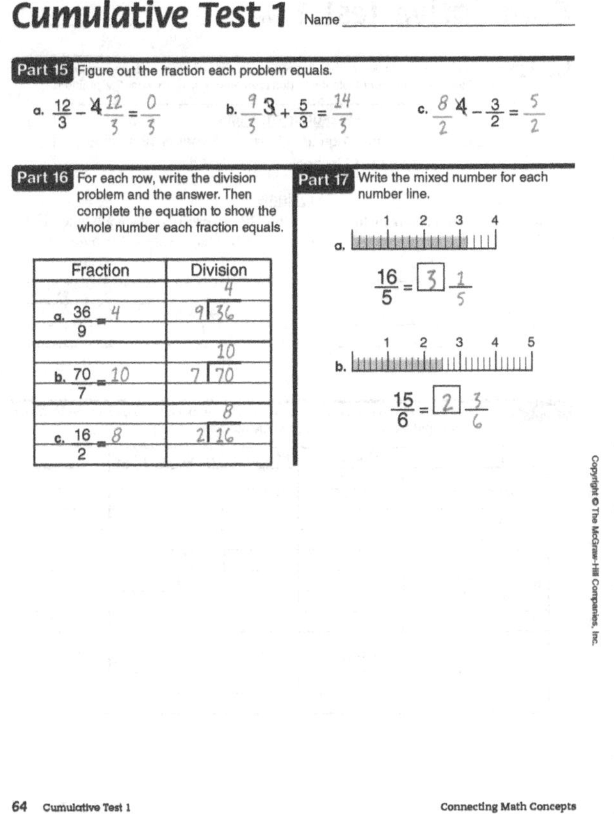

Cumulative Test 1

Name ____________

Part 15 Figure out the fraction each problem equals.

a. $\frac{12}{3} - 4\,\frac{12}{3} = \frac{0}{3}$ b. $\frac{9}{3}\,3 + \frac{5}{3} = \frac{14}{3}$ c. $\frac{8}{2}\,4 - \frac{3}{2} = \frac{5}{2}$

Part 16 For each row, write the division problem and the answer. Then complete the equation to show the whole number each fraction equals.

Fraction	Division
a. $\frac{36}{9} = 4$	9) 36 → 4
b. $\frac{70}{7} = 10$	7) 70 → 10
c. $\frac{16}{2} = 8$	2) 16 → 8

Part 17 Write the mixed number for each number line.

a. 1 2 3 4

$\frac{16}{5} = 3\frac{1}{5}$

b. 1 2 3 4 5

$\frac{15}{6} = 2\frac{3}{6}$

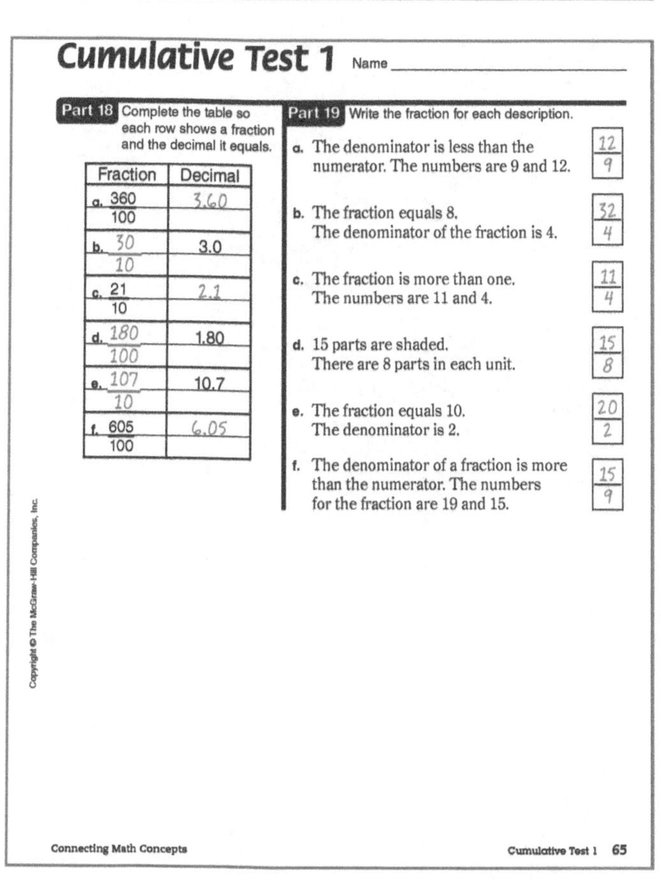

Cumulative Test 1

Name ____________

Part 18 Complete the table so each row shows a fraction and the decimal it equals.

Fraction	Decimal
a. $\frac{360}{100}$	3.60
b. $\frac{30}{10}$	3.0
c. $\frac{21}{10}$	2.1
d. $\frac{180}{100}$	1.80
e. $\frac{107}{10}$	10.7
f. $\frac{605}{100}$	6.05

Part 19 Write the fraction for each description.

a. The denominator is less than the numerator. The numbers are 9 and 12. $\frac{12}{9}$

b. The fraction equals 8. The denominator of the fraction is 4. $\frac{32}{4}$

c. The fraction is more than one. The numbers are 11 and 4. $\frac{11}{4}$

d. 15 parts are shaded. There are 8 parts in each unit. $\frac{15}{8}$

e. The fraction equals 10. The denominator is 2. $\frac{20}{2}$

f. The denominator of a fraction is more than the numerator. The numbers for the fraction are 19 and 15. $\frac{15}{9}$

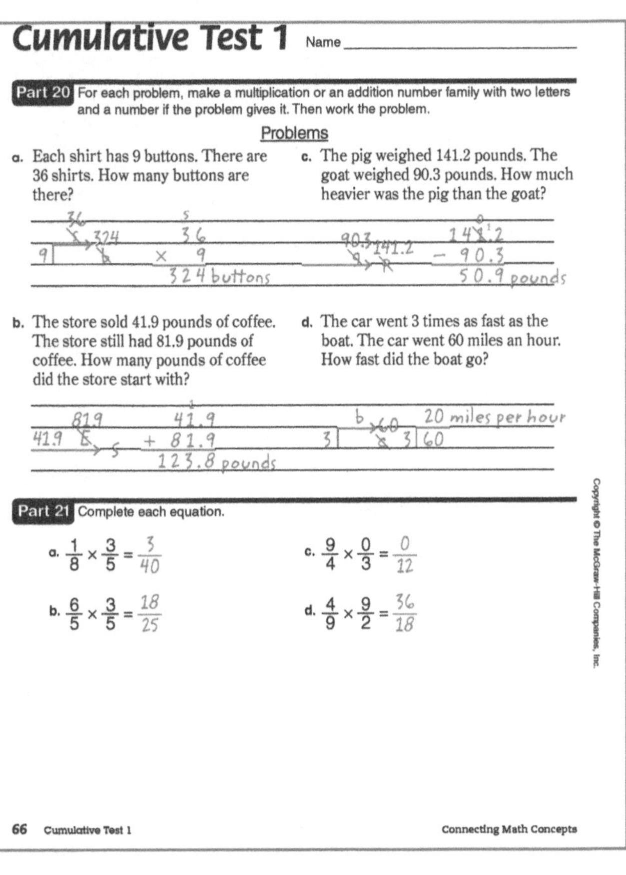

Cumulative Test 1

Name ____________

Part 20 For each problem, make a multiplication or an addition number family with two letters and a number if the problem gives it. Then work the problem.

Problems

a. Each shirt has 9 buttons. There are 36 shirts. How many buttons are there?

36 × 9 = 324 buttons

b. The store sold 41.9 pounds of coffee. The store still had 81.9 pounds of coffee. How many pounds of coffee did the store start with?

41.9 + 81.9 = 123.8 pounds

c. The pig weighed 141.2 pounds. The goat weighed 90.3 pounds. How much heavier was the pig than the goat?

141.2 − 90.3 = 50.9 pounds

d. The car went 3 times as fast as the boat. The car went 60 miles an hour. How fast did the boat go?

3) 60 → 20 miles per hour

Part 21 Complete each equation.

a. $\frac{1}{8} \times \frac{3}{5} = \frac{3}{40}$

b. $\frac{6}{5} \times \frac{3}{5} = \frac{18}{25}$

c. $\frac{9}{4} \times \frac{0}{3} = \frac{0}{12}$

d. $\frac{4}{9} \times \frac{9}{2} = \frac{36}{18}$

Cumulative Test 1 Name ____________

Part 22 For each problem, make a multiplication family with two letters and a number. Use one of the measurement facts if you don't remember it. Then work the problem.

Measurement Facts

1 gallon is 4 quarts. 1 quart is 2 pints. 1 year is 12 months.
1 foot is 12 inches. 1 week is 7 days.

Problems

a. Jane worked for a company for 12 years. How many months did Jane work for the company?

12 → M 12 Y

12 × 12 = 24 + 120 = 144 months

b. There were 16 pints in a bucket. How many quarts were in the bucket?

2 → q 16 R

8 quarts; 2 | 16

Part 23 Complete the table so each row shows the fraction equation, the division problem and the answer, and the mixed number or whole number.

Fraction	Division	Mixed or Whole Number
a. $\frac{255}{5} = 51$	51; 5 | 255; 25	51
b. $\frac{123}{10} = 12\frac{3}{10}$	12 3; 10 | 123 10	$12\frac{3}{10}$; $\frac{120}{10} + \frac{3}{10} = \frac{123}{10}$
c. $\frac{72}{2} = 36$	36; 2 | 72; 6	36

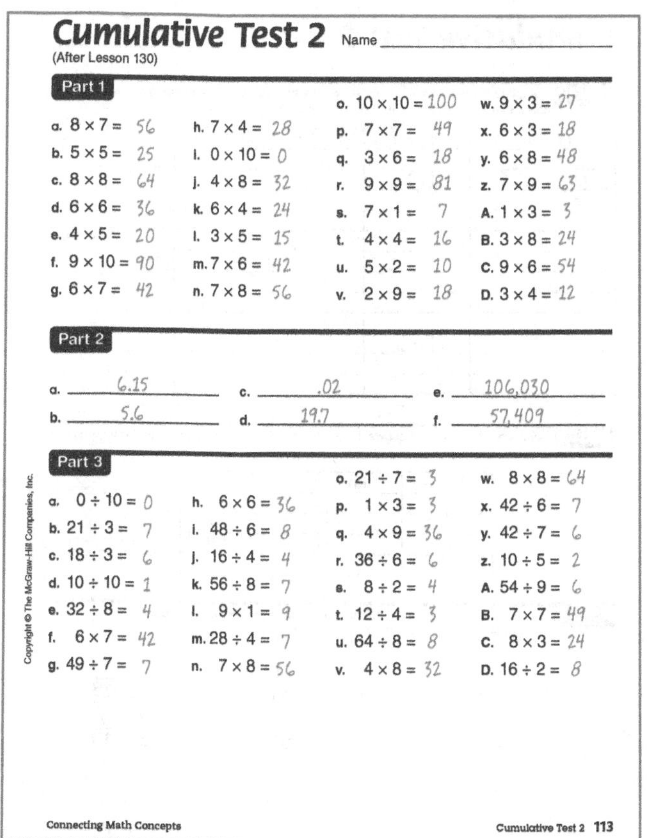

Cumulative Test 2

Name ________

(After Lesson 130)

Part 1

a. 8 × 7 = 56
b. 5 × 5 = 25
c. 8 × 8 = 64
d. 6 × 6 = 36
e. 4 × 5 = 20
f. 9 × 10 = 90
g. 6 × 7 = 42
h. 7 × 4 = 28
i. 0 × 10 = 0
j. 4 × 8 = 32
k. 6 × 4 = 24
l. 3 × 5 = 15
m. 7 × 6 = 42
n. 7 × 8 = 56
o. 10 × 10 = 100
p. 7 × 7 = 49
q. 3 × 6 = 18
r. 9 × 9 = 81
s. 7 × 1 = 7
t. 4 × 4 = 16
u. 5 × 2 = 10
v. 2 × 9 = 18
w. 9 × 3 = 27
x. 6 × 3 = 18
y. 6 × 8 = 48
z. 7 × 9 = 63
A. 1 × 3 = 3
B. 3 × 8 = 24
C. 9 × 6 = 54
D. 3 × 4 = 12

Part 2

a. 6.15
b. 5.6
c. .02
d. 19.7
e. 106,030
f. 57,409

Part 3

a. 0 ÷ 10 = 0
b. 21 ÷ 3 = 7
c. 18 ÷ 3 = 6
d. 10 ÷ 10 = 1
e. 32 ÷ 8 = 4
f. 6 × 7 = 42
g. 49 ÷ 7 = 7
h. 6 × 6 = 36
i. 48 ÷ 6 = 8
j. 16 ÷ 4 = 4
k. 56 ÷ 8 = 7
l. 9 × 1 = 9
m. 28 ÷ 4 = 7
n. 7 × 8 = 56
o. 21 ÷ 7 = 3
p. 1 × 3 = 3
q. 4 × 9 = 36
r. 36 ÷ 6 = 6
s. 8 ÷ 2 = 4
t. 12 ÷ 4 = 3
u. 64 ÷ 8 = 8
v. 4 × 8 = 32
w. 8 × 8 = 64
x. 42 ÷ 6 = 7
y. 42 ÷ 7 = 6
z. 10 ÷ 5 = 2
A. 54 ÷ 9 = 6
B. 7 × 7 = 49
C. 8 × 3 = 24
D. 16 ÷ 2 = 8

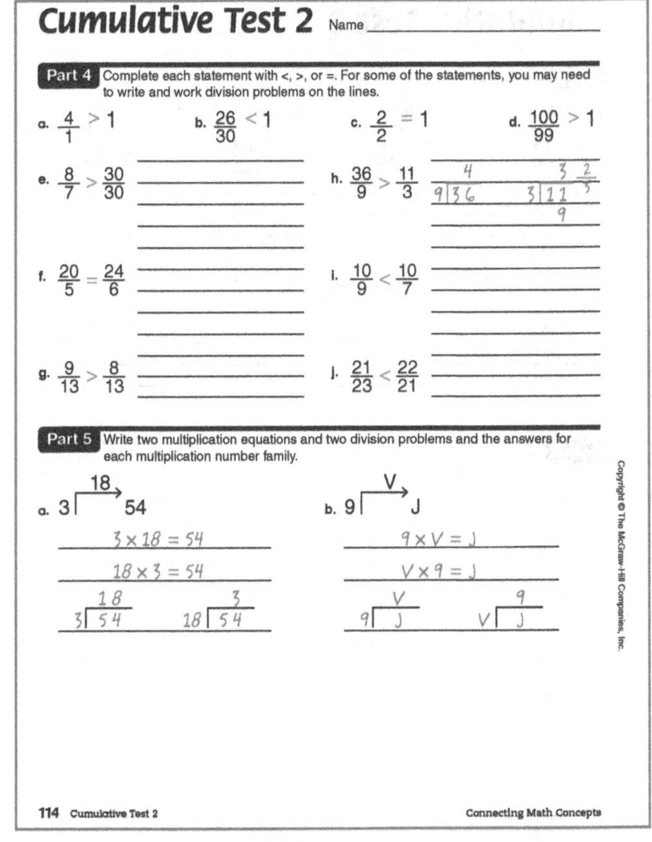

Cumulative Test 2

Name ________

Part 4 Complete each statement with <, >, or =. For some of the statements, you may need to write and work division problems on the lines.

a. $\frac{4}{1} > 1$
b. $\frac{26}{30} < 1$
c. $\frac{2}{2} = 1$
d. $\frac{100}{99} > 1$
e. $\frac{8}{7} > \frac{30}{30}$
f. $\frac{20}{5} = \frac{24}{6}$
g. $\frac{9}{13} > \frac{8}{13}$
h. $\frac{36}{9} > \frac{11}{3}$ (9⟌36 = 4; 3⟌11 = 3 r 2)
i. $\frac{10}{9} < \frac{10}{7}$
j. $\frac{21}{23} < \frac{22}{21}$

Part 5 Write two multiplication equations and two division problems and the answers for each multiplication number family.

a. 3 ⟶ 18 ⟶ 54
3 × 18 = 54
18 × 3 = 54
3⟌54 = 18; 18⟌54 = 3

b. 9 ⟶ V ⟶ J
9 × V = J
V × 9 = J
9⟌J = V; V⟌J = 9

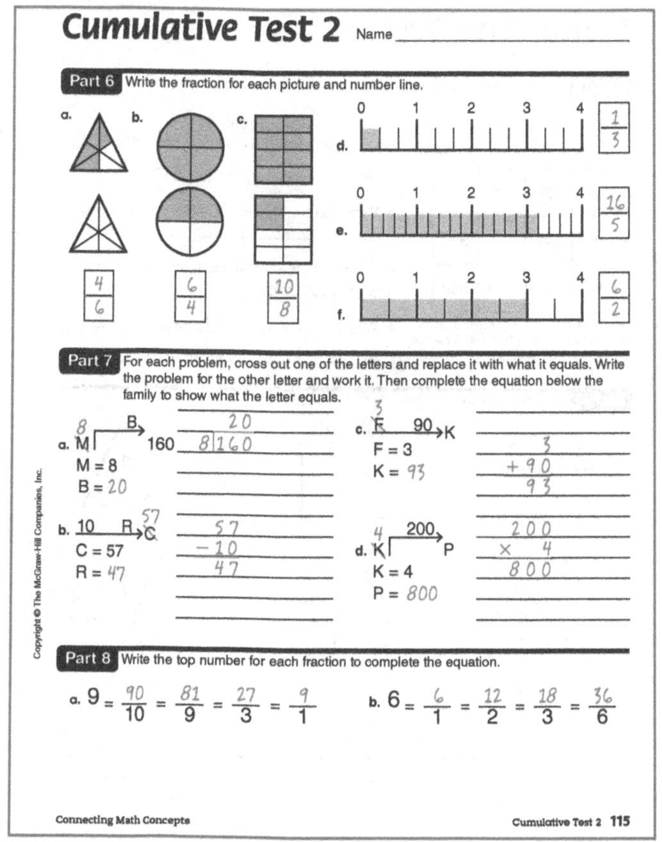

Cumulative Test 2

Name ________

Part 6 Write the fraction for each picture and number line.

a. 4/6
b. 6/4
c. 10/8
d. 1/3
e. 16/5
f. 6/2

Part 7 For each problem, cross out one of the letters and replace it with what it equals. Write the problem for the other letter and work it. Then complete the equation below the family to show what the letter equals.

a. M ⟶ B ⟶ 160 (M = 8): 8⟌160 = 20
M = 8
B = 20

b. 10 ⟶ R ⟶ C (C = 57): 57 − 10 = 47
C = 57
R = 47

c. F ⟶ 90 ⟶ K (F = 3): 3 + 90 = 93
F = 3
K = 93

d. K ⟶ 200 ⟶ P (K = 4): 200 × 4 = 800
K = 4
P = 800

Part 8 Write the top number for each fraction to complete the equation.

a. $9 = \frac{90}{10} = \frac{81}{9} = \frac{27}{3} = \frac{9}{1}$

b. $6 = \frac{6}{1} = \frac{12}{2} = \frac{18}{3} = \frac{36}{6}$

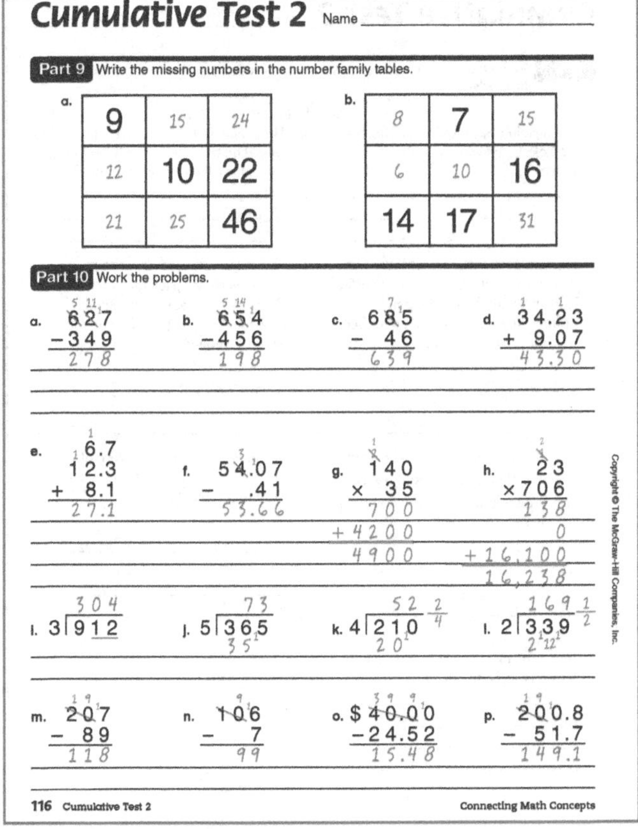

Cumulative Test 2

Name ________

Part 9 Write the missing numbers in the number family tables.

a.

9	15	24
12	10	22
21	25	46

b.

8	7	15
6	10	16
14	17	31

Part 10 Work the problems.

a. 627 − 349 = 278
b. 654 − 456 = 198
c. 685 − 46 = 639
d. 34.23 + 9.07 = 43.30
e. 6.7 + 12.3 + 8.1 = 27.1
f. 54.07 − .41 = 53.66
g. 140 × 35: 700 + 4200 = 4900
h. 23 × 706: 138 + 0 + 16,100 = 16,238
i. 3⟌912 = 304
j. 5⟌365 = 73
k. 4⟌210 = 52 r 2
l. 2⟌339 = 169 r 1
m. 207 − 89 = 118
n. 106 − 7 = 99
o. $40.00 − 24.52 = 15.48
p. 200.8 − 51.7 = 149.1

Cumulative Test 2 Name

Part 10 Work the problems. (continued)

q. $21\frac{2}{6} + 9\frac{5}{6}$ = $30\frac{7}{6}$

r. $26\frac{10}{10} - 5\frac{3}{10}$ = $20\frac{7}{10}$

s. $8\frac{4}{7} - 4\frac{4}{7}$ = 4

t. $2\frac{4}{9} \times 6$ = $12\frac{24}{9}$

u. $4\frac{2}{3} \times 5$ = $20\frac{10}{3}$

v. $\frac{41}{100}00 + \frac{5}{10}\,0 + \frac{28}{100} = \frac{4178}{100}$

w. $\frac{509}{100} - \frac{3}{100}00 = \frac{209}{100}$

Part 11 Complete the equations for the problems you can work. Cross out the problems you can't work

a. ~~$\frac{2}{7} \times \frac{10}{8}$~~ =

b. ~~$\frac{6}{9} \times \frac{8}{8}$~~ =

c. $\frac{6}{9} \times \frac{5}{1} = \frac{30}{9}$

d. $\frac{12}{3} - \frac{10}{3} = \frac{2}{3}$

e. $\frac{9}{8} \times \frac{1}{8} = \frac{9}{64}$

f. ~~$\frac{5}{9} \times \frac{4}{7}$~~ =

g. $\frac{3}{4} + 6\ \frac{24}{4} = \frac{27}{4}$

h. $\frac{4}{3} \times \frac{2}{1} \times \frac{5}{9} = \frac{40}{27}$

i. $\frac{5}{L} \times \frac{3M}{2} = \frac{15M}{2L}$

j. ~~$\frac{5L}{M} \times \frac{3L}{T}$~~ =

k. $\frac{12N}{F} - \frac{10N}{F} = \frac{2N}{F}$

Cumulative Test 2 Name

Part 12 Complete the table so each row shows the fraction equation, the division problem and the answer, and the mixed number or whole number.

Fraction	Division	Mixed or Whole Number
a. $\frac{121}{3} = 40\frac{1}{3}$	$3\overline{)121}$ = $40\frac{1}{3}$	$40\frac{1}{3}$
b. $\frac{101}{5} = 20\frac{1}{5}$	$5\overline{)101}$ = $20\frac{1}{5}$	$20\frac{1}{5}$; $\frac{100}{5} + \frac{1}{5} = \frac{101}{5}$; $20 \times 5 = 100$
c. $\frac{96}{4} = 24$	$4\overline{)96}$ = 24	24

Part 13 Complete each row so it shows the tenths, hundredths, and percent that are equal.

	Tenths	Hundredths	%
a.	.3	.30	30%
b.	2.0	2.00	200%
c.	16.5	16.50	1650%
d.	.6	.60	60%

Part 14 For each problem, complete the fraction multiplication problem equation. Then write = or ≠ to complete the statements below.

a. $\frac{3}{5} \times \frac{9}{7} = \frac{27}{35}$ $\qquad \frac{3}{5} \neq \frac{27}{35}$

b. $\frac{4}{9} \times \frac{9}{9} = \frac{36}{81}$ $\qquad \frac{4}{9} = \frac{36}{81}$

c. $\frac{7}{6} \times \frac{1}{3} = \frac{7}{18}$ $\qquad \frac{7}{6} \neq \frac{7}{18}$

Cumulative Test 2 Name

Part 15 For each problem, make a multiplication or an addition number family and work the problem.

Problems

a. Each jacket has 6 buttons. There are 72 jackets. How many buttons are there?

72 × 6 = 432 buttons

b. The store sold 43.2 pounds of coffee. The store still had 145.6 pounds of coffee. How many pounds of coffee did the store start with?

145.6 + 43.2 = 188.8 pounds

c. The sheep weighed $39\frac{1}{4}$ kilograms. The dog weighed $21\frac{3}{4}$ kilograms. How much lighter was the dog than the sheep?

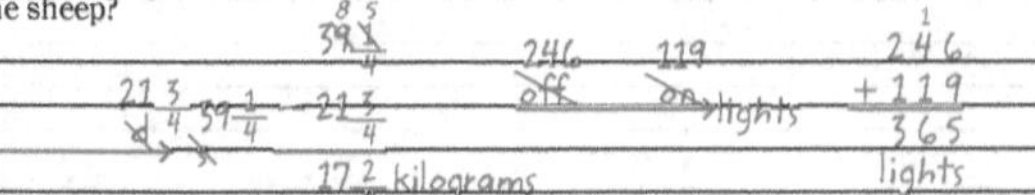

$39\frac{1}{4} - 21\frac{3}{4} = 17\frac{2}{4}$ kilograms

d. The car went 5 times as fast as the boat. The car went 75 miles an hour. How fast did the boat go?

$5\overline{)75}$ = 15 miles an hour

e. The fence around the park was 3 miles long. The fence was 15,840 feet long. How many feet are in each mile?

$3\overline{)15,840}$ = 5 280 feet

f. There were 584 packages on the plane. 420 packages were small. How many packages on the plane were not small?

584 − 420 = 164 packages

g. In a building, 246 lights were off. 119 lights were on. How many lights were in the building?

246 + 119 = 365 lights

Cumulative Test 2 Name

Part 16 Answer questions a through d. Complete sentences e and f.

	market	park	both places
men	138	201	339
women	142	178	320
adults	280	379	659

The table shows the number of men and women in two places. The places are the market and the park.

Questions

a. How many adults were in the market? 280

b. How many men were in the park? 201

c. Were there more men or women in the park? men

d. Were there fewer women in the market or the park? market

e. 201 men were in the park.

f. 659 adults were in both places.

Part 17 Write the angle for each letter.

a. 180°

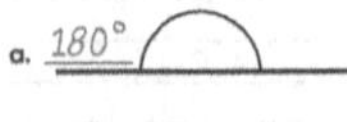

b. 270°

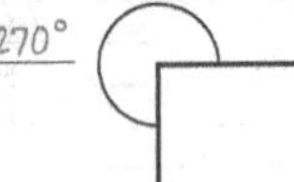

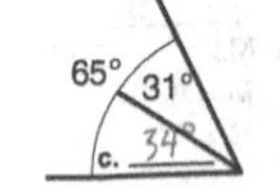

d. 107°

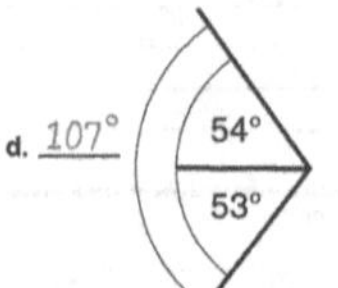

e. 190°

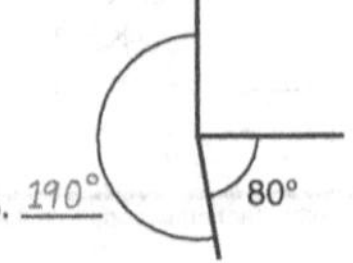

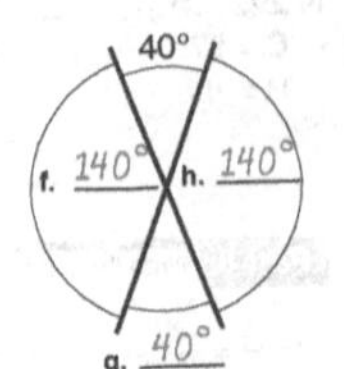

65 − 31 = 34°; 54 + 53 = 107°; 90 + 80 = 170°; 360 − 170 = 190°; 180 − 40 = 140°

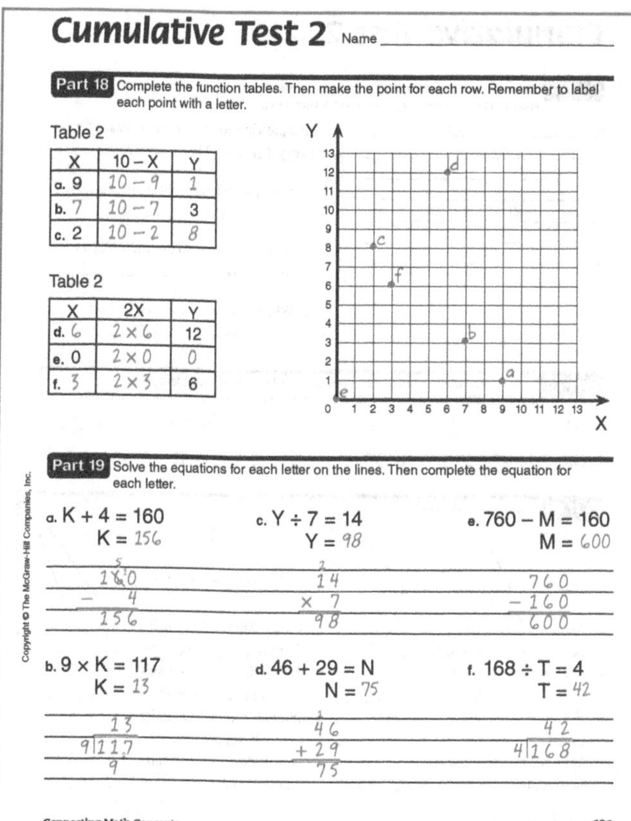

Cumulative Test 2 Name ____

Part 18 Complete the function tables. Then make the point for each row. Remember to label each point with a letter.

Table 2

X	10 − X	Y
a. 9	10 − 9	1
b. 7	10 − 7	3
c. 2	10 − 2	8

Table 2

X	2X	Y
d. 6	2 × 6	12
e. 0	2 × 0	0
f. 3	2 × 3	6

Part 19 Solve the equations for each letter on the lines. Then complete the equation for each letter.

a. K + 4 = 160
K = 156
160 − 4 = 156

c. Y ÷ 7 = 14
Y = 98
14 × 7 = 98

e. 760 − M = 160
M = 600
760 − 160 = 600

b. 9 × K = 117
K = 13
9⟌117 = 13

d. 46 + 29 = N
N = 75
46 + 29 = 75

f. 168 ÷ T = 4
T = 42
4⟌168 = 42

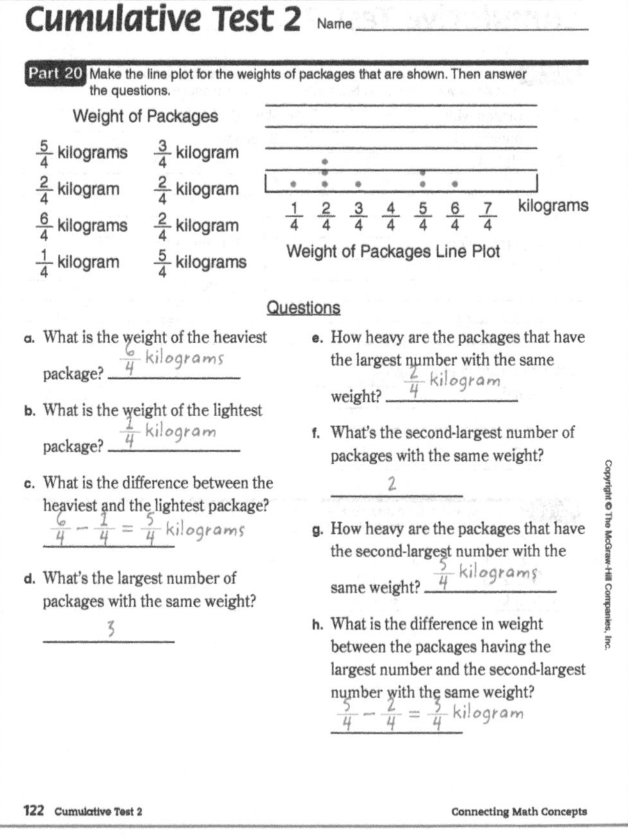

Cumulative Test 2 Name ____

Part 20 Make the line plot for the weights of packages that are shown. Then answer the questions.

Weight of Packages

$\frac{5}{4}$ kilograms $\frac{3}{4}$ kilogram
$\frac{2}{4}$ kilogram $\frac{2}{4}$ kilogram
$\frac{6}{4}$ kilograms $\frac{2}{4}$ kilogram
$\frac{1}{4}$ kilogram $\frac{5}{4}$ kilograms

Weight of Packages Line Plot

Questions

a. What is the weight of the heaviest package? $\frac{6}{4}$ kilograms

b. What is the weight of the lightest package? $\frac{1}{4}$ kilogram

c. What is the difference between the heaviest and the lightest package? $\frac{6}{4} - \frac{1}{4} = \frac{5}{4}$ kilograms

d. What's the largest number of packages with the same weight? 3

e. How heavy are the packages that have the largest number with the same weight? $\frac{2}{4}$ kilogram

f. What's the second-largest number of packages with the same weight? 2

g. How heavy are the packages that have the second-largest number with the same weight? $\frac{5}{4}$ kilograms

h. What is the difference in weight between the packages having the largest number and the second-largest number with the same weight? $\frac{5}{4} - \frac{2}{4} = \frac{3}{4}$ kilogram

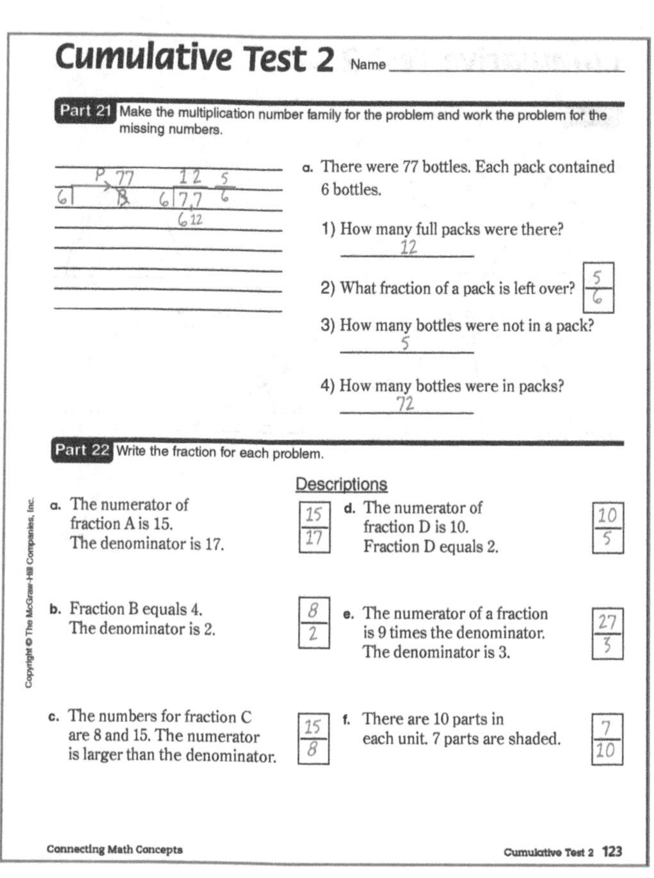

Cumulative Test 2 Name ____

Part 21 Make the multiplication number family for the problem and work the problem for the missing numbers.

6 → B → P 77; 12 r 5: 6⟌77 = 6 12

a. There were 77 bottles. Each pack contained 6 bottles.

1) How many full packs were there? 12

2) What fraction of a pack is left over? $\frac{5}{6}$

3) How many bottles were not in a pack? 5

4) How many bottles were in packs? 72

Part 22 Write the fraction for each problem.

Descriptions

a. The numerator of fraction A is 15. The denominator is 17. $\frac{15}{17}$

b. Fraction B equals 4. The denominator is 2. $\frac{8}{2}$

c. The numbers for fraction C are 8 and 15. The numerator is larger than the denominator. $\frac{15}{8}$

d. The numerator of fraction D is 10. Fraction D equals 2. $\frac{10}{5}$

e. The numerator of a fraction is 9 times the denominator. The denominator is 3. $\frac{27}{3}$

f. There are 10 parts in each unit. 7 parts are shaded. $\frac{7}{10}$

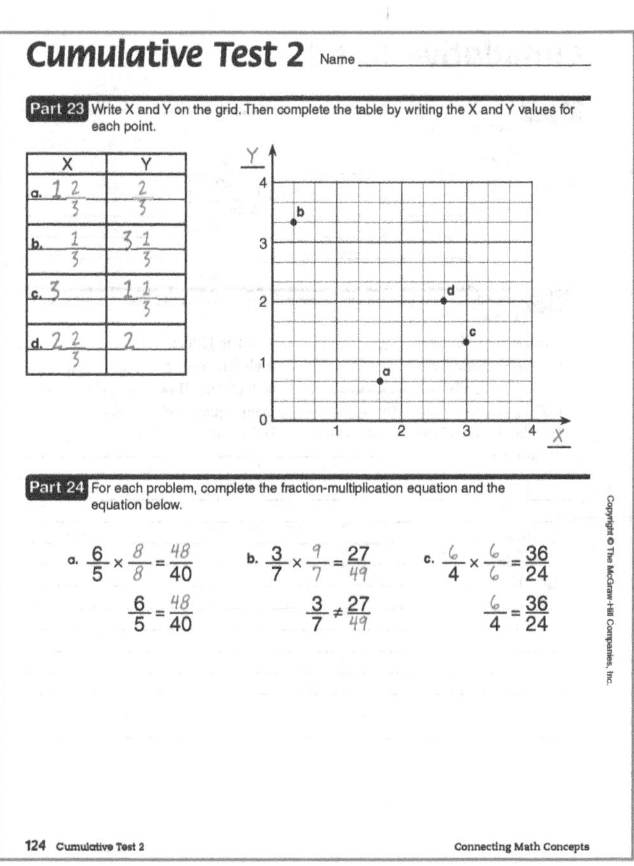

Cumulative Test 2 Name ____

Part 23 Write X and Y on the grid. Then complete the table by writing the X and Y values for each point.

	X	Y
a.	$1\frac{2}{3}$	$\frac{2}{3}$
b.	$\frac{1}{3}$	$3\frac{1}{3}$
c.	3	$1\frac{1}{3}$
d.	$2\frac{2}{3}$	2

Part 24 For each problem, complete the fraction-multiplication equation and the equation below.

a. $\frac{6}{5} \times \frac{8}{8} = \frac{48}{40}$ $\quad \frac{6}{5} = \frac{48}{40}$

b. $\frac{3}{7} \times \frac{9}{7} = \frac{27}{49}$ $\quad \frac{3}{7} \neq \frac{27}{49}$

c. $\frac{6}{4} \times \frac{6}{6} = \frac{36}{24}$ $\quad \frac{6}{4} = \frac{36}{24}$

Cumulative Test 2 Name ____________

Part 25 Write the fraction, decimal, or percent that answers each question. For some of the questions you'll have to write a fraction multiplication equation.

a. What fraction with a denominator of 20 equals 95%? $\frac{19}{20}$

$\frac{19}{20} \times \frac{5}{5} = \frac{95}{100}$ $5\overline{)95}$ = 19

d. What decimal does 8% equal? .08

b. What percent does $\frac{9}{4}$ equal? 225%

$\frac{9}{4} \times \frac{25}{25} = \frac{225}{100}$ $25 \times 9 = 225$

e. What fraction with a numerator of 7 equals 350%? $\frac{7}{2}$

$\frac{7}{2} \times \frac{50}{50} = \frac{350}{100}$ $7\overline{)350}$ = 50

c. What percent does 7.2 equal? 720%

Part 26 For a, b, and c, write and work estimation problems. Round problem a to the tens. Round b to the hundreds. Round c to the thousands.

	a.	b.	c.
851	850	900	1000
3028	3030	3000	3000
+5367	+5370	+5400	+5000
	9250	9300	9000

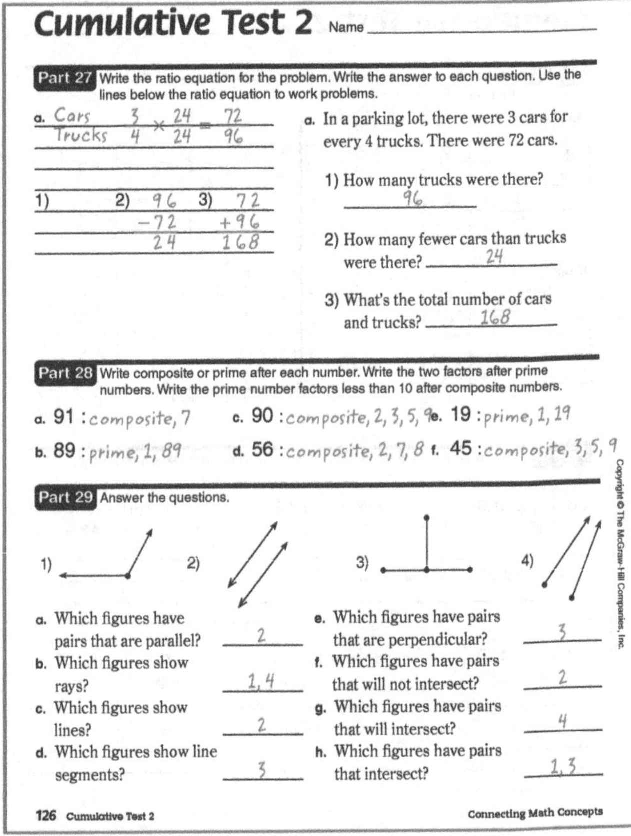

Cumulative Test 2 Name ____________

Part 27 Write the ratio equation for the problem. Write the answer to each question. Use the lines below the ratio equation to work problems.

a. $\frac{\text{Cars}}{\text{Trucks}} \; \frac{3}{4} \times \frac{24}{24} = \frac{72}{96}$

1)   2) $96 - 72 = 24$ 3) $72 + 96 = 168$

a. In a parking lot, there were 3 cars for every 4 trucks. There were 72 cars.

1) How many trucks were there? 96

2) How many fewer cars than trucks were there? 24

3) What's the total number of cars and trucks? 168

Part 28 Write composite or prime after each number. Write the two factors after prime numbers. Write the prime number factors less than 10 after composite numbers.

a. 91 : composite, 7   c. 90 : composite, 2, 3, 5, 9   e. 19 : prime, 1, 19

b. 89 : prime, 1, 89   d. 56 : composite, 2, 7, 8   f. 45 : composite, 3, 5, 9

Part 29 Answer the questions.

1)   2)   3)   4)

a. Which figures have pairs that are parallel? 2

b. Which figures show rays? 1, 4

c. Which figures show lines? 2

d. Which figures show line segments? 3

e. Which figures have pairs that are perpendicular? 3

f. Which figures have pairs that will not intersect? 2

g. Which figures have pairs that will intersect? 4

h. Which figures have pairs that intersect? 1, 3

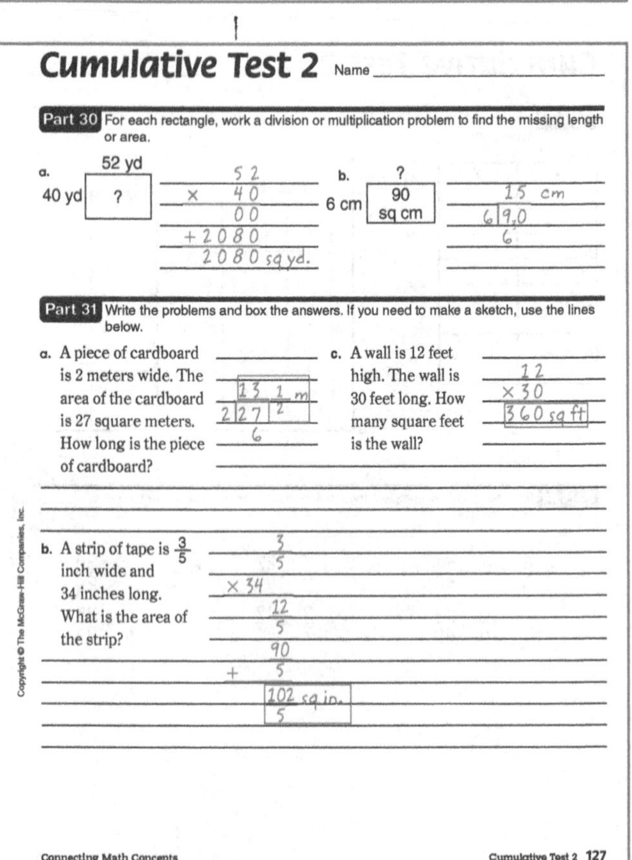

Cumulative Test 2 Name ____________

Part 30 For each rectangle, work a division or multiplication problem to find the missing length or area.

a. 52 yd × 40 yd = ? 52×40: 00 + 2080 = 2080 sq yd.

b. ? × 6 cm = 90 sq cm $6\overline{)90}$ = 15 cm

Part 31 Write the problems and box the answers. If you need to make a sketch, use the lines below.

a. A piece of cardboard is 2 meters wide. The area of the cardboard is 27 square meters. How long is the piece of cardboard? $2\overline{)27}$ = $13\frac{1}{2}$ m

c. A wall is 12 feet high. The wall is 30 feet long. How many square feet is the wall? 12×30 = 360 sq ft

b. A strip of tape is $\frac{3}{5}$ inch wide and 34 inches long. What is the area of the strip? $\frac{3}{5} \times 34 = \frac{12}{5} + \frac{90}{5} = \frac{102}{5}$ sq in.

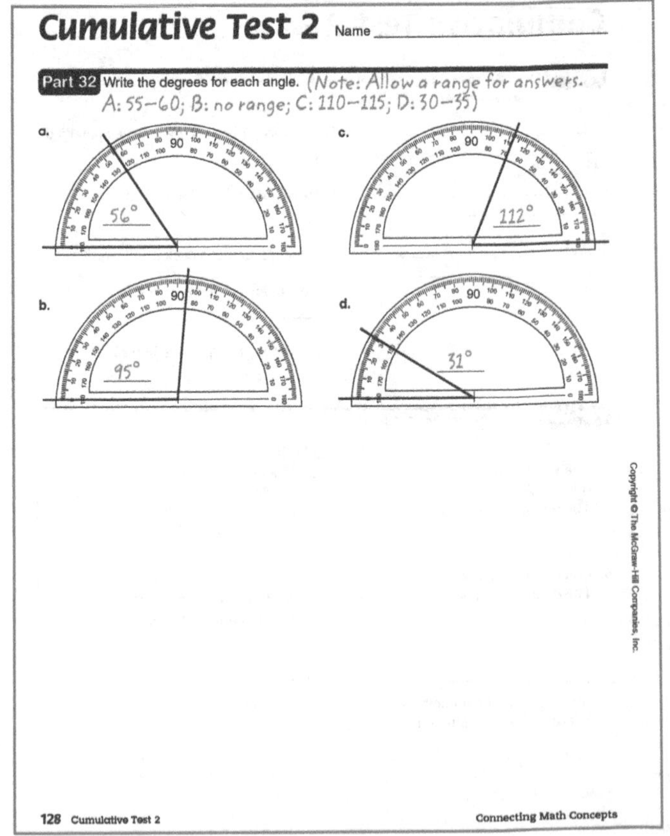

Cumulative Test 2 Name ____________

Part 32 Write the degrees for each angle. (Note: Allow a range for answers. A: 55–60; B: no range; C: 110–115; D: 30–35)

a. 56°   c. 112°

b. 95°   d. 31°

Cumulative Test 2 Name ____________

Part 33 Use the symbols <, >, and = to write the statement for each problem.

217, 2167, 267, 2617

Write the values from smallest to largest.

a. 217 < 267 < 2167 < 2617

7.9 $\frac{25}{4}$ 843%

(6 $\frac{1}{4}$ 8.43)

Write the values from largest to smallest.

b. 843% > 7.9 > $\frac{25}{4}$

400% .51 $\frac{9}{2}$

(4.00 4 $\frac{1}{2}$)

Write the values from largest to smallest.

c. $\frac{9}{2}$ > 400% > .51

Cumulative Test 2 Name ____________

Part 34 For each problem, make the number family with three names. Then work the column problem for the numerators. Complete the fraction number family and box the answer.

a. In a TV store, $\frac{24}{90}$ of the sets were turned off. What fraction of the sets were turned on?

off $\frac{24}{90}$ on $\boxed{\frac{66}{90}}$ → sets $\frac{90}{90}$ 90 − 24 = 66

b. There were 84 customers in the store. There were 24 workers in the store. What fraction of the people in the store were customers?

customers $\boxed{\frac{84}{108}}$ 24 workers $\frac{24}{108}$ → people $\frac{108}{108}$ 84 + 24 = 108

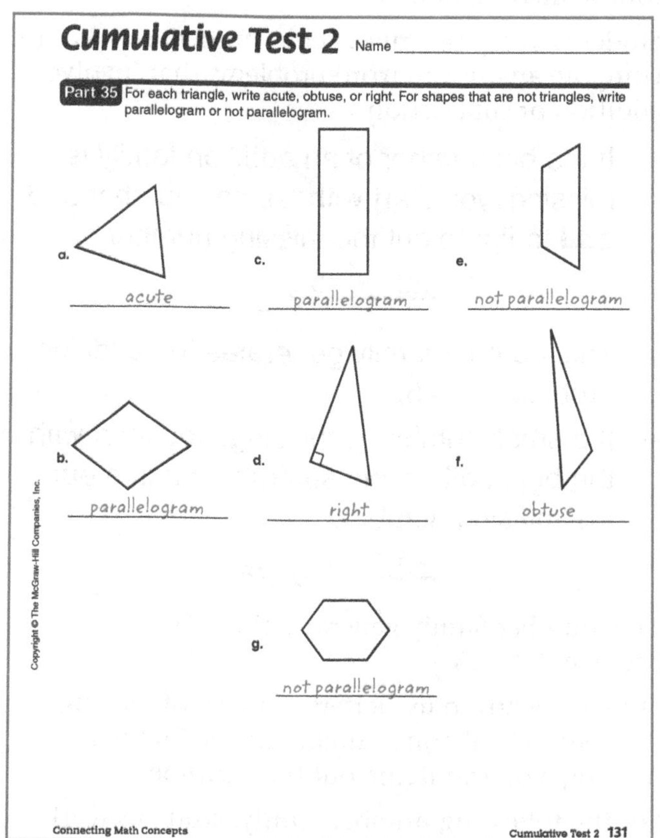

Cumulative Test 2 Name ____________

Part 35 For each triangle, write acute, obtuse, or right. For shapes that are not triangles, write parallelogram or not parallelogram.

a. acute

b. parallelogram

c. parallelogram

d. right

e. not parallelogram

f. obtuse

g. not parallelogram

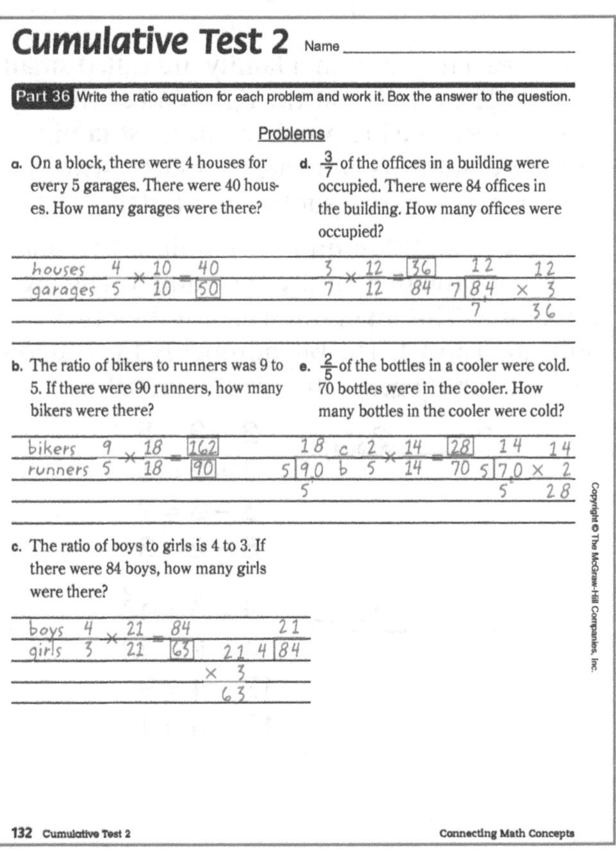

Cumulative Test 2 Name ____________

Part 36 Write the ratio equation for each problem and work it. Box the answer to the question.

Problems

a. On a block, there were 4 houses for every 5 garages. There were 40 houses. How many garages were there?

houses $\frac{4}{5} \times \frac{10}{10} = \frac{40}{\boxed{50}}$ garages

b. The ratio of bikers to runners was 9 to 5. If there were 90 runners, how many bikers were there?

bikers $\frac{9}{5} \times \frac{18}{18} = \frac{\boxed{162}}{90}$ runners (90 ÷ 5 = 18)

c. The ratio of boys to girls is 4 to 3. If there were 84 boys, how many girls were there?

boys $\frac{4}{3} \times \frac{21}{21} = \frac{84}{\boxed{63}}$ girls (84 ÷ 4 = 21; 21 × 3 = 63)

d. $\frac{3}{7}$ of the offices in a building were occupied. There were 84 offices in the building. How many offices were occupied?

$\frac{3}{7} \times \frac{12}{12} = \frac{\boxed{36}}{84}$ (84 ÷ 7 = 12; 12 × 3 = 36)

e. $\frac{2}{5}$ of the bottles in a cooler were cold. 70 bottles were in the cooler. How many bottles in the cooler were cold?

c $\frac{2}{5}$ b $\times \frac{14}{14} = \frac{\boxed{28}}{70}$ (70 ÷ 5 = 14; 14 × 2 = 28)

Tracks

Addition Number Families (Lessons 1–130)

Overview

Work with number families begins on Lesson 1 and continues throughout the program. Number families act as an organizational tool for fact memorization and for solving story problems. Each family shows three related values. There are two types of number families—addition families and multiplication families. Addition families have three values that are related through addition and subtraction. Multiplication families have three values that are related through multiplication and division.

Addition families are shown on a straight arrow:

Multiplication-division families are shown on an arrow that looks like a division sign:

The first two numbers in a family are called small numbers. The number at the right end of the arrow is called the big number. For most families, these three numbers generate four facts—two addition facts and two subtraction facts.

Here are two addition number families: The small numbers for the first family are 2 and 3. The big number is 5. The small numbers for the other family are 9 and 4. The big number is 13. The four facts each family generates are shown below:

2 —3→ 5

$2 + 3 = 5$
$3 + 2 = 5$
$5 - 3 = 2$
$5 - 2 = 3$

9 —4→ 13

$9 + 4 = 13$
$4 + 9 = 13$
$13 - 4 = 9$
$13 - 9 = 4$

Multiplication number families generate multiplication and division facts. Here's a family with the small numbers of 5 and 3. The big number is 15. The four facts that are generated are shown below:

5 ⌐3→ 15

$5 \times 3 = 15$
$3 \times 5 = 15$
$15 \div 3 = 5$
$15 \div 5 = 3$

The main rationale for using number families to teach facts is that if students learn the arrangement of the three numbers in each family, they have a model for learning and remembering four facts. So the number families reduce the memory load required for students to become proficient with facts.

Another application of number families is solving word problems. Many students struggle solving word problems because they don't have a strategy for separating the computation that's required from the relationship of the values described. In *Connecting Math Concepts,* students learn to create number families for word problems and to solve for the missing number.

Students apply two rules about number families to compute answers to word problems that involve addition or subtraction:

- If the big number of an addition family is missing, you start with a small number and **add** to figure out the missing number:

  23 —5→ _

  This number family generates the addition problem 23 + 5.

- If a small number is missing, you start with the big number and subtract to figure out the missing number:

  23 —→ 28

This number family generates the subtraction problem 28 – 23.

Students learn to use letters in place of missing numbers. If only one number in the family is missing, you can figure out that number.

For the following number family, students work the problem 95 – 32 to figure out J:

32 —J→ 95

For the following number family, students work the problem 19 + 45 to figure out Z:

$$\underrightarrow{19 \quad 45} Z$$

For families with two letters, students learn to replace a letter with what it equals and solve for the remaining letter:

$$\underrightarrow{35 \quad M} J$$

$$J = 94$$

Students cross out J and replace it with what it equals:

$$\underrightarrow{35 \quad M} \overset{94}{\not{J}}$$

Students work the problem 94 – 35 to figure out M.

Working with number families that have letters prepares students for word problems. Here's a word problem:

A sheep was 27 pounds heavier than a goat.

The sheep weighed 189 pounds.

How heavy was the goat?

Students might intuitively add to solve this comparison word problem because the problem indicates that the sheep was heavier than the goat. Students might reason that heavier is more, so you add. Using number families, students write the letters for the bigger value and the smaller value. The sheep is heavier so S is the letter for the big number. The goat is lighter so G is the letter for a small number. 27 pounds tells how much more or less, so 27 is the other small number.

Here's the family with two letters and a number for the problem:

$$\underrightarrow{27 \quad G} S$$

The problem gives a number for one of the letters. The sheep weighed 189 pounds, so S is replaced with 189:

$$\underrightarrow{27 \quad G} \overset{189}{\not{S}}$$

To find the answer to G, students subtract to find the missing small number:

$$\begin{array}{r} 189 \\ -\ 27 \\ \hline 162 \end{array} \text{ pounds}$$

The goat weighed 162 pounds.

Students who don't have a strategy for separating the operation required from the relationship of the values often choose the wrong operation and add, concluding that the goat weighed 216 pounds.

NUMBER FAMILY FOUNDATIONS

On Lesson 1, Exercise 1 students learn basic number-family conventions.

Here's the first part of the exercise:

a. You're going to learn a lot of math.
- You'll learn how to solve hard problems, and you'll learn a lot about numbers and how they work.
- Here's a rule about this program: You'll use everything we work on. Sometimes, we'll work on things that seem simple. Remember how to do them because you'll use them when we do things in the next lesson and the lesson after that.
- Remember, everything we work on, we'll use. So learn to do everything perfectly.
- Here's another rule: When you get a problem wrong, do not erase your answer. Just make a line through your answer and write the correct answer next to it. That rule is very important.
- Listen again: When you get a problem wrong, do not erase your answer. Just make a line through your answer and write the correct answer next to it.

b. (Display:) [1:1A]

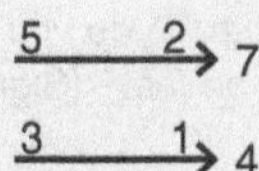

These are number families.
Each number family has three numbers.
These are the three numbers that always go together in addition and subtraction facts.
The first two numbers in the family are small numbers.
- **(Point to 5.)** What are the small numbers in this family? **(Signal.)** *5 and 2.*
- **(Point to 3.)** What are the small numbers in this family? **(Signal.)** *3 and 1.*

c. **(Point to 7.)** The number at the end of the arrow is the big number.
- **(Point to 7.)** What's the big number in this family? **(Signal.)** *7.*
- **(Point to 4.)** What's the big number in this family? **(Signal.)** *4.*
- Once more: How many numbers are in a number family? **(Signal.)** *3.*
- How many are small numbers? **(Signal.)** *2.*

from Lesson 1, Exercise 1

Teaching Notes: The brackets in the margin show parts of the exercise that are to be repeated if students make mistakes or don't respond well. It's important that students are very reliable at identifying the small numbers and the big number. Make sure that students respond together and respond on signal. Automaticity in identifying the value for the big number and the small numbers is necessary for students to become proficient in using the word problem strategies taught in *CMC Level E.*

Later in Lesson 1, students say all four equations for families and then write the equations.

a. Open your workbook to Lesson 1 and find part 1. ✔
(Teacher reference:)

a. 17 4→21 b. N 20→K

You're going to write two addition facts and two subtraction facts for each number family in part 1.

- Touch family A. ✔
- What are the small numbers in family A? (Signal.) *17 and 4.*
- What's the big number? (Signal.) *21.*
- If you start with 17, do you say an addition fact or a subtraction fact? (Signal.) *An addition fact.*
- Say the fact that starts with 17. (Signal.) *17 + 4 = 21.*
- Say the fact that starts with 4. (Signal.) *4 + 17 = 21.*
- Say the fact that starts with 21 and minuses 4. (Signal.) *21 − 4 = 17.*
- Say the other subtraction fact. (Signal.) *21 − 17 = 4.*

(Repeat until firm.)

b. You'll write all four facts for the family on the lines below it.

- Touch where you'll write 17 plus 4 equals 21. ✔ (Students touch first space.)
- Touch where you'll write 4 plus 17 equals 21. ✔ (Students touch second space.)
- Touch where you'll write 21 minus 4 equals 17. ✔ (Students touch third space.)
- Touch where you'll write 21 minus 17 equals 4. ✔ (Students touch fourth space.)

(Repeat until firm.)

c. Write all four facts for family A on the lines below it.
(Observe students and give feedback.)

- Check your work.
(Display:) [1:5A]

a. 17 4→21

17 + 4 = 21
4 + 17 = 21
21 − 4 = 17
21 − 17 = 4

Here's what you should have for family A.

d. Touch family B. ✔

- What are the small numbers for family B? (Signal.) *N and 20.*
- What is the big number for family B? (Signal.) *K.*

(Repeat step d until firm.)

e. If you start with N, do you say an addition fact or a subtraction fact? (Signal.) *An addition fact.*

- If you start with K, do you say an addition fact or a subtraction fact? (Signal.) *A subtraction fact.*

(Repeat step e until firm.)

f. Start with N and say the fact. (Signal.) *N + 20 = K.*

- Start with 20 and say the fact. (Signal.) *20 + N = K.*
- Start with K and say the fact that minuses 20. (Signal.) *K − 20 = N.*
- Say the other subtraction fact. (Signal.) *K − N = 20.*

(Repeat step f until firm.)

g. Write all four facts for family B on the lines below it.
(Observe students and give feedback.)

- Check your work.
(Display:) [1:5B]

b. N 20→K

N + 20 = K
20 + N = K
K − 20 = N
K − N = 20

Here's what you should have for family B.

Lesson 1, Exercise 5

In step A, students say the four facts for family A. In step B, they touch where they'll write each fact. In step C, they write the facts for family A.

In the rest of the exercise, students go through similar steps with a family that has a number and two letters.

Teaching Note: Make sure students are firm on the steps that you "Repeat until firm." For some students, saying equations that have letters is a new and strange experience. Tell students they will soon work very difficult problems that have letters. Remind students that the number and letters follow the same rules.

On Lesson 2, students review the work done on Lesson 1 and say problems for families that have a missing number.

For example, this family has a missing small number:

7 ——→ 8

A small number is missing so students subtract to find it. Students say the problem 8 minus 7 to find the missing number.

On Lesson 3, students write problems for number families that have missing numbers.

Here's the first part of the exercise from Lesson 3:

a. Find part 4 in your workbook. ✔
(Teacher reference:)

a. ___ 5→ 7 ______
b. 3 ___→ 13 ______
c. 12 1→ ___ ______
d. 9 3→ ___ ______
e. ___ 8→ 15 ______
f. 5 ___→ 9 ______

- Family A. Is a small number missing in the family? (Signal.) *Yes.*
- So do you subtract to find the missing number? (Signal.) *Yes.*
- Say the problem for family A. (Signal.) *7 − 5.*

b. (Repeat the following tasks for families B through F:)

Family ___. Is a small number missing?		So do you subtract to find the missing number?	Say the problem for family ___.	
B	*Yes*	*Yes*	B	*13 − 3*
C	*No*	*No*	C	*12 + 1*
D	*No*	*No*	D	*9 + 3*
E	*Yes*	*Yes*	E	*15 − 8*
F	*Yes*	*Yes*	F	*9 − 5*

(Repeat families that were not firm.)

c. Below family A, write the problem for finding the missing number. Then complete the equation.
(Observe students and give feedback.)

d. Check your work.

- Touch and read the equation you wrote below family A. (Signal.) *7 − 5 = 2.*

(Display:) [3:8A]

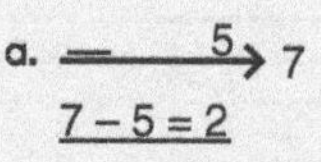

Here's what you should have for A.

from Lesson 3, Exercise 8

In steps A and B you present three tasks for each family:

- Is a small number missing?
- So do you subtract to find the missing number?
- Say the problem for family ___.

In step C, students write the problem and the answer for finding the missing number.

Teaching Notes: The questions presented in steps A and B establish the basis for the operation students use to find the missing number in a family. The questions are parallel. If the answer to the first question is yes, the answer to the follow-up question is yes. Students subtract, so they know what problem to say. If the answer to the first question is no, the answer to the follow-up question is no. Students do not subtract, so, by default, they add. Presenting these series of questions as they are written is important for students adopting a solid strategy for finding the missing number in number families. Repeat the series of questions for any family that is not firm. In steps C and D remind students to write the equation that they say. (Sometimes, students try to keep the numbers in the same order they appear in the family.)

Starting with Lesson 4, students write problems for families that have a letter. Students treat the letter the same way they treat missing numbers in families. If the letter is a small number, they subtract to find it. If the letter is the big number, they add to find it.

Here's a set of problems from Lesson 4:

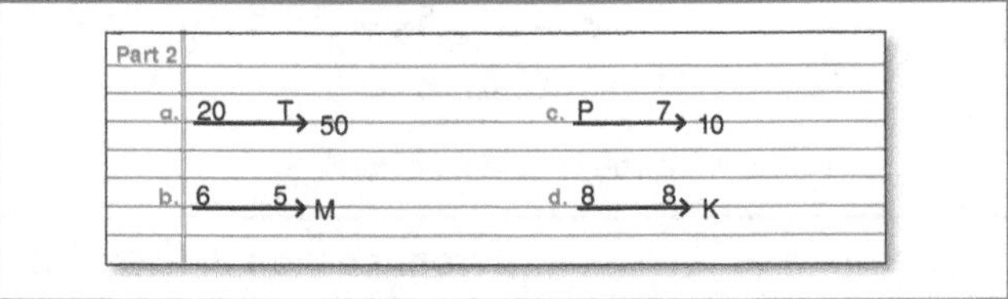

The series of questions for figuring out what the letter in each family equals is the same as families with a missing number, except the first question asks, "Is the letter a small number?"

In Lesson 5, students work problems that require computation:

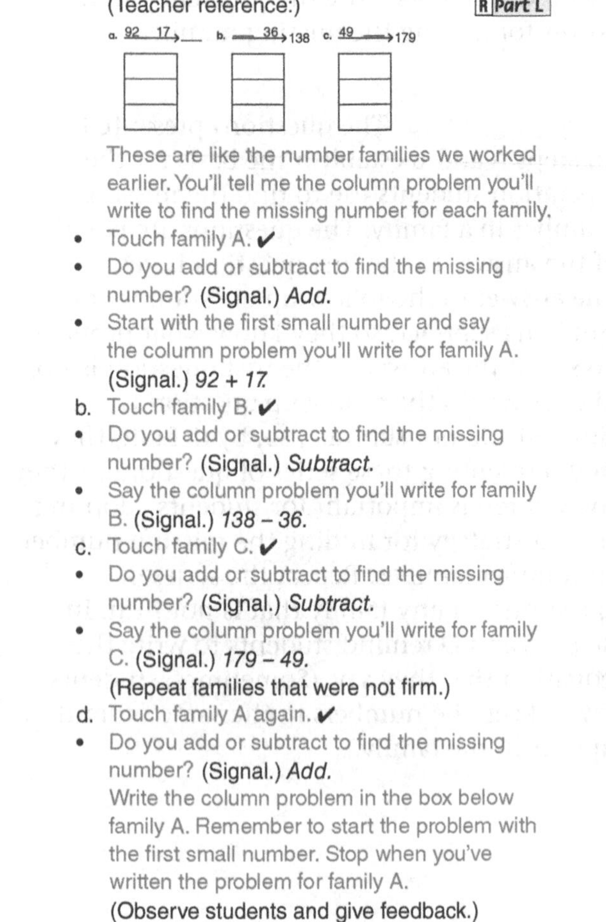

(Teacher reference:) **R Part L**

a. 92 17 → ___ b. ___ 36 → 138 c. 49 ___ → 179

These are like the number families we worked earlier. You'll tell me the column problem you'll write to find the missing number for each family.

- Touch family A. ✔
- Do you add or subtract to find the missing number? (Signal.) *Add.*
- Start with the first small number and say the column problem you'll write for family A. (Signal.) *92 + 17.*

b. Touch family B. ✔
- Do you add or subtract to find the missing number? (Signal.) *Subtract.*
- Say the column problem you'll write for family B. (Signal.) *138 – 36.*

c. Touch family C. ✔
- Do you add or subtract to find the missing number? (Signal.) *Subtract.*
- Say the column problem you'll write for family C. (Signal.) *179 – 49.*

(Repeat families that were not firm.)

d. Touch family A again. ✔
- Do you add or subtract to find the missing number? (Signal.) *Add.* Write the column problem in the box below family A. Remember to start the problem with the first small number. Stop when you've written the problem for family A. (Observe students and give feedback.)

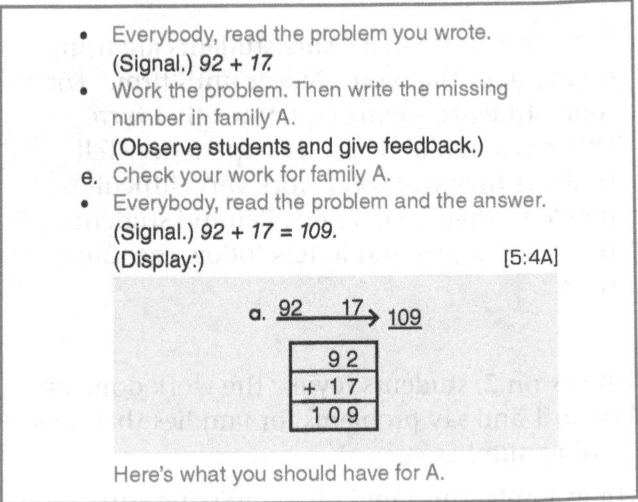

- Everybody, read the problem you wrote. (Signal.) *92 + 17.*
- Work the problem. Then write the missing number in family A. (Observe students and give feedback.)

e. Check your work for family A.
- Everybody, read the problem and the answer. (Signal.) *92 + 17 = 109.* (Display:) [5:4A]

a. 92 17 → 109

92
+ 17
109

Here's what you should have for A.

from Lesson 5, Exercise 4

In steps A through C, students indicate whether they add or subtract to find the missing number and say the column problem they will write. In steps D and E, students work problem A and check their work.

Teaching Note: The exercise provides directions that are more abbreviated than those presented earlier. The first question they answer is, "Do you add or subtract to find the missing number?" If students don't answer on time, pause a second or two before you ask the question. If some students answer incorrectly, stop and prompt them to respond correctly using the same series of questions presented in Lesson 3. After students respond correctly, go back to the first family and present the tasks in steps A through C. Provide a little more think time and prompt students to determine for each family if a small number is missing.

On Lesson 6, students learn to write families with two letters and a number from statements that compare. For example: J is 11 less than P.

Students first identify the number in the family. The number tells how many more or less. Students learn to write it as the first small number in the family. Students then determine from the statement which letter is the big number. J is less than P, so P is the big number. J is a small number.

Here's the first part of the exercise from Lesson 6:

(Teacher reference:)

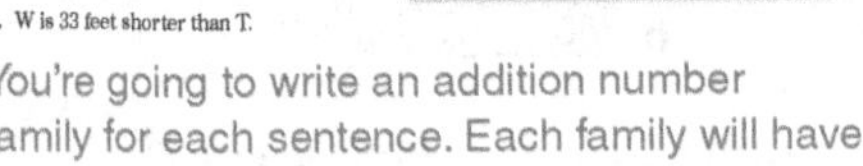

Sentences

a. T is 16 less than J.

b. P weighs 8 pounds more than V.

c. Z weighs 96 pounds less than G.

d. W is 33 feet shorter than T.

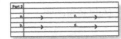

You're going to write an addition number family for each sentence. Each family will have a number and two letters.

- The number in each sentence tells about how much more or less. So will you write the number as the first small number or the big number? **(Signal.)** *The first small number.*

b. Sentence A: T is 16 less than J.
- What's the first small number in the family? **(Signal.)** *16.*
- Listen again: T is 16 less than J. What's the big number—T or J? **(Signal.)** *J.* Yes, J is the big number.

c. Sentence B: P weighs 8 pounds more than V. What's the first small number? **(Signal.)** *8.*
- Listen again: P weighs 8 pounds more than V. What's the big number—P or V? **(Signal.)** *P.* Yes, P is the big number.

d. Sentence C: Z weighs 96 pounds less than G. What's the first small number? **(Signal.)** *96.*
- Listen again: Z weighs 96 pounds less than G. What's the big number—Z or G? **(Signal.)** *G.* Yes, G is the big number.

e. Sentence D: W is 33 feet shorter than T. What's the first small number? **(Signal.)** *33.*
- Listen again: W is 33 feet shorter than T. What's the big number—W or T? **(Signal.)** *T.* Yes, T is the big number.

(Repeat sentences that were not firm.)

from Lesson 6, Exercise 6

In this exercise, students don't solve a problem but become familiar with the conventions of writing a family with two letters and a number.

Teaching Note: Practice presenting steps D and E before teaching the lesson. The teaching goal of this exercise and the goals that follow on Lesson 7 are for students to become well practiced in identifying the first small number in the family and indicating which of the two letters is the big number. The sentences are simple. They use the words *more* or *less,* and the sentences do not have distracting information in them. Firm students on making families for these sentences before presenting more complicated comparison sentences, which appear later in the program.

The introduction on Lesson 7 involves sentences that give information about how to configure the number family. For instance, J is 11 years younger than T. Students write the letters J and T. They write 11 as the first small number.

Also on Lesson 7, students begin work with problems that give a number and two letters. The problems give a value for one of the letters. Students replace that letter with the number it equals and solve for the remaining letter in the family.

Here's part of the exercise from Lesson 7.

a. **(Display:)** [7:5A]

$\xrightarrow{45 \quad B} C$

C = 55

This is a family with two letters and a number. Below the family is an equation that tells what one of the letters equals.
- Tell me the number in the family. Get ready. **(Signal.)** *45.*
- Tell me the letter for the big number. Get ready. **(Signal.)** *C.*
- Tell me the other letter. Get ready. **(Signal.)** *B.*

b. **(Point to C =.)** Read the equation below the family. **(Signal.)** *C = 55.* That tells you that you can cross out C in the family and write 55 above it.

(Add to show:) [7:5B]

$\xrightarrow{45 \quad B} \overset{55}{\cancel{C}}$

C = 55

Now the family has two numbers. So we can figure out what B equals.
- Say the problem for figuring out B. Get ready. **(Signal.)** *55 – 45.*
- What's 55 – 45? **(Signal.)** *10.* So I cross out B and write 10 above it and circle it to show that it's the number we figured out.

(Add to show:) [7:5C]

$\xrightarrow{45 \quad \overset{⑩}{\cancel{B}}} \overset{55}{\cancel{C}}$

C = 55

Here's what we have for the family.

from Lesson 7, Exercise 5

Word Problems

Comparison Word Problems

If students are firm on what has been taught, they should have no problems with this extension. The purpose of the exercise is to provide practice in working with families that are like the ones students will write when solving word problems.

On Lesson 8, students write number families from sentences that tell about two names.

For example: Tim was 12 inches shorter than Dan. The first small number is 12. The names are Tim and Dan, so students write T and D. Tim was shorter, so Dan is the big number.

Here's the family: 12 ——T——> D

Here are the sentences from Lesson 8:

Sentences

a. Tim was 12 inches shorter than Dan.
b. Mr. King was 13 years older than Mr. Brown.
c. Ann weighed 34 pounds less than Bob.
d. The Snake was 36 inches longer than the Worm.

Part 1
a. ——> c. ——>
b. ——> d. ——>

On Lesson 10, students work complete word problems.

Here's part of the exercise:

e. I'll read problem B: T is 80 miles longer than Q. T is 220 miles long. How long is Q?
The sentence T is 80 miles longer than Q tells how to make the number family.
- What do I write first in the family? (Signal.) *80.*
- Raise your hand when you know the letter for the big number. ✔
- What letter do we write for the big number? (Signal.) *T.*
- What letter do we write for the other small number? (Signal.) *Q.*
(Display:) [10:8E]

b. 80 ——Q——> T

f. The sentence T is 220 miles long tells what one of the letters equals. Which letter does that sentence tell about? (Signal.) *T.*
- What's the number for T? (Signal.) *220.*
- Which letter in the family do I cross out? (Signal.) *T.*
- What do I write above it? (Signal.) *220.*
(Add to show:) [10:8F]

b. 80 ——Q——> ~~T~~ (220)

g. The problem asks: How long is Q?
- Say the problem for figuring out Q. (Signal.) *220 – 80.*
(Add to show:) [10:8G]

b. 80 ——Q——> ~~T~~ (220)

$$\begin{array}{r} \overset{1}{\cancel{2}}{}^{1}20 \\ -\ 80 \\ \hline 140 \end{array}$$

Here's the problem and the answer for B.
- The answer is 140. So how long is Q? (Signal.) *140 miles.*
(Add to show:) [10:8H]

b. 80 ——Q——> ~~T~~ (220)

$$\begin{array}{r} \overset{1}{\cancel{2}}{}^{1}20 \\ -\ 80 \\ \hline \boxed{140\text{ miles}} \end{array}$$

from Lesson 10, Exercise 8

Teaching Note: The teacher indicates that one of the sentences tells how to make the number family. In later exercises, students will identify the sentence.

Make sure that students are firm on responding to the question, "What do I write first in the family?" and on responding to the direction, "Raise your hand when you know the letter for the big number."

On the following lessons, students continue to work on word problems that have letters. In the same lesson range, students also work with problems that give names. Students identify the letters and indicate where they go in the family.

Here's part of the exercise from Lesson 13:

e. You'll work the next problem on your lined paper.
- Write part 4 on your lined paper with the letter B below. Then make an addition number family arrow.
(Observe students and give feedback.)
- I'll read problem B. Follow along: The Truck was 218 inches long. The Boat was 126 inches shorter than the Truck. How long was the Boat?
- One sentence in the problem tells how many more or less. Raise your hand when you know which sentence. ✔
- Read the sentence that tells how many more or less. (Call on a student.) *The boat was 126 inches shorter than the truck.*
- Write a number and two letters in the family for B.
(Observe students and give feedback.)

f. Check your work.
- What did you write for the first small number? (Signal.) *126.*
- What's the letter for the big number? (Signal.) *T.*
- What's the other small number? (Signal.) *B.*
(Display:) [13:9E]

Part 4

b. 126 B → T

Here's the family you should have for B.

g. A sentence gives a number for the truck or the boat. What's the number? (Signal.) *218.*
- What letter do you cross out? (Signal.) *T.*
- Cross out T and write what it equals. Then write the column problem next to the family and work it. Make sure to keep your columns lined up. Remember the abbreviation for the units in the answer.
(Observe students and give feedback.)

h. Check your work.
- Read the number problem and the answer. (Signal.) *218 – 126 = 92.*
- The problem asks: How long was the boat? Say the answer with the unit name. (Signal.) *92 inches.*
- What did you write for inches? (Signal.) *I N.*
(Add to show:) [13:9F]

Part 4

b. 126 B → ~~T~~ 218

$$\begin{array}{r} \overset{1}{\cancel{2}}18 \\ -126 \\ \hline 92 \text{ in.} \end{array}$$

Here's what you should have for B. Make sure what you wrote looks like this.

from Lesson 13, Exercise 9

Teaching Note: The sentence that tells how to make the family is: "The boat was 126 inches shorter than the truck." The first letter of *boat* and *truck* are underlined. Students follow the familiar routine of writing the number and the two letters conveyed by the sentence. Students are to work the problem and write the answer as a number with a unit name.

Work with **comparison problems** continues through Lesson 19, with less structure, which means that students assume responsibility for finding the sentence that compares, writing a number and two letters in the family, replacing one letter with a number, working the problem, and writing the answer as a number with a unit name.

Sequence Word Problems

On Lesson 19, students work **sequence problems.** These are different from problems that compare two things. The procedures that students follow are also different. For sequence problems, students do not write initials for names. Instead, they write the same pair of letters in all problems—**S** and **E.** S represents the value the problem **starts** with. E represents the value the problem **ends** with.

For example:

Alex started out with 18 dollars. He ended up with 4 dollars. How much money did he spend?

He ended with less, so E is a small number. He started with more, so S is the big number. The problem gives numbers for both letters, so students cross out the letters and write the numbers:

→ ~~E~~ (4) → ~~S~~ (18)

Students work the problem 18 – 4 to find the amount spent.

On Lesson 19, students are introduced to conventions for writing the letters S and E in families. On Lesson 20, students work from sentences to make families with one letter—E. The sentences give information about whether E is the big number or a small number. Students will build on this discrimination throughout the start-end sequence.

Here's part of the exercise from Lesson 20:

TEXTBOOK PRACTICE

a. Open your textbook to Lesson 20 and find part 1. ✔
(Teacher reference:)

Sentences

a. Bob started out with 15 chips and ate some of the chips.

b. Jerry started out with some chips and bought another bag of chips.

c. Amy had 3 dimes and earned 8 dimes.

d. Mr. Jones started out with some trucks and sold 11 trucks.

e. Ms. Green had 8 peaches in a bowl and picked 4 peaches from a tree.

These sentences tell about how much more or less a person got. You're going to make a family for each sentence with the letter E.

- Write part 1 on your lined paper with the letters A through E below. Make an addition number family arrow after each letter.
(Observe students and give feedback.)

b. I'll read sentence A. Follow along.
Listen: Bob started out with 15 chips and ate some of the chips.

- Did Bob end up with more or less than he started with? (Signal.) *Less.*
- So is E the big number or a small number? (Signal.) *A small number.*
- Write E in family A. ✔

c. Sentence B: Jerry started out with some chips and bought another bag of chips.

- Is E the big number or a small number? (Signal.) *The big number.*
- Write E in family B. ✔

d. Write E as the big number or a small number in the rest of the families.
(Observe students and give feedback.)

e. Check your work.
I'll read the rest of the sentences. You'll tell me if you wrote E as the big number or a small number for each sentence.

- Sentence C: Amy had 3 dimes and earned 8 dimes. Did you write E as the big number or a small number for sentence C? (Signal.) *The big number.*
- Sentence D: Mr. Jones started out with some trucks and sold 11 trucks. Did you write E as the big number or a small number for sentence D? (Signal.) *A small number.*
- Sentence E: Ms. Green had 8 peaches in a bowl and picked 4 peaches from a tree. Did you write E as the big number or a small number for sentence E? (Signal.) *The big number.*

(Display:) [20:7G]

Part 1

a. —E→ d. —E→

b. ——→E e. ——→E

c. ——→E

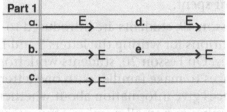

Here's E in the families for the sentences.

f. The families don't show S. Put S in each family. Remember, if E is a small number, S is the big number. If E is the big number, S is a small number.
(Observe students and give feedback.)
(Add to show:) [20:7H]

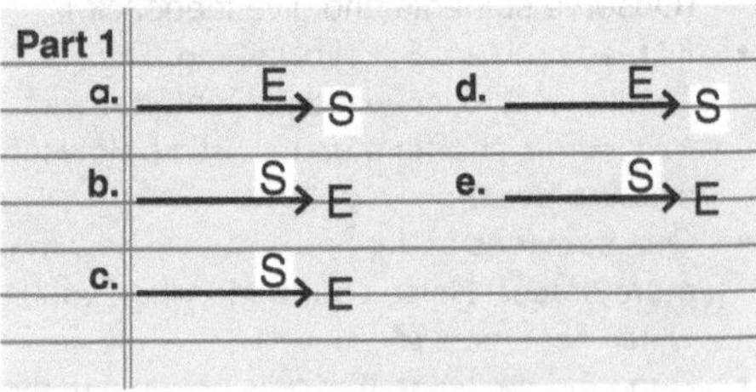

Here are E and S in the families for all of the sentences.

from Lesson 20, Exercise 7

Students make a series of families that have the letter E. At the end of the exercise, students go back and write S in each family.

On Lesson 23, students make number families with S, E, and a number.

Here's an example that shows the structure the teacher provides for students to make the families:

h. 5 birds flew into the field.

- Raise your hand when you know if E is the big number or a small number for the field. ✔
- Is E the big number or a small number? (Signal.) *The big number.*
- What is S? (Signal.) *A small number.*
- Listen: 5 birds flew into the field. What's the first small number? (Signal.) *5.*

from Lesson 23, Exercise 7

Note that the 5 is the first small number in the family above.

On Lesson 26, full sequence word problems are introduced. The strategy students follow is to find the sentence in the problem that tells about getting more or less. That sentence has information about whether *end* is the big number or a small number. The sentence also gives the first small number in the family.

Here's the first part of the exercise from Lesson 26:

(Teacher reference:)

Problems

a. Jan started out with some eggs. Then she gave away 22 eggs. Jan ended up with 37 eggs. How many eggs did Jan start with?

b. There were some fish in a pond. Then another 63 fish were put into the pond. The pond ended up with 175 fish. How many fish started out in the pond?

c. Jill had 210 dollars. She spent 130 dollars. How much did Jill end up with?

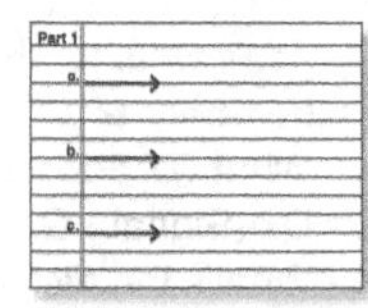

You're going to make number families for problems that tell about Start or End and work them.

- Write part 1 on your lined paper with the letters A, B, and C below. Make an addition number family arrow after each letter. **(Observe students and give feedback.)**

b. I'll read problem A. Follow along and get ready to tell me the sentence that tells about getting more or less.

- Listen: Jan started out with some eggs. Then she gave away 22 eggs. Jan ended up with 37 eggs. How many eggs did Jan start with? Here's the sentence that tells about getting more or less: Then she gave away 22 eggs.
- Say the sentence that tells about getting more or less. Get ready. **(Signal.)** *Then she gave away 22 eggs.*

c. Raise your hand when you know if E is the big number or a small number for Jan. ✔

- What is E? **(Signal.)** *A small number.*
- So what's S? **(Signal.)** *The big number.*
- What's the number that tells how many more or less? **(Signal.)** *22.*
- Make the family for problem A with a number and E and S. **(Observe students and give feedback.)** **(Display:)** [26:7A]

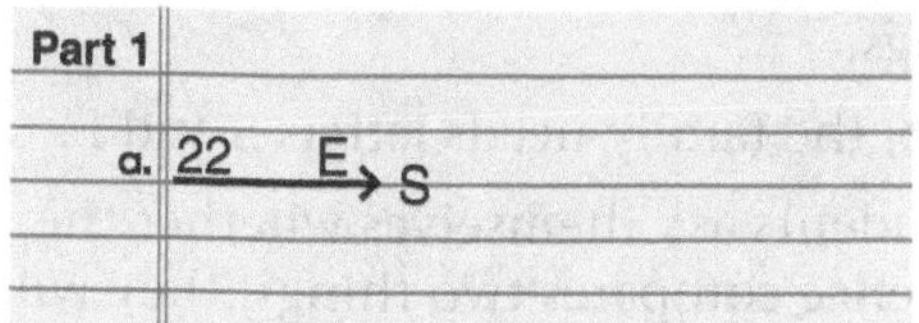

Here's the family for A.

d. The problem gives a number for start or end. Raise your hand when you know that number. ✔

- Is the number the problem gives for start or end? **(Signal.)** *End.*
- What's the number for end? **(Signal.)** *37.*
- Cross out E and write what it equals. ✔ **(Add to show:)** [26:7B]

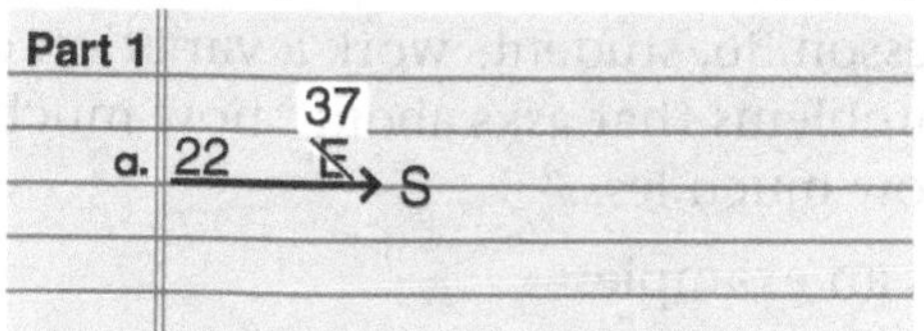

e. Say the problem you'll work to find S. **(Signal.)** *22 + 37.*

- Write the column problem next to family A and work it. Write the units in the answer. **(Observe students and give feedback.)**
- Read the number problem and the whole answer for A. **(Signal.)** *22 + 37 = 59 eggs.*
- How many eggs did Jan start with? **(Signal.)** *59.* **(Add to show:)** [26:7C]

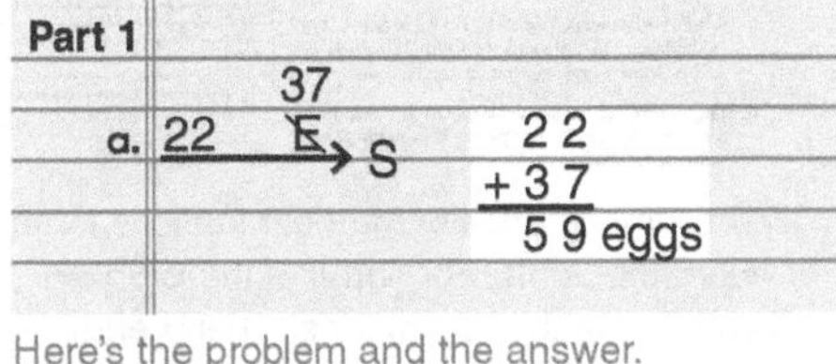

Here's the problem and the answer.

from Lesson 26, Exercise 7

Teaching Note: Students have worked all parts of the problem. This is the first time, however, that all parts are combined. There are a lot of steps, but if students start with the sentence that tells about getting more or less, they have a framework for putting in a number for start or end and a number for getting more or less. When these numbers are in the family, there's only one problem that can be worked. This problem is confirmed by the question the word problem asks.

If students get stuck or confused, remind them of the steps.

- Did you make a family for the sentence that tells about getting more or less?
- Did you put in a number for getting more or less? What number?
- Did you put in a number for start or end? What number?
- Say the problem for the missing letter.

On the following lessons, the structure you provide is reduced progressively until students work problems with minimal guidance.

Beginning on Lesson 33, students work problem sets composed of both comparison problems and start-end problems.

Here's the first part of the exercise from Lesson 34:

a. Open your textbook to Lesson 34 and find part 1. ✔

(Teacher reference:)

Problems

a. A bicycle costs $86 more than a skateboard. The bicycle costs $192. How much does the skateboard cost?

b. A vine had grapes on it. People picked 269 grapes off of the vine. The vine ended up with 377 grapes. How many grapes did the vine start out with?

c. A train had some people on it. Then 106 people got on the train. The train ended up with 211 people on it. How many people started out on the train?

d. There were 125 ducks on a lake. On the lake, there were 49 fewer ducks than geese. How many geese were on the lake?

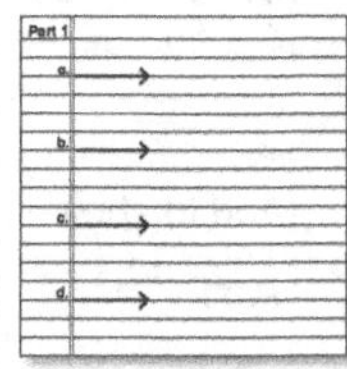

You'll make addition number families for these word problems. For some of the problems, you'll write the letters for start and end. I'll call on a student to read each problem. Then you'll tell me about the family.

- Read problem A. (Call on a student.) *A bicycle costs $86 more than a skateboard. The bicycle costs $192. How much does the skateboard cost?*
- Will you write the letters E and S in family A? (Signal.) *No.*
- Raise your hand when you know what you'll write for the big number and the small numbers. ✔
- What will you write for the big number in family A? (Signal.) *B.*
- What's the first small number? (Signal.) *86.*
- What's the other small number? (Signal.) *S.*

b. Read problem B. (Call on a student.) *A vine had grapes on it. People picked 269 grapes off of the vine. The vine ended up with 377 grapes. How many grapes did the vine start out with?*

- Will you write the letters E and S in family B? (Signal.) *Yes.*
- Raise your hand when you know if E is the big number or a small number for the grapes. ✔
- Is E the big number or a small number? (Signal.) *A small number.*
- What's the first small number in family B? (Signal.) *269.*

(If students are 100%, skip to step e.)

c. Read problem C. (Call on a student.) *A train had some people on it. Then 106 people got on the train. The train ended up with 211 people on it. How many people started out on the train?*

- Will you write the letters E and S in family C? (Signal.) *Yes.*
- Raise your hand when you know if E is the big number or a small number for the train. ✔
- Is E the big number or a small number? (Signal.) *The big number.*
- What's the first small number in family D? (Signal.) *106.*

d. Read problem D. (Call on a student.) *There were 125 ducks on a lake. The lake had 49 fewer ducks than geese. How many geese were on the lake?*

- Will you write the letters E and S in family D? (Signal.) *No.*
- Raise your hand when you know what you'll write for the big number and the small numbers. ✔
- What will you write for the big number in family D? (Signal.) *G.*
- What's the first small number? (Signal.) *49.*
- What's the other small number? (Signal.) *D.*

(Repeat problems that were not firm.)

from Lesson 34, Exercise 7

This part provides students with practice in identifying the letters and number in the family. If students have trouble with any items, repeat those items after going through item D.

After doing the part of the exercise shown, students make the families for the problems, check them, and then work the problems.

Teaching Note: The biggest problem that students have is not being sure about whether to write S and E in the family or to write letters for two things that are compared.

The simplest correction is to tell them to find the sentence that tells them how to make the number family. Read the sentence and ask, "Does that sentence compare two things?" (or, "Does that sentence tell about two things?")

If yes, the family needs letters for those two things.

If no, the family needs letters S and E.

If students ask themselves whether the sentence compares two things, they will always discriminate between start-end problems and comparison problems.

Structured exercises that have a mix of comparison and start-end problems continue on each lesson through Lesson 35.

On Lesson 36, students work a variation of start-end problems that asks about "how much more" or "how much less."

Here's an example:

A truck started out with 44 boxes, picked up some more boxes and ended with 91 boxes. How many boxes did the truck pick up?

The question indicates that the missing number is the number for getting more or less and is the first small number in the family. If students put in the numbers the problem gives, they will see that the number for getting more or less is missing.

To work any problem of this type, students subtract the small number that is shown from the big number.

Here's part of the exercise from Lesson 38. The teacher first asks about the letters in each family and then directs students to work the problems:

(Teacher reference:)

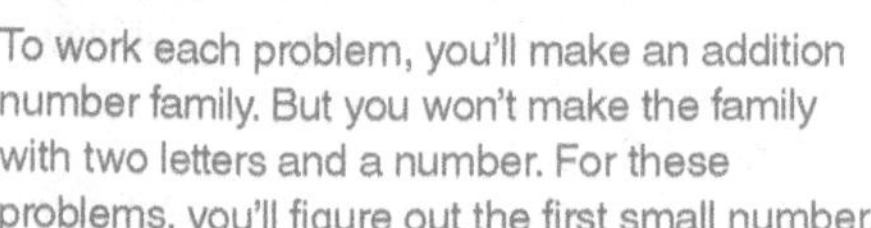
Problems

a. There were 114 students in the gym. There were 181 students in the auditorium. How many more students were in the auditorium than the gym?

b. 84 people were on a train. Then some of those people got off of the train. The train ended up with 25 people on it. How many people got off of the train?

c. A tank had 217 gallons in it. Then some more gallons were poured into the tank. The tank ended up with 307 gallons in it. How many gallons were poured into the tank?

d. A truck was lighter than a boat. The boat weighed 2510 kilograms. The truck weighed 1910 kilograms. How much lighter was the truck than the boat?

To work each problem, you'll make an addition number family. But you won't make the family with two letters and a number. For these problems, you'll figure out the first small number.

- Write part 2 on your lined paper with the letters A through D below. Make an addition number family after each letter.
  (Observe students and give feedback.)

b. Now I'll call on one of you to read each problem. Then you'll tell me about the number family you'll make.
- Read problem A. (Call on a student.) *There were 114 students in the gym. There were 181 students in the auditorium. How many more students were in the auditorium than the gym?*
- Does this problem tell about start and end? (Signal.) *No.*
- What letter will you write for the big number? (Signal.) *A.*
- What letter will you write for a small number? (Signal.) *G.*
- Tell me the number you'll write for A. Get ready. (Signal.) *181.*
- Tell me the number you'll write for G. Get ready. (Signal.) *114.*

c. Read problem B. (Call on a student.) *84 people were on a train. Then some of those people got off of the train. The train ended up with 25 people on it. How many people got off of the train?*
- Does this problem tell about start and end? (Signal.) *Yes.*
- What letter will you write for the big number? (Signal.) *S.*
- What letter will you write for a small number? (Signal.) *E.*
- Tell me the number you'll write for S. Get ready. (Signal.) *84.*
- Tell me the number you'll write for E. Get ready. (Signal.) *25.*

(If students have performed perfectly on word problems for at least two lessons, skip to step f.)

d. Read problem C. (Call on a student.) *A tank had 217 gallons in it. Then some more gallons were poured into the tank. The tank ended up with 307 gallons in it. How many gallons were poured into the tank?*
- Does this problem tell about start and end? (Signal.) *Yes.*
- What letter will you write for the big number? (Signal.) *E.*
- What letter will you write for a small number? (Signal.) *S.*
- Tell me the number you'll write for S. Get ready. (Signal.) *217.*
- Tell me the number you'll write for E. Get ready. (Signal.) *307.*

e. Read problem D. (Call on a student.) *A truck was lighter than a boat. The boat weighed 2510 kilograms. The truck weighed 1910 kilograms. How much lighter was the truck than the boat?*
- Does this problem tell about start and end? (Signal.) *No.*
- What letter will you write for the big number? (Signal.) *B.*
- What letter will you write for a small number? (Signal.) *T.*
- Tell me the number you'll write for B. Get ready. (Signal.) *2510.*
- Tell me the number you'll write for T. Get ready. (Signal.) *1910.*

(Repeat problems that were not firm.)

f. Work all the problems. Put your pencil down when you've worked the problems in part 2.
(Observe students and give feedback.)

g. Check your work.
- Problem A. Read the column problem and the answer. (Signal.) *181 − 114 = 67.*
- How many more students were in the auditorium than the gym? (Signal.) *67.*

(Display:) [38:7A]

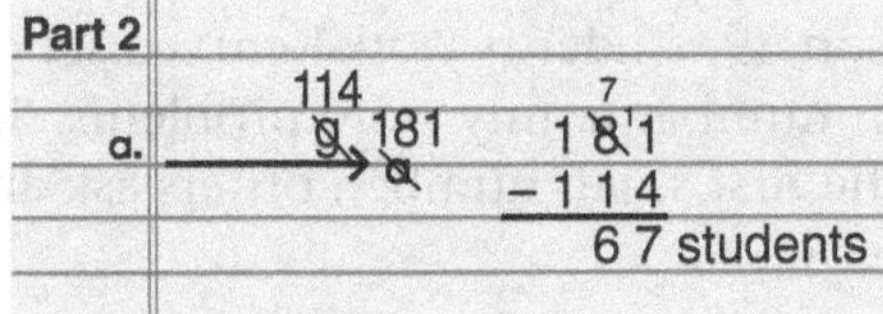

Here's what you should have for problem A.

h. Problem B. Read the column problem and the answer. (Signal.) *84 − 25 = 59.*
- How many people got off of the train? (Signal.) *59.*

(Display:) [38:7B]

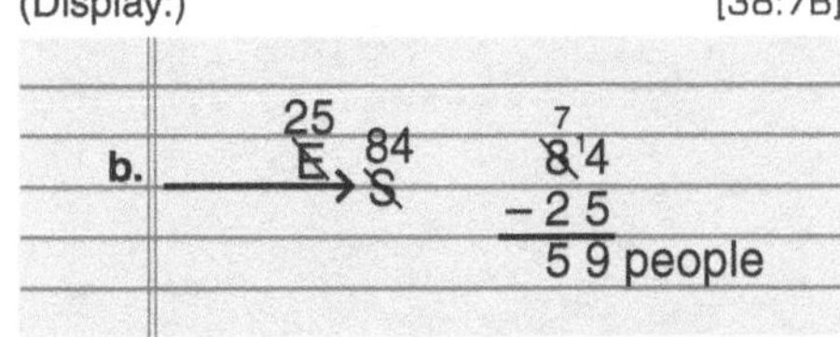

Here's what you should have for problem B.

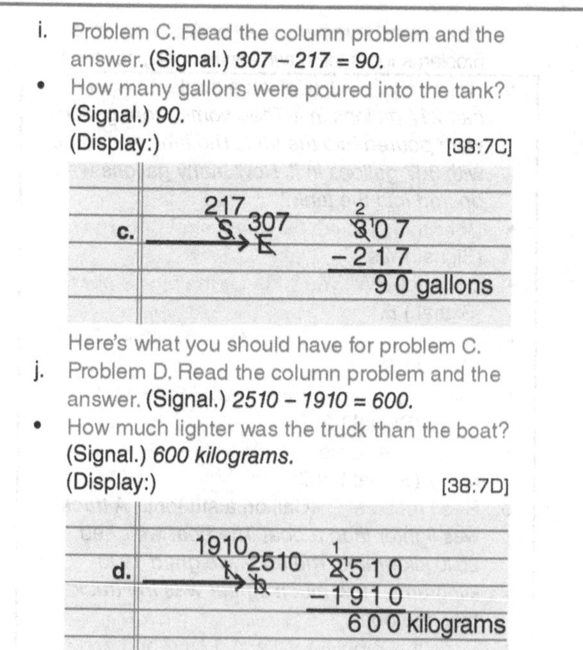

i. Problem C. Read the column problem and the answer. (Signal.) *307 – 217 = 90.*

- How many gallons were poured into the tank? (Signal.) *90.*

(Display:) [38:7C]

Here's what you should have for problem C.

j. Problem D. Read the column problem and the answer. (Signal.) *2510 – 1910 = 600.*

- How much lighter was the truck than the boat? (Signal.) *600 kilograms.*

(Display:) [38:7D]

Here's what you should have for problem D.

Lesson 38, Exercise 7

Teaching Note: In steps B through E, make sure that students respond correctly when identifying the letters for each family and the numbers for the letters. Repeat any verbal items that are not firm.

When students work problems (starting in step F), make sure they show the family with two numbers, work the column problem, and write the unit name in the answer.

On Lesson 39, students work a mixed set of addition number family word problems. Some ask about the first small number; others ask about one of the letters.

Here's the set of problems from Lesson 40 and the first part of the exercise:

a. Open your textbook to Lesson 40 and find part 1. ✔
(Teacher reference:)

Problems

a. A tub of water started out at 85 degrees. The water cooled off. The tub of water ended up at 59 degrees. How much did the water cool off?

b. A trailer was 231 centimeters wide. A sofa was 39 centimeters narrower than the trailer. How wide was the sofa?

c. In the morning, there were some cows in a field. Then 195 cows walked into the field. The field ended up with 608 cows in it. How many cows were in the field in the morning?

d. Fran read 89 books. Greg read 46 books. How many more books did Fran read than Greg?

Part 1
a. →
b. →
c. →
d. →

To work each problem, you'll make an addition number family. For some of the problems, you'll make a family with two letters and a number. For the other problems, you'll make a family with two letters. For those problems, you'll figure out the first small number.

- Read problem A. (Call on a student.) *A tub of water started out at 85 degrees. The water cooled off. The tub of water ended up at 59 degrees. How much did the water cool off?*
- Will you make an addition number family with the letters for start and end? (Signal.) *Yes.*
- What letter will you write for the big number? (Signal.) *S.*
- What's the other letter you'll write? (Signal.) *E.*
- Does the problem give a number for how much the temperature changed? (Signal.) *No.*
- So will you write a number for the first small number? (Signal.) *No.*

from Lesson 40, Exercise 5

Students read and answer the same set of questions for all the problems. Then they work the problems. The key discriminations are addressed by the last two questions:

> Does the problem give a number for how different two things are or how much something has changed?

If the answer is yes, the students write the first small number in the family. If the answer is no, students will have to figure out the first small number.

Classification Word Problems

Work on classification word problems begins on Lesson 107. Classification word problems are different from comparison problems or sequence word problems because classification number families have three names, not two. The names for classification number families always refer to a class and two subsets of that class. For a classification word problem that deals with bald men and men who are not bald, the name of the class is men. So, men is the name for the big number in the family. The subsets of men are bald and not bald. Those are the names for the small numbers in the family. Here's the family with three names:

For the classification word problem that deals with adults, men, and women, the name of the class is adults. So adults is the name for the big number in the family. The subsets of adults are men and women. Those are the names for the small numbers in the family. Here's the family with three names:

men ——— women → adults

On Lesson 107, students verbally review names of several classes and identify names for the small numbers and the big number.

Here's the part of the exercise that follows the verbal examples:

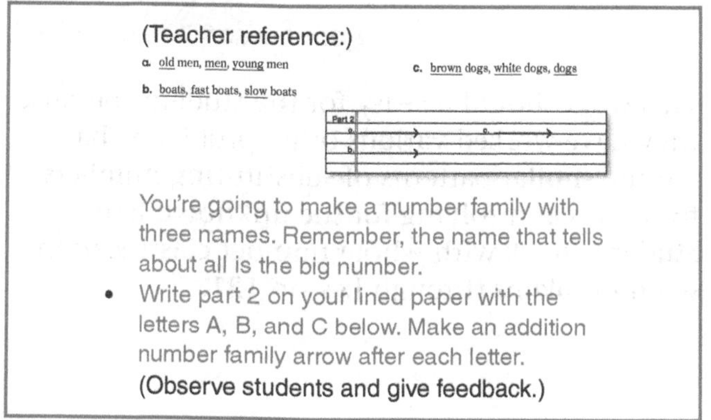

(Teacher reference:)

a. old men, men, young men
b. boats, fast boats, slow boats
c. brown dogs, white dogs, dogs

You're going to make a number family with three names. Remember, the name that tells about all is the big number.

- Write part 2 on your lined paper with the letters A, B, and C below. Make an addition number family arrow after each letter. (Observe students and give feedback.)

b. The names for A are old men, men, young men.
- Which name tells about all? (Signal.) *Men.*
- What's the name for the big number? (Signal.) *Men.*

c. The names for B are boats, fast boats, slow boats.
- What's the name for the big number? (Signal.) *Boats.*

d. The names for C are brown dogs, white dogs, dogs.
- What's the name for the big number? (Signal.) *Dogs.*

(Repeat steps b through d until firm.)

e. Make the number family for A. Use the underlined words. They are old, men, young. (Observe students and give feedback.)

f. Check your work for family A.
- What are the names for the small numbers? (Signal.) *Old (and) young.*
- What's the name for the big number? (Signal.) *Men.*

(Display:) [107:5I]

Part 2
a. old ——— young → men

Here's what you should have for A.

g. Make the number families for B and C. Remember, use the underlined words. (Observe students and give feedback.)

h. Check your work for family B.
- What are the names for the small numbers? (Signal.) *Fast (and) slow.*
- What's the name for the big number? (Signal.) *Boats.*

(Display:) [107:5J]

b. fast ——— slow → boats

Here's what you should have for B.

i. Family C. What are the names for the small numbers? (Signal.) *Brown (and) white.*
- What's the name for the big number? (Signal.) *Dogs.*

(Display:) [107:5K]

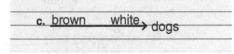

c. brown ——— white → dogs

Here's what you should have for C.

from Lesson 107, Exercise 5

Teaching Note: If students have trouble identifying the name for the big number, show them how to test the names. Say each name and ask if it tells about the other two names.

For example:

The names are old men, men, and young men.

Listen: Does old men tell about men and young men?

So is it the big number?

Listen: Does men tell about old men and young men?

So is men the big number?

Students continue to make families for three names in Lessons 108 and 109.

On Lesson 110, students work complete problems. They make the number family with letters for the three classes. They put in the numbers the problem gives, and solve for the name that has no number.

Here's part of the exercise that follows verbal examples:

e Read problem A. (Call on a student.) *There were 89 men in the store. 22 of the men wore hats. How many men did not wear hats?*

- Which name tells about the big number? (Signal.) *Men.*
- What are the names you'll write for the small numbers? (Signal.) *Hats (and) not hats.*
- Make the family with the names men, hats, and not hats. Stop when you've done that much.
  (Observe students and give feedback.)
  (Display:) [110:7E]

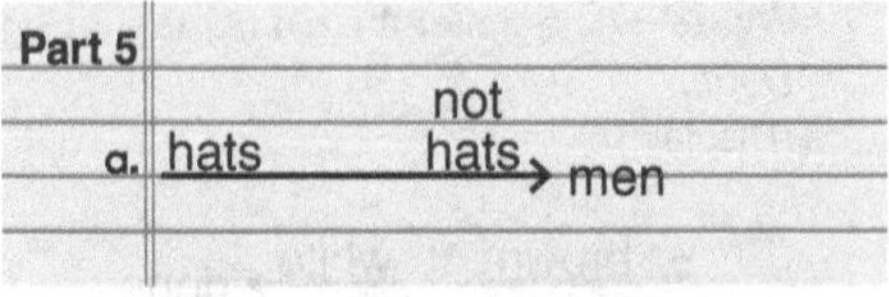

Here's what you should have for family A.

(If students are firm, skip to step g.)

f. Read problem B. (Call on a student.) *In a store, 59 women wore dresses. 47 women did not wear dresses. How many women were in the store?*

- Which name tells about the big number? (Signal.) *Women.*
- What are the names you'll write for the small numbers? (Signal.) *Dresses (and) not dresses.*
- Make the family with the names dresses, not dresses, and women. Stop when you've done that much.
  (Observe students and give feedback.)
  (Display:) [110:7F]

b. dresses — not dresses → women

Here's what you should have for family B.

g. Work problem A. Replace the names with the numbers the problem gives. Work the column problem for the missing number. Put your pencil down when you know how many men did not wear hats.
(Observe students and give feedback.)

h. Check your work for problem A.

- Everybody, read the column problem and the answer. (Signal.) *89 – 22 = 67.*
- How many men did not wear hats? (Signal.) *67.*
  (Display:) [110:7G]

Part 5
a. 22 ~~hats~~ — not hats → 89 ~~men~~

$$\begin{array}{r} 89 \\ -\,22 \\ \hline 67 \end{array}$$ men

Here's what you should have for problem A.

from Lesson 110, Exercise 7

This work should be easy for the students because they have worked various other problems that require similar patterns of substituting numbers for letters and solving for the unknown letter. Students work with whole-number classification word problems through Lesson 121.

NUMBER FAMILIES WITH FRACTIONS

On Lesson 121, students work with number families that have fractions. Students are shown that a number family with fractions that have the same denominators works the same way as a family with whole numbers.

Here's part of the introduction on Lesson 121:

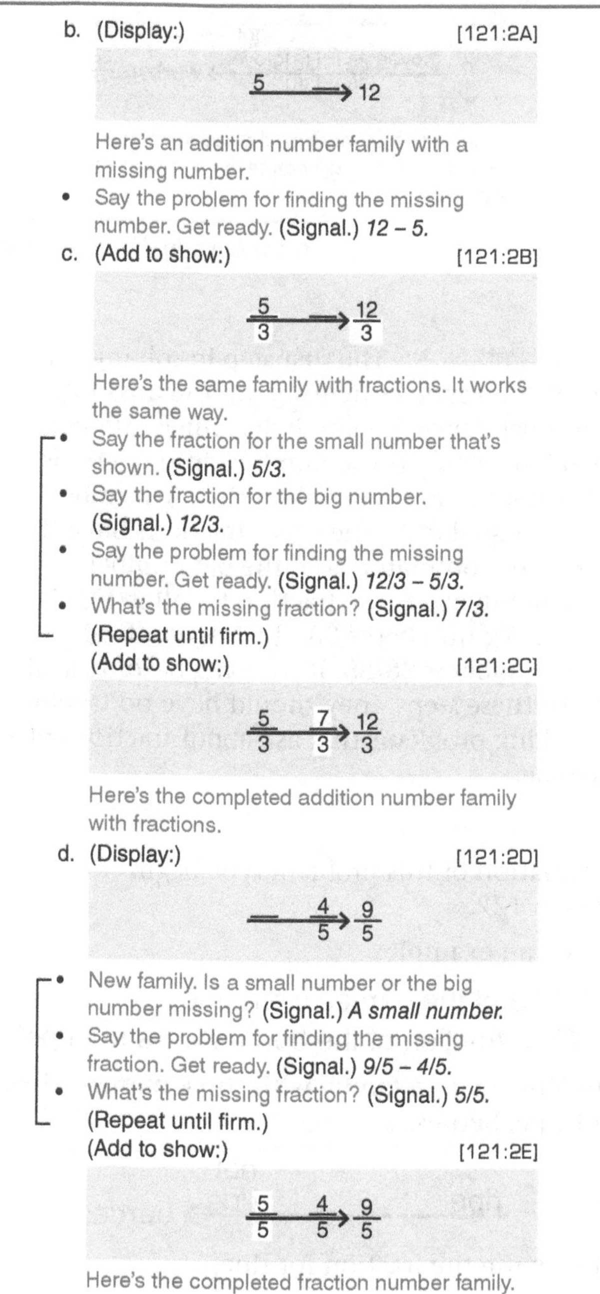

b. (Display:) [121:2A]

$5 \longrightarrow 12$

Here's an addition number family with a missing number.

- Say the problem for finding the missing number. Get ready. (Signal.) *12 – 5.*

c. (Add to show:) [121:2B]

$\frac{5}{3} \longrightarrow \frac{12}{3}$

Here's the same family with fractions. It works the same way.

- Say the fraction for the small number that's shown. (Signal.) *5/3.*
- Say the fraction for the big number. (Signal.) *12/3.*
- Say the problem for finding the missing number. Get ready. (Signal.) *12/3 – 5/3.*
- What's the missing fraction? (Signal.) *7/3.*

(Repeat until firm.)

(Add to show:) [121:2C]

$\frac{5}{3} \quad \frac{7}{3} \longrightarrow \frac{12}{3}$

Here's the completed addition number family with fractions.

d. (Display:) [121:2D]

$\underline{\quad} \quad \frac{4}{5} \longrightarrow \frac{9}{5}$

- New family. Is a small number or the big number missing? (Signal.) *A small number.*
- Say the problem for finding the missing fraction. Get ready. (Signal.) *9/5 – 4/5.*
- What's the missing fraction? (Signal.) *5/5.*

(Repeat until firm.)

(Add to show:) [121:2E]

$\frac{5}{5} \quad \frac{4}{5} \longrightarrow \frac{9}{5}$

Here's the completed fraction number family.

from Lesson 121, Exercise 2

The work should be easy for students because they know how to solve for whole numbers in a family, and they know the properties of fractions that have the same denominator.

On Lesson 123, students work with families that have a fraction and a whole number. Students figure out the missing value.

Here are the items from the Textbook Lesson 123:

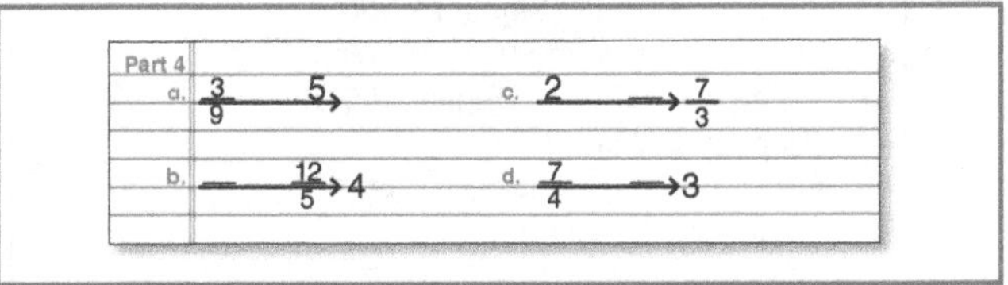

Part 4

a. $\frac{3}{9} \quad 5 \longrightarrow$

b. $\underline{\quad} \quad \frac{12}{5} \longrightarrow 4$

c. $2 \quad \underline{\quad} \longrightarrow \frac{7}{3}$

d. $\frac{7}{4} \quad \underline{\quad} \longrightarrow 3$

To work each problem, students first convert the whole number into a fraction that has the same denominator as the fraction in the family. For family A, they change 5 into 9ths—45/9. Students apply familiar rules for adding or subtracting to find the missing value.

On Lesson 125, students work classification word problems that have fractions. The procedure for identifying the big number and small numbers of families are the same as they are for whole number families.

Students work problems with fractions. For each problem students create number families that have three fractions.

For example: In a factory, 25 of the workers wore gloves. There were 38 workers in the factory. What fraction of the workers did not wear gloves?

The first step students take is to write the missing number in the family.

$$25 \quad 13 \longrightarrow 38$$

The problem asks about a fraction, so the students change the whole numbers into fractions. The number for *all* equals 1. So the fraction for all workers is 38/38, and all of the fractions have the denominator of 38.

$$\frac{25}{38} \quad \frac{13}{38} \longrightarrow \frac{38}{38}$$

Students are now able to answer the question about the fraction of workers that did not wear gloves.

The key discrimination is identifying the fraction for the whole group. In step A of the exercise from Lesson 125, students learn how to do this:

a. You're going to write an addition number family with three names and a fraction for each name. Listen: There were 12 boys. The fraction of boys who were boys was 12/12. What fraction of the boys were boys? (Signal.) *12/12.*
- Listen: There were 30 boys. What fraction of boys were boys? (Signal.) *30/30.*
- Listen: There were 89 sheep. What fraction of sheep were sheep? (Signal.) *89/89.*

(Repeat until firm.)

b. Find part 3 in your textbook. ✔

(Teacher reference:)

a. In a factory, 25 of the workers wore gloves. There were 38 workers in the factory. What fraction of the workers in the factory did not wear gloves?

b. There were 36 students in a class. 21 of the students were boys. What fraction of the students were girls?

c. There were 14 men in a swimming pool. There were 6 women in the swimming pool. What fraction of the adults in the swimming pool were men?

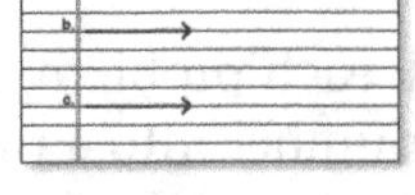

These problems are a lot like problems you've worked before. This time you'll solve the family for a missing fraction.
- Read problem A. (Call on a student.) *In a factory, 25 of the workers wore gloves. There were 38 workers in the factory. What fraction of the workers in the factory did not wear gloves?*

The problem asks about the fraction of workers who did not wear gloves. Before we can answer that question, we have to know the number of workers who did not wear gloves.

(Display:) [125:5A]

- Look at the family and say the problem for finding the number of workers who did not wear gloves. Get ready. (Signal.) *38 – 25.*

The answer is 13.

c. (Add to show:) [125:5B]

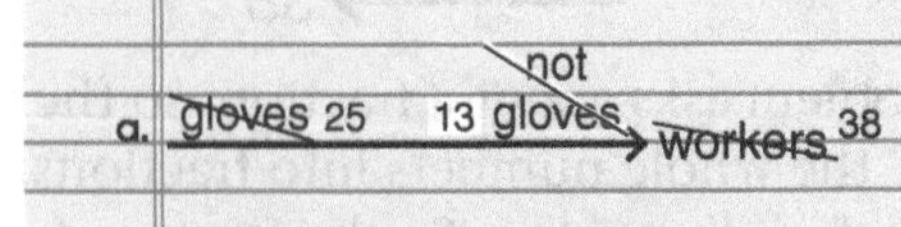

Now we have a number for each of the names. We can change the numbers into fractions.
- There were 38 workers. So what fraction of the workers were workers? (Signal.) *38/38.*
- (Point to **25.**) What's the fraction for workers who wore gloves? (Signal.) *25/38.*
- (Point to **13.**) What's the fraction for workers who didn't wear gloves? (Signal.) *13/38.*

(Repeat until firm.)

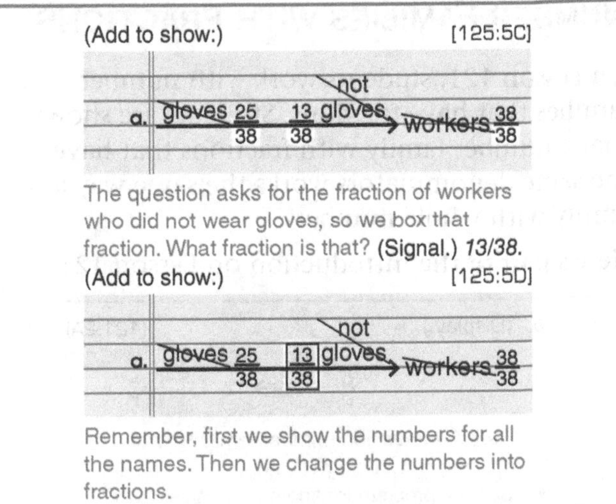

(Add to show:) [125:5C]

a. gloves 25/38 13/38 not gloves → workers 38/38

The question asks for the fraction of workers who did not wear gloves, so we box that fraction. What fraction is that? (Signal.) *13/38.*

(Add to show:) [125:5D]

a. gloves 25/38 [13/38] not gloves → workers 38/38

Remember, first we show the numbers for all the names. Then we change the numbers into fractions.

from Lesson 125, Exercise 5

Teaching Note: The first step in solving the problem involves figuring out the missing number. Once the family has three numbers, students convert the numbers into fractions. The conversion is based on the big number being equal to 1. The other fractions have the same denominator. So if the big number in the family is 17, the fraction for all is 17/17. If the big number is 88, the fraction for the big number is 88/88. If students become facile with these steps, they should have no trouble working problems that ask about fractions of a whole.

A variation of this problem type begins on Lesson 127.

Here's an example:

24/73 of the berries were ripe.
What fraction of the berries were not ripe?

Students make a family with three names—ripe, not ripe, berries:

ripe — not ripe → berries

They write the fraction for ripe:

ripe 24/73 — not ripe → berries

The fraction shows the denominator for the other fractions. The fraction for the big number equals 1:

$$\text{ripe } \frac{24}{73} \longrightarrow \text{not ripe} \quad \text{berries } \frac{73}{73}$$

Students subtract to find the numerator for not ripe, 73 – 24. Then they convert the answer into the fraction 49/73.

$$\text{ripe } \frac{24}{73} \quad \frac{49}{73} \text{ not ripe} \longrightarrow \text{berries } \frac{73}{73}$$

Here's part of the exercise from Lesson 127:

a. Listen: There were 58 lights. 25 of the lights were on. What's the fraction for the lights that were on? (Signal.) *25 fifty-eighths.*
- What's the fraction for all of the lights? (Signal.) *58/58.*

b. There were 75 boys. 59 of the boys played football. What's the fraction for the boys that played football? (Signal.) *59 seventy-fifths.*
- What's the fraction for all of the boys? (Signal.) *75/75.*

c. Listen: 12 thirty-sevenths of the dogs wore collars. Tell me what fraction of the dogs were dogs. Get ready. (Signal.) *37/37.*
- Listen: 24 fiftieths of the berries were ripe. Tell me the fraction for all of the berries. (Signal.) *50/50.*
- Listen: 39 eighty-sixths of the lights were off. Tell me the fraction for all of the lights. Get ready. (Signal.) *86/86.*

(Repeat steps a through c until firm.)

d. Find part 5 in your textbook. ✔

(Teacher reference:)

a. 175 of the lights in a building were off. There were 281 lights in the building. What fraction of the lights in the building were on?

b. On a tree, there were 166 apples that were not ripe. There were 72 ripe apples on the tree. What fraction of the apples were ripe?

c. $\frac{24}{73}$ of the berries were ripe. What fraction of the berries were not ripe?

d. In a classroom, $\frac{19}{30}$ of the students were girls. What fraction of the students were boys?

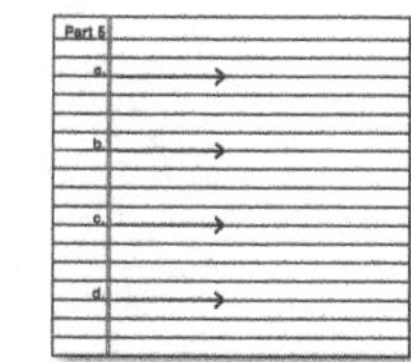

To work these problems, you're going to write addition number families with three names. The values in the families will be fractions.
- Write part 5 on your lined paper with the letters A through D below. After each letter, make an addition number family arrow.

(Observe students and give feedback.)

e. Read problem A to yourself. Make the family with three names. Figure out the missing number. Change the numbers into fractions. Box the fraction that answers the question the problem asks.

(Observe students and give feedback.)

f. Check your work for problem A.
- Tell me the names for the small numbers. (Signal.) *Off (and) on.*
- Everybody, what's the name for the big number? (Signal.) *Lights.*
- You figured out the number of lights in the building that were turned on. What number? (Signal.) *106.*
- What's the fraction for lights in the building that were on? (Signal.) *106/281.*
- What's the fraction for lights in the building that were off? (Signal.) *175/281.*
- What's the fraction for all of the lights? (Signal.) *281/281.*
- Which fraction did you box? (Signal.) *106/281.*

(Display:) [127:6A]

Here's what you should have for family A.

g. Read problem B to yourself. Make the family with three names. Figure out the missing number. Change the numbers into fractions. Box the fraction that answers the question the problem asks.

(Observe students and give feedback.)

h. Check your work for problem B.
- Tell me the names for the small numbers. Get ready. (Signal.) *Ripe (and) not ripe.*
- Everybody, what's the name for the big number? (Signal.) *Apples.*
- You figure out the number of apples. What number? (Signal.) *238.*
- What's the fraction for the apples that were not ripe? (Signal.) *166/238.*
- What's the fraction for the apples that were ripe? (Signal.) *72/238.*
- What's the fraction for all of the apples? (Signal.) *238/238.*
- Which fraction did you box? (Signal.) *72/238.*

(Display:) [127:6B]

Here's what you should have for family B.

i. Read problem C. (Call on a student.) *24/73 of the berries were ripe. What fraction of the berries were not ripe?*
- There are three names. Everybody, what's the name for the big number? (Signal.) *Berries.*
- Tell me the names for the small numbers. (Signal.) *Ripe (and) not ripe.*
- The problem tells about a fraction for ripe berries. If the fraction for ripe berries is 24/73, what's the fraction for all of the berries? (Signal.) *73/73.*
- Make the family for problem C. Replace ripe and berries with the fractions they equal. Write the column problem for the numerators and work it. Put your pencil down when you've completed the fraction number family and boxed the answer.

(Observe students and give feedback.)

j. Check your work for problem C.

- Read the column problem and the answer for the numerators. Get ready. **(Signal.)** *73 – 24 = 49.*
- What fraction of the berries were not ripe? **(Signal.)** *49/73.*

(Display:) [127:6C]

c. ripe $\frac{24}{73}$ — $\boxed{\frac{49}{73}}$ not ripe → berries $\frac{73}{73}$ $73 - 24 = 49$ (berries)

Here's what you should have for family C.

k. Read problem D. **(Call on a student.)** ***In a classroom, 19/30 of the students were girls. What fraction of the students were boys?***

- Tell me the fraction of students that were girls. Get ready. **(Signal.)** *19/30.*
- If the fraction for girls is 19/30, what's the fraction for all of the students? **(Signal.)** *30/30.*
- Make the family for problem D. Replace girls and students with the fractions they equal. Write the column problem for the numerators and work it. Put your pencil down when you've completed the fraction number family and boxed the answer.

(Observe students and give feedback.)

l. Check your work for problem D.

- Read the column problem and the answer for the numerators. Get ready. **(Signal.)** *30 – 19 = 11.*
- What fraction of the students were boys? **(Signal.)** *11/30.*

(Display:) [127:6D]

d. girls $\frac{19}{30}$ — $\boxed{\frac{11}{30}}$ boys → students $\frac{30}{30}$ $30 - 19 = 11$ (boys)

Here's what you should have for family D.

Lesson 127, Exercise 6

Multiplication Number Families (Lessons 9–130)

Multiplication number families look different than addition families, but they are parallel in the relationships they show. Here are three multiplication number families:

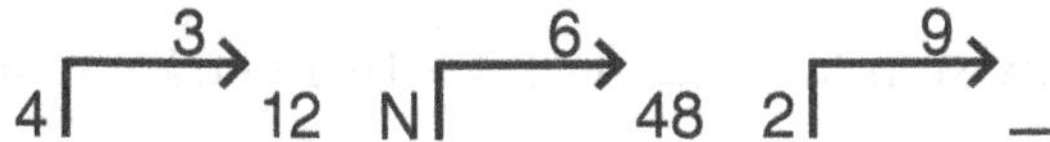

The configuration of the families resembles a division problem. Each family consists of two small numbers and a big number. The first two numbers are small numbers. The big number is at the end of the arrow.

To find a missing big number, you multiply the small numbers. To find a missing small number, you divide the big number by the small number shown.

Word problems that involve multiplication or division can be represented with multiplication number families.

For example:

> The boat was 4 times the length of the car. The car was 12 feet long. What was the length of the boat?

Here's the family with the times number and the letters for boat and car:

4 C B

The problem gives a number for car (12). With that number in the family, students work the problem for finding B: 4 × 12. So the big number in the family is 48. The boat was 48 feet long.

4 ~~C~~ 12 B

$$\begin{array}{r} 12 \\ \times\ 4 \\ \hline 48 \end{array}$$ feet

Writing Four Facts

The work with multiplication number families begins on Lesson 9 and continues throughout the remainder of the program. This means that while students are working on problems involving addition and subtraction, they are also working with multiplication number families in the same lessons. This work begins early so that students who did not go through *CMC Level D* will learn at least the basics before students who have completed *CMC Level D* start on Lesson 31 of *Level E.*

Here's part of an exercise that follows the introduction of multiplication number families:

a. Find part 4 in your workbook. ✔
(Teacher reference:) **R Part I**

a. 16 4 64 b. N 20 K

You're going to write two multiplication facts and two division facts for each family.

- Touch family A. ✔
- What are the small numbers in this family? (Signal.) *16 and 4.*
- What's the big number? (Signal.) *64.*
- Is 16 a small number or the big number? (Signal.) *A small number.*

b. If you start with 16, do you say a multiplication fact or a division fact? (Signal.) *A multiplication fact.*

- Say the fact that starts with 16. (Signal.) *16 × 4 = 64.*
- Say the fact that starts with 4. (Signal.) *4 × 16 = 64.*

c. Say the fact that starts with 64 and divides by 4. (Signal.) *64 ÷ 4 = 16.*
The first division fact is 64 ÷ 4 = 16.

- Say the other division fact. (Signal.) *64 ÷ 16 = 4.*

(Repeat steps b and c until firm.)

d. Touch where you'll write the multiplication fact that starts with the first small number. ✔

- Touch where you'll write the other multiplication fact. ✔

(Repeat until firm.)

- Write both multiplication facts below family A.

(Observe students and give feedback.)
(Display:) [9:6B]

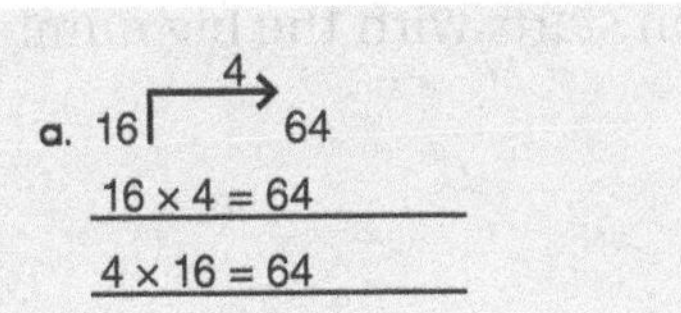

Here's what you should have for family A.

e. The signs are shown for the division facts.

- Touch where you'll write the first division fact. ✔
  The first division fact is 64 ÷ 4 = 16.
- Say that fact. (Signal.) *64 ÷ 4 = 16.*
  Remember, you'll write the big number under the division sign.
- Touch where you'll write 64 for that fact. ✔

f. Write the first division fact for family A.
(Observe students and give feedback.)
(Display:) [9:6C]

$4\overline{)64}$ with 16 above; empty division sign

Here's the division fact you should have.

g. Say the other division fact. (Signal.)
64 ÷ 16 = 4.

- Write that fact for family A.
  (Observe students and give feedback.)
  (Display:) [9:6D]

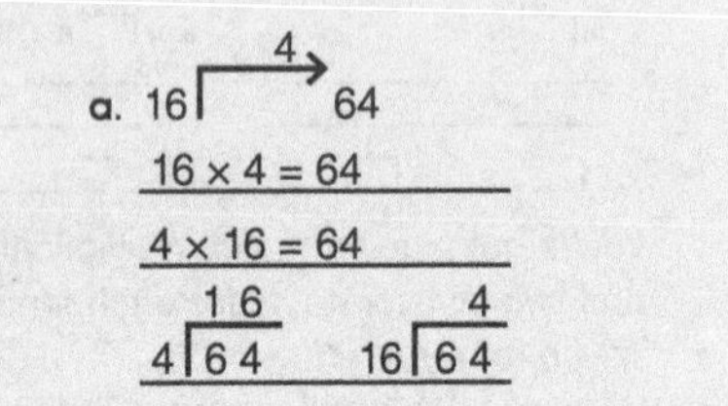
a. 16 —4→ 64
16 × 4 = 64
4 × 16 = 64
$4\overline{)64}$ = 16 $16\overline{)64}$ = 4

Here are the four facts you should have for family A.

h. Touch family B. ✔

- What are the small numbers? (Signal.)
  N and 20.
- What's the big number? (Signal.) *K.*
  My turn to start with N and say the fact:
  N × 20 = K.
- Say that fact. (Signal.) *N × 20 = K.*
- Say the other multiplication fact. (Signal.)
  20 × N = K.

i. My turn to say the fact that divides by 20:
K ÷ 20 = N.

- Say the fact that divides by 20. (Signal.)
  K ÷ 20 = N.
- Say the other division fact. (Touch.)
  K ÷ N = 20.
  (Repeat steps h and i until firm.)

from Lesson 9, Exercise 6

Before students do the Workbook activity, they identify the multiplication arrow and indicate that multiplying starts with a small number; division starts with the big number.

The Workbook activity requires students to first say the four facts for a family and then write the four facts. The family is:

16 —4→ 64

The multiplication facts are 16 × 4 = 64 and 4 × 16 = 64.

The division facts are 64 ÷ 4 = 16 and 64 ÷ 16 = 4.

In the same exercise, students also write four facts for a family that has two letters and a number.

20 —N→ K

The work with multiplication number families roughly follows the same progression that occurs with addition families. On Lesson 10, students work with families that have missing numbers. Students indicate whether the missing number is a small number or the big number. They say the problem for finding the missing number and complete the family.

Here are the problems for Textbook Lesson 11, Part 3:

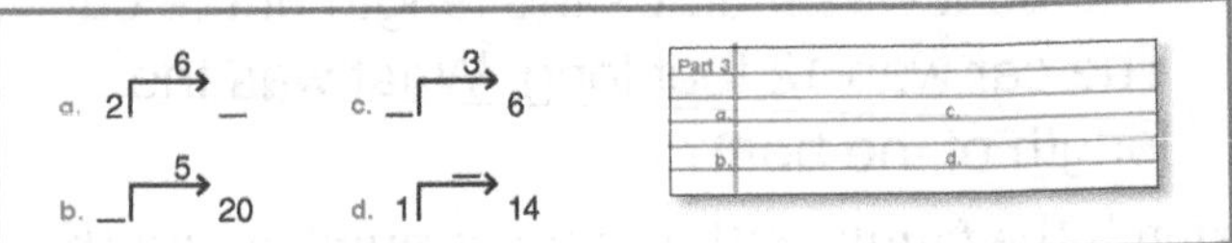

On Lesson 12, students figure out the missing number in families that have two numbers and a letter.

T —10→ 50   5 —6→ M

Students identify whether the letter is a small number or the big number, indicate the problem that they work to figure out what the letter equals, and work the problems.

On Lesson 14, students work with problems that have two letters. Some problems require addition-subtraction, others require multiplication-division.

Here's part of the exercise:

f. Touch problem A in your textbook. It gives a number for one of the letters. What letter? (Signal.) *A.*
- Put the number in the family and then stop. ✔
- Now the family has one letter. Say the problem for figuring out that letter. (Signal.) *5 × 30.*

g. Problem B gives a number for one of the letters. What letter? (Signal.) *A.*
- Put that number in family B. ✔

h. The rest of the problems show a number for one of the letters. Put those numbers in the families.
(Observe students and give feedback.)

i. Check your work.
- Family A. What kind of family is it? (Signal.) *Multiplication family.*
- Say the problem for the numbers that are shown. (Signal.) *5 × 30.*

j. Family B. What kind of family is it? (Signal.) *Addition family.*
- Say the problem for the numbers that are shown. (Signal.) *5 + 30.*

k. Family C. What kind of family is it? (Signal.) *Addition family.*
- Say the problem for the numbers that are shown. (Signal.) *180 – 18.*

l. Family D. What kind of family is it? (Signal.) *Multiplication family.*
- Say the problem for the numbers that are shown. (Signal.) *180 ÷ 18.*
(Repeat families that were not firm.)
(Display:) [14:6A]

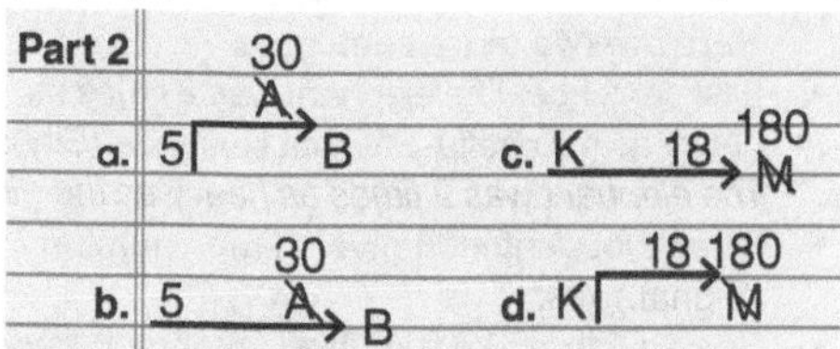

Here's what you should have for part 2.

m. Write all the problems for part 2 and work them.
(Observe students and give feedback.)

n. Check your work.
- Family A. Read the problem you worked and the answer. (Signal.) *5 × 30 = 150.*
- What does B equal? (Signal.) *150.*

o. (Repeat the following tasks for problems B through D:)

Family __. Read the problem you worked and the answer.		What does __ equal?	
B	*5 + 30 = 35*	B	*35*
C	*180 – 18 = 162*	K	*162*
D	*180 ÷ 18 = 10*	K	*10*

from Lesson 14, Exercise 6

Before working this part, students identify whether the family is an addition family or multiplication family. Students copy the problems on their lined paper.

In steps F through H, students replace a letter in each family. Then students find the missing number in each family. They first identify the kind of family it is and say the problem for the missing letter. They work the problem and write the missing number.

Students should not have serious problems with the mixed set of problems because the decisions that they have to make about each missing number are parallel for addition families and multiplication families.

If they make the mistake of confusing addition with multiplication or subtraction with division, ask them:

"Is this an addition family?"
"So do you add or subtract?"
"What do you do?"

Students continue to solve problems based on addition and multiplication families through Lesson 18. These problems appear as part of students' Independent Work until structured work begins on multiplication word problems (Lesson 43).

MULTIPLICATION-FAMILY WORD PROBLEMS

Work with multiplication-family word problems continues through the end of the level. The program introduces three types of multiplication-family word problems—times problems, problems that refer to *each* or *every,* and problems that refer to measurement facts.

Here are examples of the problem types:

Times problem: The boat was 3 times as heavy as the car. The car weighed 2600 pounds. How much did the boat weigh?

Each-every problem: There were 9 players on each team. The league had 18 teams. How many players were in the league?

Measurement-fact problem: There were 116 quarts of fuel in a tank. How many gallons of fuel were in the tank?

This track parallels that of addition-family problems (sequence and comparison). Students first work with sentences that tell how to make the number families. Students make families that have a number and two letters.

After this practice, students work entire problems. They make the family, replace a letter with a number, work the problem for the remaining letter, and answer the question the problem asks.

Times Word Problems

The work with times problems begins on Lesson 43. At this point of the program, students have worked addition number-family problems that compare. The first exercises for times problems give students practice in distinguishing between sentences that refer to times and sentences from comparison problems that refer to how many more or less.

Here's the first part of the exercise from Lesson 43:

a. (Display:) [43:6A]

Tim was 6 inches taller than Billy.

Tim was 2 times as tall as Billy.

You're going to learn about sentences that give a times number. For those sentences, you can write a multiplication number family with two letters and a number.

- I'll read the first sentence: Tim was 6 inches taller than Billy. Say that sentence. (Signal.) *Tim was 6 inches taller than Billy.*
  That sentence does not give a times number. So you can't make a multiplication number family for it.

b. Does "Tim was 6 inches taller than Billy" give a times number? (Signal.) *No.*

- So can you make a multiplication number family for it? (Signal.) *No.*

c. Next sentence: Tim was 2 times as tall as Billy. Say that sentence. (Signal.) *Tim was 2 times as tall as Billly.*

- That sentence gives a times number. Tim was how many **times** as tall as Billy? (Signal.) *2.*
- The sentence gives a times number, so can you make a multiplication number family for it? (Signal.) *Yes.*

d. (Display:) [43:6B]

The elephant was 2 times as heavy as the giraffe.

The elephant was 3 tons heavier than the giraffe.

Here are two more sentences.

- First sentence: The elephant was 2 times as heavy as the giraffe. Say that sentence. (Signal.) *The elephant was 2 times as heavy as the giraffe.*
- Does that sentence give a times number? (Signal.) *Yes.*
- So can you make a multiplication number family for it? (Signal.) *Yes.*
- The elephant was how many times as heavy as the giraffe? (Signal.) *2.*
  (Repeat until firm.)

e. Next sentence: The elephant was 3 tons heavier than the giraffe. Say that sentence. (Signal.) *The elephant was 3 tons heavier than the giraffe.*

- Does that sentence give a times number? (Signal.) *No.*
- So can you make a multiplication number family for it? (Signal.) *No.*

from Lesson 43, Exercise 6

Students continue to make number families from sentences through Lesson 45. On Lesson 46, students work complete problems. The steps they follow parallel steps for addition number-family problems. Students first identify the sentence that tells how to make the family. They make the family with two letters and a "times" number. They replace one of the letters with a number.

Here's part of the exercise from Lesson 46:

b. I'll read problem A: The white bird weighed 36 ounces. The blue bird weighed 4 times as much as the white bird. How much did the blue bird weigh?
- Raise your hand when you know which sentence tells the times number. ✔
- Read the sentence that tells the times number. (Call on a student.) *The blue bird weighed 4 times as much as the white bird.*
- Everybody, what's the times number? (Signal.) *4.*
- So what number will you write for the first small number? (Signal.) *4.*
- Did the blue bird or the white bird weigh more? (Signal.) *The blue bird.*
- So what letter will you write for the big number? (Signal.) *B.*
- What letter will you write for the other small number? (Signal.) *W.*

c. Make the multiplication number family with two letters and a number for problem A. Put your pencil down when you've made the family. (Observe students and give feedback.) (Display:) [46:8A]

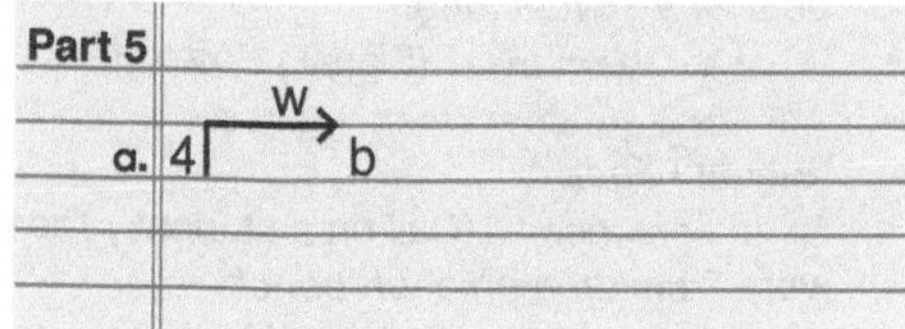

Here's what you should have.

d. The problem gives a number for the blue bird or the white bird. Raise your hand when you know that number. ✔
- What's the number? (Signal.) *36.*
- Is that the number for the blue bird or the white bird? (Signal.) *The white bird.*
- So what letter will you cross out in the family? (Signal.) *W.*
- Cross out W and replace it with what it equals. Write the problem and figure out how much the blue bird weighs. Write the answer with a unit name. (Observe students and give feedback.)

e. Read the number problem you worked and the answer. (Signal.) *36 × 4 = 144.*
- Everybody, how much did the blue bird weigh? (Signal.) *144 ounces.* (Add to show:) [46:8B]

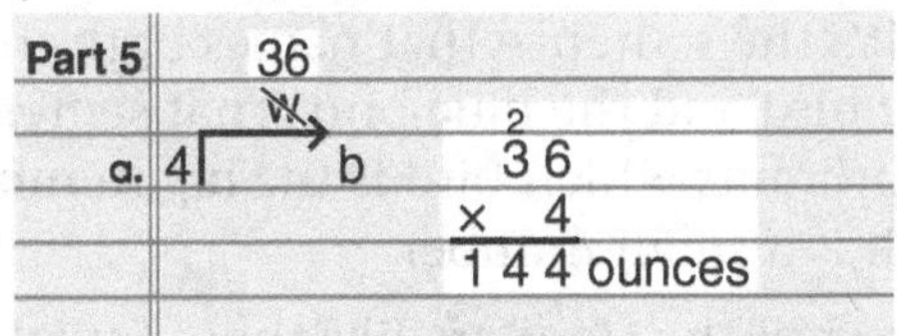

Here's what you should have for problem A.

f. I'll read problem B. Follow along: There were 2 times as many green cars as blue cars. There were 46 green cars. How many blue cars were there?
- Touch the sentence that gives the times number. ✔
- Make the family for B. Then stop. (Observe students and give feedback.)

g. Family B. What did you write for the times number? (Signal.) *2.*
- What letter did you write for the other small number? (Signal.) *B.*
- What letter did you write for the big number? (Signal.) *G.* (Display:) [46:8C]

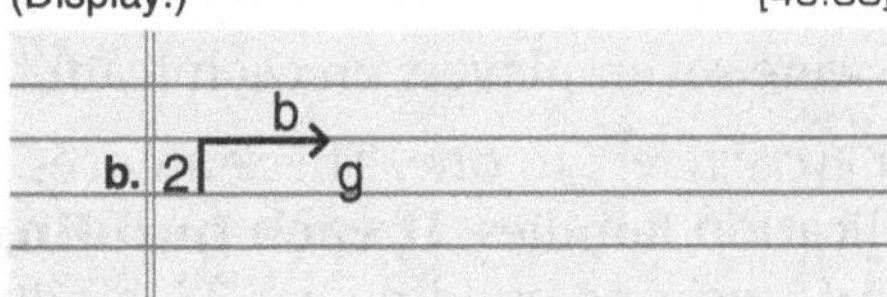

Here's what you should have for family B.

h. The problem gives a number for blue cars or green cars. Raise your hand when you know that number. ✔
- What's the number? (Signal.) *46.*
- What letter will you cross out and replace with 46? (Signal.) *G.*

i. Replace G with what it equals. Write the problem and figure out how many blue cars there were. Remember the unit name. (Observe students and give feedback.)

j. Check your work for problem B.
- Read the number problem and the answer. (Signal.) *46 ÷ 2 = 23.*
- How many blue cars were there? (Signal.) *23.* (Add to show:) [46:8D]

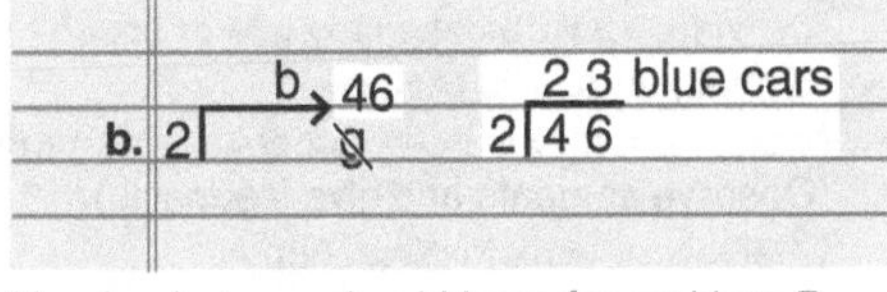

Here's what you should have for problem B.

from Lesson 46, Exercise 8

Teaching Note: Students should not have trouble identifying the sentence that tells how to make the number family. If students have trouble with the first example, remind them that it's the sentence that names both the white bird and the blue bird. That sentence also prompts which bird is the big number and which is a small number.

The rest of the steps are like those for addition-family problems.

Students continue making families from sentences and working times problems through Lesson 50.

Each-Every Word Problems

On Lesson 52, students work with sentences that tell about *each* or *every.* The word *each* or *every* may appear in the subject or predicate of the key sentence.

"Each building had 24 windows."

"There were seven players on each team."

Each or *every* refers to one. The sentences generate multiplication families. **If each building has 24 windows,** one building has 24 windows.

There are more windows than buildings, so *buildings* is a small number and *windows* is the big number. 24 is the times number.

Here's the family:

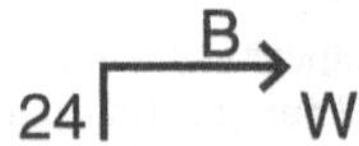

Here's part of the exercise from Lesson 51:

b. Read sentence A. (Call on a student.) *There were 12 jars in each box.*
- What are the names? (Signal.) *Jars (and) box(es).*
- Are there 12 times as many jars or boxes? (Signal.) *Jars.*
- Make the family with two letters and a number. (Observe students and give feedback.)

c. Check your work for sentence A.
- Everybody, what's the letter for the big number? (Signal.) *J.*
- What's the first small number? (Signal.) *12.*
- What's the other small number? (Signal.) *B.*

(Display:) [51:4G]

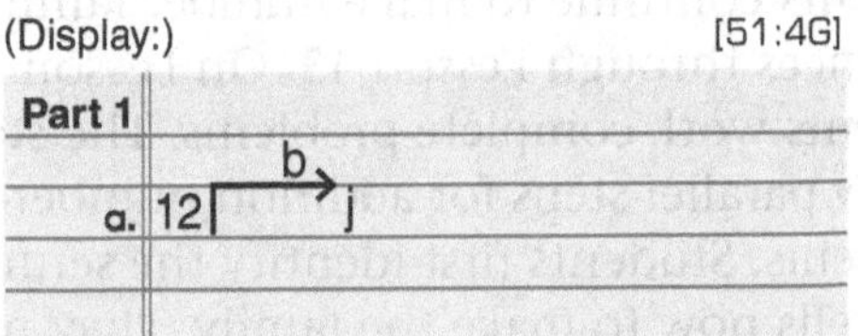

Here's what you should have for sentence A.

d. Read sentence B. (Call on a student.) *Every jar held 30 beans.*
- What are the names? (Signal.) *Jar(s) (and) beans.*
- Are there 30 times as many jars or beans? (Signal.) *Beans.*
- Make the family with two letters and a number. (Observe students and give feedback.)

e. Check your work for family B.
- Everybody, what's the letter for the big number? (Signal.) *B.*
- What's the first small number? (Signal.) *30.*
- What's the other small number? (Signal.) *J.*

(Display:) [51:4H]

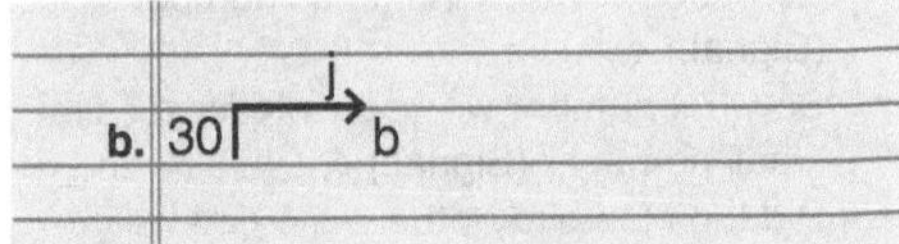

Here's what you should have for sentence B.
(If students are 100%, skip to step i.)

f. Read sentence C. (Call on a student.) *There were 24 students in each classroom.*
- What are the names? (Signal.) *Students (and) classroom(s).*
- Are there 24 times as many students or classrooms? (Signal.) *Students.*

g. Read sentence D. (Call on a student.) *Each boat was 16 feet long.*
- What are the names? (Signal.) *Boat(s) (and) feet.*
- Are there 16 times as many boats or feet? (Signal.) *Feet.*

h. Read sentence E. (Call on a student.) *There were 3 children on each bench.*
- What are the names? (Signal.) *Children (and) bench(es).*
- Are there 3 times as many children or benches? (Signal.) *Children.*

i. Make families for sentences C, D, and E. (Observe students and give feedback.)

j. Check your work.
- Sentence C. What's the letter for the big number? (Signal.) *S.*
- What's the first small number? (Signal.) *24.*
- What's the other small number? (Signal.) *C.*

(Display:) [51:4I]

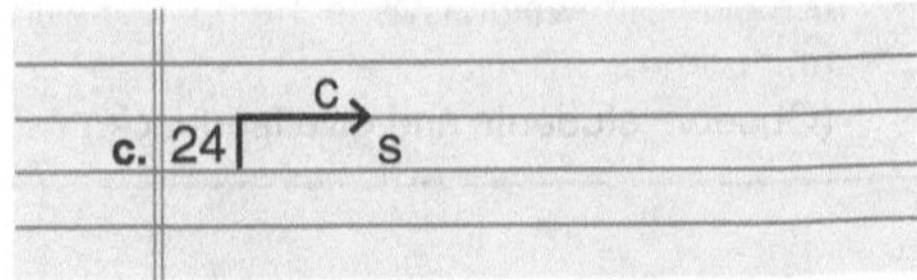

Here's what you should have for sentence C.

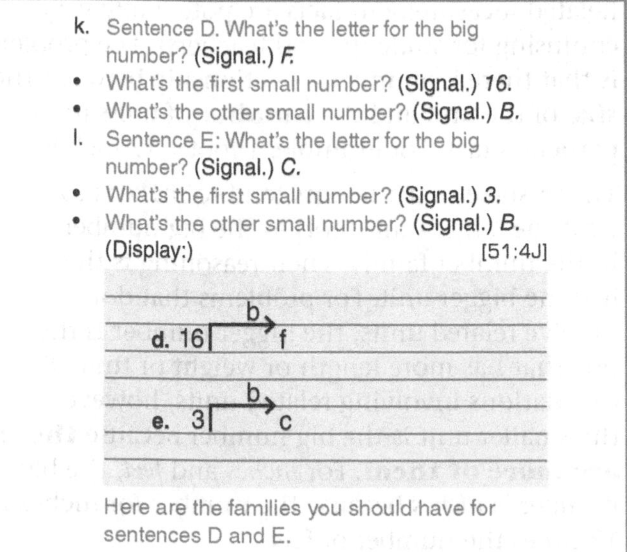

k. Sentence D. What's the letter for the big number? (Signal.) *F.*
- What's the first small number? (Signal.) *16.*
- What's the other small number? (Signal.) *B.*

l. Sentence E: What's the letter for the big number? (Signal.) *C.*
- What's the first small number? (Signal.) *3.*
- What's the other small number? (Signal.) *B.*

(Display:) [51:4J]

Here are the families you should have for sentences D and E.

from Lesson 51, Exercise 4

Teaching Note: If students have trouble with the question, "Are there 30 times as many jars or beans?" read the sentence, "Every jar held 30 beans." Then ask students, "How many beans are in one jar?"

"So are there 30 times as many beans or jars?"

On Lesson 52, students work complete problems. Here are the problems from Lesson 52:

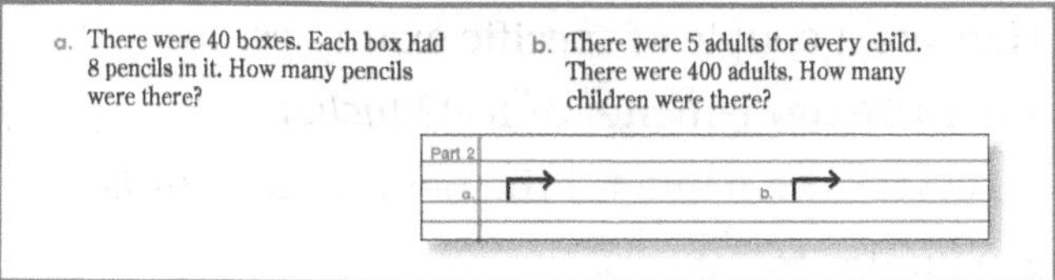

a. There were 40 boxes. Each box had 8 pencils in it. How many pencils were there?

b. There were 5 adults for every child. There were 400 adults. How many children were there?

The most critical step is for students to find the sentence that tells how to make the number family. If students make the family correctly, working the problem is a familiar operation.

Starting on Lesson 55, students work problem sets that have addition-subtraction problems and multiplication-division problems. The underlined sentence in each problem tells how to make the family. Students have to decide whether the problem tells about times or more and less.

Here's part of the exercise from Lesson 55:

(Teacher reference:)

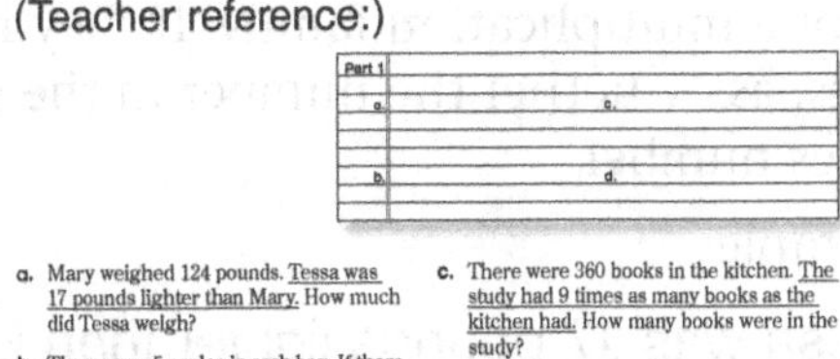

a. Mary weighed 124 pounds. Tessa was 17 pounds lighter than Mary. How much did Tessa weigh?

b. There were 5 apples in each bag. If there were 155 apples, how many bags were there?

c. There were 360 books in the kitchen. The study had 9 times as many books as the kitchen had. How many books were in the study?

d. The building was 25 feet taller than the tree. The building was 84 feet tall. How tall was the tree?

You'll make number families to solve these word problems. For some of the problems, you'll make an addition number family. For the other problems, you'll make a multiplication family. You'll read the underlined sentence for each problem and make the number family with two letters and a number. You'll work the rest of the problems as part of your independent work.
- Write part 1 on your lined paper with the letters A through D below.
(Observe students and give feedback.)

b. Read the underlined sentence in problem A. (Call on a student.) *Tessa was 17 pounds lighter than Mary.*
- Do you make a multiplication or addition family for sentence A? (Signal.) *(An) addition (family).*
- What number will you write for the first small number? (Signal.) *17.*
- Raise your hand when you can tell me the letters you'll write for the big number and the other small number. ✔
- What letter will you write for the big number? (Signal.) *M.*
- What letter will you write for the other small number? (Signal.) *T.*

c. Read the underlined sentence in problem B. (Call on a student.) *There were 5 apples in each bag.*
- What kind of family do you make for sentence B? (Signal.) *(A) multiplication (family).*
- What number will you write for the first small number? (Signal.) *5.*
- Are there 5 times as many apples or bags? (Signal.) *Apples.*
- So what letter will you write for the big number? (Signal.) *A.*
- What letter will you write for the other small number? (Signal.) *B.*

from Lesson 55, Exercise 6

After students read the underlined sentence, they indicate whether they will make an addition family or a multiplication family. If they make mistakes, ask whether the number in the sentence is a times number.

For example:

Tessa was 17 pounds lighter than Mary.

If students indicate that they are to make a multiplication family, say, "The sentence tells how many **pounds** lighter she is. That's not the same as how many **times** lighter she is." Point out that **the only families students write for times problems are multiplication families.**

Unit Conversion

Unit conversion is built around relationships of two systems of measurements. For instance, both feet and inches can be used to measure distance. Both years and days can be used to measure time. Both ounces and pounds can be used to measure weight.

Here's the table of measurement facts that students use in *CMC Level E:*

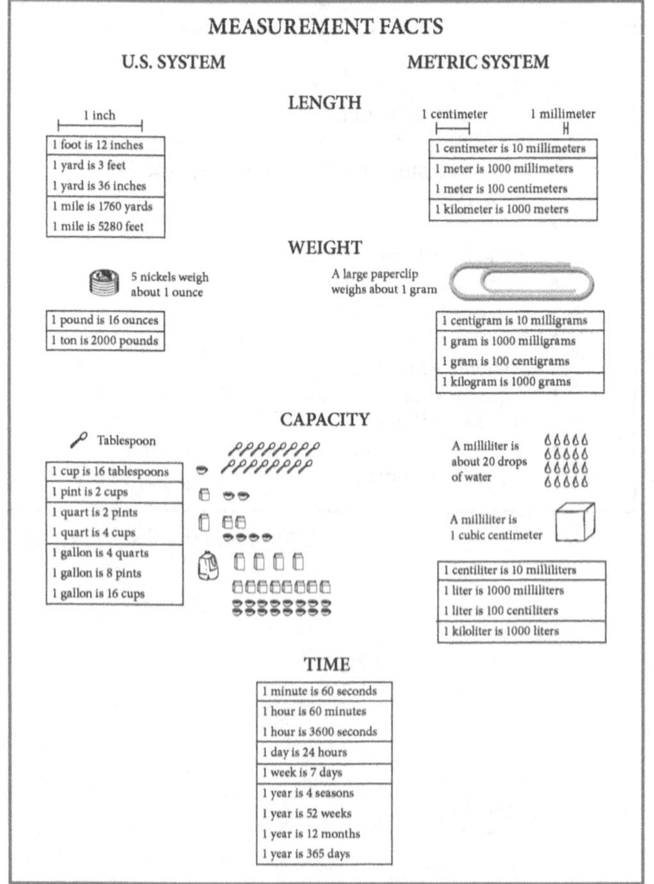

Related measurement facts are potentially very confusing for students (and teachers). The problem is that there is an inverse relationship between the **size** of the unit and the **number** of units for a particular fact. For example, 1 foot = 12 inches.

The mistake that students tend to make is to treat the larger unit (foot) as the big number in the number family. Their reasoning is that it is the bigger unit. For problems that don't involve related units, the bigger number is the one that has more length or weight or time. For calculations involving related units, however, the smaller unit is the big number because **there are more of them.** For *inches* and *feet,* the big number is *inches* because the number for inches is 12 times the number of feet.

Here's the family:

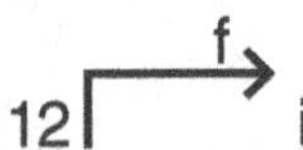

Because of this inverse relationship, we have to be very careful in keying on the **number,** not the **size** of the units.

Here is the pair of questions to ask when creating number families as measurement facts. The form of the first question is:

Are there more ____ or ____?

The second question is always:

So what's the big number?

Here are a couple of specific examples:

Are there more inches or feet? *Inches.*

So what's the name for the big number? *Inches.*

1 pound = 16 ounces.

Are there more pounds or ounces? *Ounces.*

So what's the name for the big number? *Ounces.*

If students remember the question to ask about numbers, they won't have trouble with measurement relationships.

On Lesson 56, students are introduced to measurement facts.

Here's part of the exercise:

a. We're going to make multiplication number families for measurement facts.
(Display:) [56:5A]

1 foot is 12 inches.

Here's a measurement fact: 1 foot is 12 inches. This fact tells you that **each** foot is 12 inches.

- Say the sentence with each. (Signal.) *Each foot is 12 inches.*
- What's the times number? (Signal.) *12.*
- Are there 12 times as many feet or inches? (Signal.) *Inches.*
- So what's the letter for the big number? (Signal.) *I.*
- What's the letter for the other small number? (Signal.) *F.*

(Repeat until firm.)
(Add to show:) [56:5B]

1 foot is 12 inches.

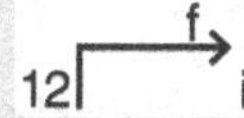

Here's the family with a times number and two letters for 1 foot is 12 inches.

b. (Display:) [56:5C]

1 year is 52 weeks.

New fact: 1 year is 52 weeks. That tells you that **each** year is 52 weeks.

- Say the sentence with each. (Signal.) *Each year is 52 weeks.*
- What's the times number? (Signal.) *52.*
- Are there 52 times as many years or weeks? (Signal.) *Weeks.*
- So what's the letter for the big number? (Signal.) *W.*
- What's the letter for the other small number? (Signal.) *Y.*

(Repeat until firm.)
(Add to show:) [56:5D]

1 year is 52 weeks.

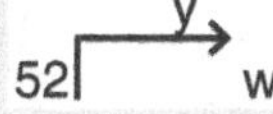

Here's the family with a number and two letters for 1 year is 52 weeks.

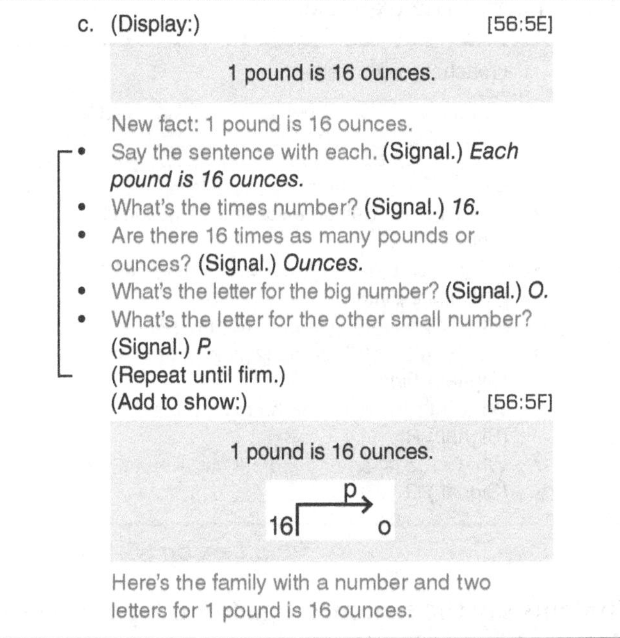

c. (Display:) [56:5E]

1 pound is 16 ounces.

New fact: 1 pound is 16 ounces.

- Say the sentence with each. (Signal.) *Each pound is 16 ounces.*
- What's the times number? (Signal.) *16.*
- Are there 16 times as many pounds or ounces? (Signal.) *Ounces.*
- What's the letter for the big number? (Signal.) *O.*
- What's the letter for the other small number? (Signal.) *P.*

(Repeat until firm.)
(Add to show:) [56:5F]

1 pound is 16 ounces.

p
16 o

Here's the family with a number and two letters for 1 pound is 16 ounces.

from Lesson 56, Exercise 5

Teaching Note: The key question asks about the relationship between the two units:

In step A: Are there 12 times as many feet or inches?

In step B: Are there 52 times as many years or weeks?

In step C: Are there 16 times as many pounds or ounces?

Make sure students are firm on answering these questions. If they aren't firm, they will have trouble making number families because the name for the larger **number (not unit)** is the big number in the family.

Following the verbal work, students make multiplication number families for new measurement facts.

Here's part of the exercise:

(Teacher reference:)

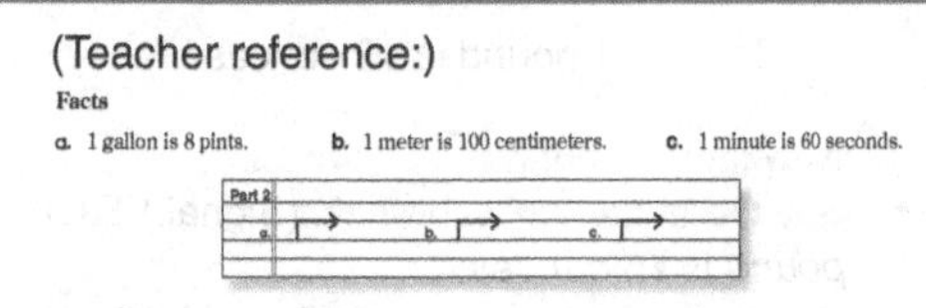

Facts
a. 1 gallon is 8 pints.
b. 1 meter is 100 centimeters.
c. 1 minute is 60 seconds.

- Read fact A. (Call on a student.) *1 gallon is 8 pints.*
- Say the sentence with each. (Signal.) *Each gallon is 8 pints.*
- What's the times number? (Signal.) *8.*
- Are there 8 times as many gallons or pints? (Signal.) *Pints.*
- So what's the letter for the big number? (Signal.) *P.*
- What's the letter for the other small number? (Signal.) *G.*

from Lesson 56, Exercise 5

Students say the sentence,1 gallon is 8 pints, with the word *each,* identify the times number (8), and identify the big number in the family.

On Lesson 59, students work complete problems.

Here's part of the exercise:

g. Touch the problems in part 3. ✔
- Read problem A. (Call on a student.) *The workers were in the mine for 6 days. How many hours were they in the mine?*
- One of the units the problem names is days. What's the other unit the problem names? (Signal.) *Hours.*
- Touch the fact that names days and hours. ✔
- Everybody, say that fact. Get ready. (Signal.) *1 day is 24 hours.*
- Make a number family for that fact for problem A. (Observe students and give feedback.)

h. Check your work.
- What's the times number for family A? (Signal.) *24.*
- What's the letter for the big number? (Signal.) *H.*
- What's the other letter? (Signal.) *D.*

(Display:) [59:6A]

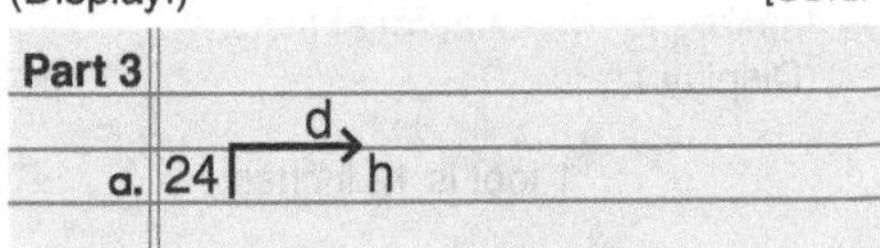

Here's the number family you should have for A.

i. Look at problem A. Raise your hand when you know the letter you'll replace in family A and the number you'll replace it with. ✔
- What letter will you replace in family A? (Signal.) *D.*
- What number will you replace D with? (Signal.) *6.*

(Add to show:) [59:6B]

Part 3
a. 24 6 (d) h

Here's the family with a number for days.

j. Say the problem to figure out the number of hours. Get ready. (Signal.) *24 × 6.*
- Write and work the problem for A. (Observe students and give feedback.)
- Everybody, what's 24 × 6? (Signal.) *144.*
- What unit did you write in the answer? (Signal.) *Hours.* Yes, the workers were in the mine for 144 hours.

from Lesson 59, Exercise 6

Students continue to work on complete problems through Lesson 88.

Starting on Lesson 61, they refer to the measurement fact table on the inside back cover of their Textbook.

Facts (Lessons 1–85)

CMC Level E teaches multiplication-division facts, but not addition-subtraction facts. Those facts should have been mastered when students exited *CMC Level D.*

Structured work on multiplication facts starts on Lesson 1 and continues through Lesson 85. Most instruction involves number families with a strong emphasis on division facts.

Families	Facts	Introduced or Reviewed on Lesson
2s, 5s, and 10s	2 × 1, 2 × 2, 2 × 3, 2 × 4, 2 × 5, 2 × 6, 2 × 7, 2 × 8, 2 × 9, 2 × 10, 5 × 1, 5 × 2, 5 × 3, 5 × 5, 5 × 6, 5 × 7, 5 × 8, 5 × 9, 5 × 10, 10 × 1, 10 × 2, 10 × 3, 10 × 4, 10 × 5, 10 × 6, 10 × 7, 10 × 8, 10 × 9, 10 × 10	1–10
Squares	1 × 1, 2 × 2, 3 × 3, 4 × 4, 5 × 5, 6 × 6, 7 × 7, 8 × 8, 9 × 9, 10 × 10	8–13
9s	9 × 1, 9 × 2, 9 × 3, 9 × 4, 9 × 5, 9 × 6, 9 × 7, 9 × 8, 9 × 9, 9 × 10	11–21
3s	3 × 1, 3 × 2, 3 × 3, 3 × 4, 3 × 5, 3 × 6, 3 × 7, 3 × 8, 3 × 9, 3 × 10	14–21
$3 \xrightarrow{3} 9$ $4 \xrightarrow{3} 12$ $6 \xrightarrow{3} 18$	3 × 3, 9 ÷ 3, 4 × 3, 3 × 4, 12 ÷ 3, 12 ÷ 4, 6 × 3, 3 × 6, 18 ÷ 3, 18 ÷ 6	21–25
$4 \xrightarrow{4} 16$ $6 \xrightarrow{4} 24$ $6 \xrightarrow{6} 36$	4 × 4, 16 ÷ 4, 6 × 4, 4 × 6, 24 ÷ 4, 24 ÷ 6, 6 × 6, 36 ÷ 6	26–30
$9 \xrightarrow{7} 63$ $9 \xrightarrow{8} 72$ $9 \xrightarrow{9} 81$	9 × 7, 7 × 9, 63 ÷ 7, 63 ÷ 9, 9 × 8, 8 × 9, 72 ÷ 8, 72 ÷ 9, 9 × 9, 81 ÷ 9	31–35
$7 \xrightarrow{3} 21$ $8 \xrightarrow{3} 24$	7 × 3, 3 × 7, 21 ÷ 3, 21 ÷ 7, 8 × 3, 3 × 8, 24 ÷ 3, 24 ÷ 8	36–45
$7 \xrightarrow{4} 28$ $8 \xrightarrow{4} 32$	7 × 4, 4 × 7, 28 ÷ 4, 28 ÷ 7, 8 × 4, 4 × 8, 32 ÷ 4, 32 ÷ 8	46–55
$7 \xrightarrow{6} 42$ $7 \xrightarrow{7} 49$	7 × 6, 6 × 7, 42 ÷ 6, 42 ÷ 7, 7 × 7, 49 ÷ 7	56–65
$8 \xrightarrow{6} 48$ $8 \xrightarrow{7} 56$ $8 \xrightarrow{8} 64$	8 × 6, 6 × 8, 48 ÷ 6, 48 ÷ 8, 8 × 7, 7 × 8, 56 ÷ 7, 56 ÷ 8, 8 × 8, 64 ÷ 8	66–75

The table shows when facts are first introduced or reviewed in *CMC Level E.*

In the first 14 lessons of *CMC Level E,* students review multiplication facts that should be familiar. These facts are not taught in the context of number families. After Lesson 14, the work with facts becomes closely related to number families, and the scope broadens from multiplication facts only to multiplication and related division facts as they are presented in families.

As the table shows, facts for 2s, 5s, 10s, and squares (2 × 2, 3 × 3 . . .) are reviewed in the first 10 lessons.

Here's the exercise from Lesson 2. It reviews multiplying by 2 and by 5:

(Teacher reference:) **R Part A**

a. 2 × 7 =	i. 6 × 5 =	q. 9 × 2 =	y. 5 × 5 =
b. 4 × 2 =	j. 5 × 3 =	r. 5 × 1 =	z. 9 × 5 =
c. 2 × 9 =	k. 8 × 5 =	s. 4 × 5 =	A. 0 × 2 =
d. 8 × 2 =	l. 5 × 4 =	t. 6 × 2 =	B. 7 × 5 =
e. 2 × 3 =	m. 5 × 7 =	u. 5 × 3 =	C. 5 × 2 =
f. 6 × 2 =	n. 5 × 5 =	v. 2 × 8 =	D. 5 × 8 =
g. 2 × 0 =	o. 5 × 1 =	w. 5 × 6 =	E. 3 × 2 =
h. 2 × 5 =	p. 5 × 9 =	x. 4 × 2 =	F. 7 × 2 =

The first column has problems that multiply by 2. I'll read each problem. You'll touch each problem and say the answer.

- 2 times 7. What's the answer? (Signal.) *14.*
- (Repeat for:) 4 × 2, *8;* 2 × 9, *18;* 8 × 2, *16;* 2 × 3, *6;* 6 × 2, *12;* 2 × 0, *0;* 2 × 5, *10.*

(Repeat problems that were not firm.)

b. Touch the first problem in the second column. ✔

This column has problems that multiply by 5. I'll read each problem. You'll touch each problem and say the answer.

- 6 times 5. What's the answer? (Signal.) *30.*
- (Repeat for:) 5 × 3, *15;* 8 × 5, *40;* 5 × 4, *20;* 5 × 7, *35;* 5 × 5, *25;* 5 × 1, *5;* 5 × 9, *45.*

(Repeat problems that were not firm.)

c. Complete the facts for problems in the first column and the second column. Then stop. Put your pencil down when you've worked through problem P.

(Observe students and give feedback.)

- Check your work.
- You'll read each problem and the answer. Make a line through each wrong answer. Write the correct answer next to it.

(Observe students and give feedback.)

- Problem A. (Signal.) *2 × 7 = 14.*
- B. (Signal.) *4 × 2 = 8.*
- (Repeat for:) C, *2 × 9 = 18;* D, *8 × 2 = 16;* E, *2 × 3 = 6;* F, *6 × 2 = 12;* G, *2 × 0 = 0;* H, *2 × 5 = 10;* I, *6 × 5 = 30;* J, *5 × 3 = 15;* K, *8 × 5 = 40;* L, *5 × 4 = 20;* M, *5 × 7 = 35;* N, *5 × 5 = 25;* O, *5 × 1 = 5;* P, *5 × 9 = 45.*

d. The problems in the last two columns multiply by 2 or by 5. You're going to write answers to those problems. You have **one** minute. Get ready. Go.

- (After 1 minute say:) Stop.
- Check your work.
- Problem Q. (Signal.) *9 × 2 = 18.*
- R. (Signal.) *5 × 1 = 5.*
- (Repeat for:) S, *4 × 5 = 20;* T, *6 × 2 = 12;* U, *5 × 3 = 15;* V, *2 × 8 = 16;* W, *5 × 6 = 30;* X, *4 × 2 = 8;* Y, *5 × 5 = 25;* Z, *9 × 5 = 45;* Capital A, *0 × 2 = 0;* Capital B, *7 × 5 = 35;* Capital C, *5 × 2 = 10;* Capital D, *5 × 8 = 40;* Capital E, *3 × 2 = 6;* Capital F, *7 × 2 = 14.*

Lesson 2, Exercise 6

Students respond orally to the problems in the first two columns. Then they write answers to those problems and check their work. Then students are timed on writing answers to problems in the last two columns. These problems are a mix of 5s and 2s. The time limit for the sixteen problems is one minute.

Teaching Note: Follow the script for presenting the oral items. For 2 × 7 = you say: "Two times seven. What's the answer?" (Signal.)

Students are to respond in unison when you signal. If students respond together, you can get good information about how solid they are. You can use this information to adjust the rate at which you present problems.

If students are very solid, speed up the presentation by reading the problems faster. If students are not responding well on signal, increase the pause between the problem and the question: "What's the answer?" (Pause. Signal.)

In the early lessons, students review the rule about the digits for multiplying by 9: The sum of the digits in the answer is 9.

Here's the exercise from Lesson 11:

(Teacher reference:) **R Part B**

a. 9 × 1 =	k. 9 × 2 =	u. 5 × 5 =	E. 4 × 9 =
b. 9 × 2 =	l. 9 × 7 =	v. 8 × 9 =	F. 8 × 8 =
c. 9 × 3 =	m. 9 × 5 =	w. 7 × 2 =	G. 8 × 10 =
d. 9 × 4 =	n. 9 × 8 =	x. 9 × 2 =	H. 4 × 4 =
e. 9 × 5 =	o. 9 × 9 =	y. 8 × 5 =	I. 4 × 2 =
f. 9 × 6 =	p. 9 × 4 =	z. 9 × 5 =	J. 7 × 7 =
g. 9 × 7 =	q. 9 × 3 =	A. 3 × 5 =	K. 3 × 9 =
h. 9 × 8 =	r. 9 × 6 =	B. 2 × 8 =	L. 7 × 9 =
i. 9 × 9 =	s. 9 × 10 =	C. 9 × 9 =	M. 10 × 10 =
j. 9 × 10 =	t. 9 × 1 =	D. 5 × 10 =	N. 6 × 9 =

The problems in the first column multiply by 9. You'll read each problem and then tell me the answer.

- Read problem A. **(Signal.)** *9 × 1.*
- What's the answer? **(Signal.)** *9.*

b. Read problem B. **(Signal.)** *9 × 2.*
- What's the answer? **(Signal.)** *18.*

c. **(Repeat the following tasks for problems C through J:)**

Read problem ___.		What's the answer?
C	9 × 3	27
D	9 × 4	36
E	9 × 5	45
F	9 × 6	54
G	9 × 7	63
H	9 × 8	72
I	9 × 9	81
J	9 × 10	90

d. Write answers to the problems in the first column.
(Observe students and give feedback.)

e. Check your work.
I'll read the problems. You'll tell me the answers.
- Problem A. What's 9 × 1? **(Signal.)** *9.*
- **(Repeat for:)** B, 9 × 2, *18;* C, 9 × 3, *27;* D, 9 × 4, *36;* E, 9 × 5, *45;* F, 9 × 6, *54;* G, 9 × 7, *63;* H, 9 × 8, *72;* I, 9 × 9, *81;* J, 9 × 10, *90.*

f. You can check your answers by adding the digits in the answer. They should always add up to 9.
- What do the digits add up to? **(Signal.)** *9.*

g. Problem A. The answer is 9.
- Problem B. The digits are 1 and 8. What's 1 + 8? **(Signal.)** *9.*
- Problem C. The digits are 2 and 7. What's 2 + 7? **(Signal.)** *9.*
- Problem D. The digits are 3 and 6. What's 3 + 6? **(Signal.)** *9.*
- Problem E. The digits are 4 and 5. What's 4 + 5? **(Signal.)** *9.*
- Problem F. The digits are 5 and 4. What's 5 + 4? **(Signal.)** *9.*
- Problem G. The digits are 6 and 3. What's 6 + 3? **(Signal.)** *9.*
- Problem H. The digits are 7 and 2. What's 7 + 2? **(Signal.)** *9.*
- Problem I. The digits are 8 and 1. What's 8 + 1? **(Signal.)** *9.*
- Problem J. The digits are 9 and zero. What's 9 + 0? **(Signal.)** *9.*
Remember, for these problems, the digits in the answer add up to 9.

h. Now you're going to write answers to the rest of the problems. You have one and a half minutes.
- Get ready. Go.
- **(After 90 seconds say:)** Stop.

i. Check your work. I'll read each problem. You'll tell me the answer.
- Problem K. What's 9 × 2? **(Signal.)** *18.*
- **(Repeat for:)** L, 9 × 7, *63;* M, 9 × 5, *45;* N, 9 × 8, *72;* O, 9 × 9, *81;* P, 9 × 4, *36;* Q, 9 × 3, *27;* R, 9 × 6, *54;* S, 9 × 10, *90;* T, 9 × 1, *9;* U, 5 × 5, *25;* V, 8 × 9, *72;* W, 7 × 2, *14;* X, 9 × 2, *18;* Y, 8 × 5, *40;* Z, 9 × 5, *45;* Capital A, 3 × 5, *15;* B, 2 × 8, *16;* C, 9 × 9, *81;* D, 5 × 10, *50;* E, *4 × 9, 36;* F, 8 × 8, *64;* G, 8 × 10, *80;* H, 4 × 4, *16;* I, 4 × 2, *8;* J, 7 × 7, *49;* K, 3 × 9, *27;* L, 7 × 9, *63;* M, 10 × 10, *100;* N, 6 × 9, *54.*

Lesson 11, Exercise 1

In steps F and G, students check their answers by adding the sum of the digits in the answer.

Teaching Note: The problems in the first column are arranged from 9 × 1 to 9 × 10. The same problems appear in random order in the second column. This arrangement increases the probability that students will make fewer mistakes on the multiplying by 9 facts when they appear in columns three and four.

After Lesson 14, all new facts are introduced through number families. Facts Table page 75 shows that facts are taught in groups. Each group consists of two to four families that generate 6–14 juxtaposed facts. Work with each fact group usually occurs in a ten-lesson block. In the block, the new facts are related to earlier-taught facts.

The same pattern of introduction and review is followed with all fact groups after Lesson 14. Not all cycles require ten lessons, but all follow the same steps:

- First, students say four facts for each new family.
- Next, students complete families with missing numbers.
- Then, students work separate exercises for multiplication and division problems.
- Finally, students work mixed sets of multiplication and division problems.

In Lessons 46–55, students learn two new families that generate eight facts: 7 × 4, 4 × 7, 28 ÷ 4, 28 ÷ 7, 8 × 4, 4 × 8, 32 ÷ 4, 32 ÷ 8.

The preceding block in Lessons 36–45 focused on two other families that have a small number of 3.

The first exercise in the cycle presents the number families. Students say four facts for each family. Then students identify the missing number in sets of families. Finally, students work a set of division problems based on the new families and families taught earlier.

Here's the introduction of two new families on Lesson 46:

a. (Display:) [46:1A]

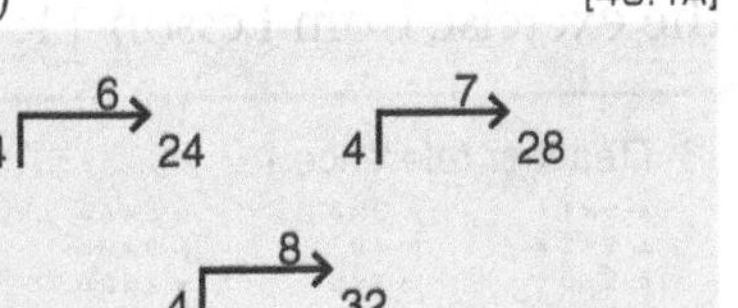

These are multiplication number families with a small number of 4.

- (Point to 4 6 24.) Say the fact that starts with the first small number. **(Signal.)** *4 × 6 = 24.*
- Say the other multiplication fact. **(Signal.)** *6 × 4 = 24.*
- Say the fact that divides by 4. **(Signal.)** *24 ÷ 4 = 6.*
- Say the other division fact. **(Signal.)** *24 ÷ 6 = 4.*

b. (Point to 4 7 28.) Say the fact that starts with the first small number. **(Signal.)** *4 × 7 = 28.*

- Say the other multiplication fact. **(Signal.)** *7 × 4 = 28.*
- Say the fact that divides by 4. **(Signal.)** *28 ÷ 4 = 7.*
- Say the other division fact. **(Signal.)** *28 ÷ 7 = 4.*

c. (Point to 4 8 32.) Say the fact that starts with the first small number. **(Signal.)** *4 × 8 = 32.*

- Say the other multiplication fact. **(Signal.)** *8 × 4 = 32.*
- Say the fact that divides by 4. **(Signal.)** *32 ÷ 4 = 8.*
- Say the other division fact. **(Signal.)** *32 ÷ 8 = 4.*

(Repeat families that were not firm.)

WORKBOOK PRACTICE

a. Open your workbook to Lesson 46 and find part 1. ✔

(Teacher reference:)

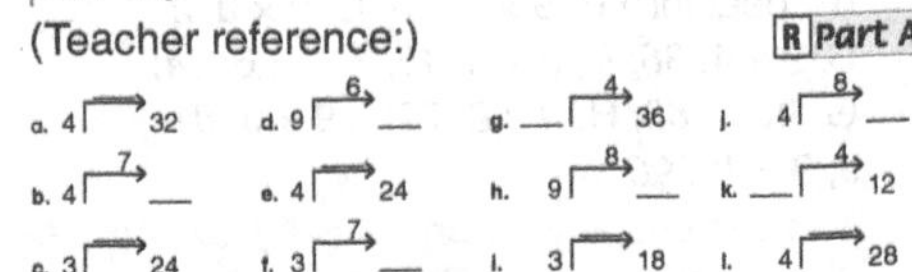

One of the numbers in each family is missing. You'll say the problem and the answer for the missing number in each family. Then you'll complete the families.

- Family A. Say the problem for the missing number. **(Signal.)** *32 ÷ 4.*
- What's 32 ÷ 4? **(Signal.)** *8.*

b. (Repeat the following tasks for families B through L:)

	Say the problem for family ___.	What's ___?	
B	*4 × 7*	4 × 7	*28*
C	*24 ÷ 3*	24 ÷ 3	*8*
D	*9 × 6*	9 × 6	*54*
E	*24 ÷ 4*	24 ÷ 4	*6*
F	*3 × 7*	3 × 7	*21*
G	*36 ÷ 4*	36 ÷ 4	*9*
H	*9 × 8*	9 × 8	*72*
I	*18 ÷ 3*	18 ÷ 3	*6*
J	*4 × 8*	4 × 8	*32*
K	*12 ÷ 4*	12 ÷ 4	*3*
L	*28 ÷ 4*	28 ÷ 4	*7*

(Repeat problems that were not firm.)

c. Now write the missing number for each family. (Observe students and give feedback.)

d. Check your work. You'll tell me the missing number you wrote for each family.

- Family A. (Signal.) *8.*
- (Repeat for:) B, *28;* C, *8;* D, *54;* E, *6;* F, *21;* G, *9;* H, *72;* I, *6;* J, *32;* K, *3;* L, *7.*

e. Find part 2 in your workbook. ✔

(Teacher reference:) **R Part E**

a. $9\overline{)81}$	g. $8\overline{)32}$	m. $6\overline{)24}$	s. $2\overline{)0}$	y. $5\overline{)5}$
b. $6\overline{)36}$	h. $8\overline{)72}$	n. $3\overline{)9}$	t. $6\overline{)6}$	z. $4\overline{)32}$
c. $9\overline{)36}$	i. $9\overline{)54}$	o. $6\overline{)18}$	u. $3\overline{)21}$	A. $9\overline{)54}$
d. $9\overline{)72}$	j. $7\overline{)28}$	p. $9\overline{)27}$	v. $5\overline{)20}$	B. $3\overline{)24}$
e. $6\overline{)6}$	k. $3\overline{)12}$	q. $10\overline{)80}$	w. $4\overline{)28}$	C. $4\overline{)24}$
f. $9\overline{)63}$	l. $5\overline{)35}$	r. $2\overline{)18}$	x. $5\overline{)30}$	D. $3\overline{)18}$

The division problems in part 2 are from number families you know. Work all of the division problems in part 2. You have two minutes.

- Get ready. Go. (Observe students and give feedback.)
- (After 2 minutes say:) Stop.

f. Check your work. You'll read the fact for each problem.

- Problem A. (Signal.) *81 ÷ 9 = 9.*
- (Repeat for:) B, *36 ÷ 6 = 6;* C, *36 ÷ 9 = 4;* D, *72 ÷ 9 = 8;* E, *6 ÷ 6 = 1;* F, *63 ÷ 9 = 7;* G, *32 ÷ 8 = 4;* H, *72 ÷ 8 = 9;* I, *54 ÷ 9 = 6;* J, *28 ÷ 7 = 4;* K, *12 ÷ 3 = 4;* L, *35 ÷ 5 = 7;* M, *24 ÷ 6 = 4;* N, *9 ÷ 3 = 3;* O, *18 ÷ 6 = 3;* P, *27 ÷ 9 = 3;* Q, *80 ÷ 10 = 8;* R, *18 ÷ 2 = 9;* S, *0 ÷ 2 = 0;* T, *6 ÷ 6 = 1;* U, *21 ÷ 3 = 7;* V, *20 ÷ 5 = 4;* W, *28 ÷ 4 = 7;* X, *30 ÷ 5 = 6;* Y, *5 ÷ 5 = 1;* Z, *32 ÷ 4 = 8;* Capital A, *54 ÷ 9 = 6;* B, *24 ÷ 3 = 8;* C, *24 ÷ 4 = 6;* D, *18 ÷ 3 = 6.*

Lesson 46, Exercise 1

Teaching Note: The exercise presents the families in the counting order. The family with small numbers of 4 and 6 had been introduced on Lesson 26.

Attend to the note at the end of the verbal practice (Repeat families that were not firm.). It's a good idea to repeat steps A through C even if students respond well the first time. Responding to the tasks in order shows how the families are related: one of the small numbers increases by one; the big number increases by four.

Students next do verbal tasks in which they say the problem for the missing number in each family and say the missing number. Some missing numbers are big numbers and some are small numbers.

This practice assures that students are aware of the arrangement of numbers in each complete number family.

If students are weak on particular missing numbers, repeat the work on them after going through all the specified tasks B through L.

The last activity presents division problems. This activity is timed. Students are to complete all thirty problems in two minutes. The problems are from the two new families and other families taught earlier in the program. Students should be facile with the facts for these problems. When the two-minute period is over, tell students who have not finished the problem set to make a vertical line after the last problem they worked. Students are not to complete problems that follow the vertical line. Direct students to make a C or an X after each problem, (C = correct; X = incorrect).

When you check the students' work, direct students who made no more than three errors to raise their hand. Praise those students.

Later in the lesson, students work a set of thirty multiplication problems.

Here's the exercise from Lesson 46:

(Teacher reference:) **R Part B**

a. 8 × 4 =	k. 6 × 5 =	u. 8 × 2 =
b. 2 × 2 =	l. 8 × 5 =	v. 9 × 6 =
c. 4 × 3 =	m. 7 × 10 =	w. 4 × 7 =
d. 6 × 4 =	n. 9 × 9 =	x. 4 × 4 =
e. 5 × 9 =	o. 4 × 6 =	y. 6 × 3 =
f. 3 × 7 =	p. 9 × 4 =	z. 9 × 7 =
g. 2 × 6 =	q. 5 × 5 =	A. 4 × 5 =
h. 8 × 3 =	r. 6 × 6 =	B. 8 × 10 =
i. 7 × 4 =	s. 5 × 7 =	C. 4 × 9 =
j. 2 × 7 =	t. 4 × 8 =	D. 10 × 10 =

Some of these multiplication problems are from families we worked with in this lesson. You'll read some of the problems and say the answer.

- Read problem A. (Signal.) *8 × 4.*
- What's the answer? (Signal.) *32.*

b. (Repeat the following tasks for problems B through J:)

Problem ___.		What's the answer?
B	2 × 2	4
C	4 × 3	12
D	6 × 4	24
E	5 × 9	45
F	3 × 7	21
G	2 × 6	12
H	8 × 3	24
I	7 × 4	28
J	2 × 7	14

(Repeat problems that were not firm.)

c. Write all of the answers to the problems in part 4. You have two minutes.

- Get ready. Go.

(Observe students and give feedback.)

- (After 2 minutes say:) Stop.

d. Check your work. You'll read the fact for each problem.

- Problem A. (Signal.) *8 × 4 = 32.*
- (Repeat for:) B, *2 × 2 = 4;* C, *4 × 3 = 12;* D, *6 × 4 = 24;* E, *5 × 9 = 45;* F, *3 × 7 = 21;* G, *2 × 6 = 12;* H, *8 × 3 = 24;* I, *7 × 4 = 28;* J, *2 × 7 = 14;* K, *6 × 5 = 30;* L, *8 × 5 = 40;* M, *7 × 10 = 70;* N, *9 × 9 = 81;* O, *4 × 6 = 24;* P, *9 × 4 = 36;* Q, *5 × 5 = 25;* R, *6 × 6 = 36;* S, *5 × 7 = 35;* T, *4 × 8 = 32;* U, *8 × 2 = 16;* V, *9 × 6 = 54;* W, *4 × 7 = 28;* X, *4 × 4 = 16;* Y, *6 × 3 = 18;* Z, *9 × 7= 63;* Capital A, *4 × 5 = 20;* B, *8 × 10 = 80;* C, *4 × 9 = 36;*D, *10 × 10 = 100.*

Lesson 46, Exercise 3

Students say the answers to the first ten problems and then write answers to all the problems in two minutes. The problem set should not be difficult for students because the new facts come from only two families.

In step B, students review the new facts. They say the problem and then say the answer.

Students may have trouble if your pacing is too fast. If they answer late or don't respond, increase the pause between the time they read the problem and you ask the question.

During the remainder of the ten-lesson cycle, students continue to do two fact exercises in each lesson. The multiplication exercise is the same in all lessons. The other exercise is the same as that in Lessons 47–50.

On Lesson 51, the presentation is shortened. Students do not say the four facts for families. They start with identifying the missing numbers in families. They complete the families. Then they do a two-minute timing.

On Lessons 52–55, students work a mixed set of multiplication and division problems. They do an oral review with the first 12 items. Then they do a two-minute timing.

Here's the exercise from Lesson 52:

(Teacher reference:)

a. $4\overline{)12}$	g. 3 × 6 =	m. $8\overline{)24}$	s. $3\overline{)18}$	y. $6\overline{)36}$
b. $3\overline{)9}$	h. $10\overline{)0}$	n. $1\overline{)5}$	t. 6 × 6 =	z. $5\overline{)30}$
c. $4\overline{)28}$	i. $7\overline{)21}$	o. $9\overline{)63}$	u. $9\overline{)72}$	A. $9\overline{)81}$
d. 9 × 7 =	j. 9 × 6 =	p. 6 × 4 =	v. 4 × 7 =	B. $5\overline{)35}$
e. 8 × 4 =	k. $4\overline{)32}$	q. 3 × 8 =	w. $9\overline{)36}$	C. $6\overline{)54}$
f. $9\overline{)54}$	l. $6\overline{)24}$	r. $4\overline{)24}$	x. $2\overline{)10}$	D. 3 × 7 =

These multiplication and division problems are from families you know. For some of the problems, you'll read the problem, say if the missing number is a small number or the big number, and say the answer.

- Read problem A. (Signal.) *12 ÷ 4.*
- Is a small number or the big number missing? (Signal.) *A small number.*
- What's the answer? (Signal.) *3.*

b. Read problem B. (Signal.) *9 ÷ 3.*

- Is a small number or the big number missing? (Signal.) *A small number.*
- What's the answer? (Signal.) *3.*

c. Read problem C. (Signal.) *28 ÷ 4.*

- What's missing? (Signal.) *A small number.*
- What's the answer? (Signal.) *7.*

d. (Repeat the following tasks for problems D through L:)

Read problem ___.		What's missing?	What's the answer?
D	9 × 7	*The big number.*	63
E	8 × 4	*The big number.*	32
F	54 ÷ 9	*A small number.*	6
G	3 × 6	*The big number*	18
H	0 ÷ 10	*A small number.*	0
I	21 ÷ 7	*A small number.*	3
J	9 × 6	*The big number.*	54
K	32 ÷ 4	*A small number.*	8
L	24 ÷ 6	*A small number.*	4

(Repeat problems that were not firm.)

e. Write all of the answers to the problems in part 1. You have two minutes.

- Get ready. Go.

(Observe students and give feedback.)

- (After 2 minutes, say:) Stop.

f. Check your work. You'll read the fact for each problem.

- Problem A. (Signal.) *12 ÷ 4 = 3.*
- (Repeat for:) B, *9 ÷ 3 = 3;* C, *28 ÷ 4 = 7;* D, *9 × 7 = 63;* E, *8 × 4 = 32;* F, *54 ÷ 9 = 6;* G, *3 × 6 = 18;* H, *0 ÷ 10 = 0;* I, *21 ÷ 7 = 3;* J, *9 × 6 = 54;* K, *32 ÷ 4 = 8;* L, *24 ÷ 6 = 4;* M, *24 ÷ 8 = 3;* N, *5 ÷ 1 = 5;* O, *63 ÷ 9 = 7;* P, *6 × 4 = 24;* Q, *3 × 8 = 24;* R, *24 ÷ 4 = 6;* S, *18 ÷ 3 = 6;* T, *6 × 6 = 36;* U, *72 ÷ 9 = 8;* V, *4 × 7 = 28;* W, *36 ÷ 9 = 4;* X, *10 ÷ 2 = 5;* Y, *36 ÷ 6 = 6;* Z, *30 ÷ 5 = 6;* Capital A, *81 ÷ 9 = 9;* B, *35 ÷ 5 = 7;* C, *54 ÷ 6 = 9;* D, *3 × 7 = 21.*

Lesson 52, Exercise 1

Teaching Note: In steps A through D, you direct students to read the problem, tell if the big number or a small number is missing, and say the answer.

The reason for calling students' attention to whether the big number or a small number is missing is that we want students to understand that the facts are related to number families. If students make this link, learning and organizing multiplication and division facts is much easier than relying on brute memory.

About two-thirds of the problems are division problems because students receive daily multiplication-fact practice in each lesson.

If students are taught to mastery, they will have a solid command of multiplication and division facts and their number families.

Optional Extra Fact Practice

If the classroom is heterogeneous, expect possibly 25 percent of the students to need extra practice. *Level E* has two provisions for more fact practice.

1. Additional fact practice worksheets are available for each series of facts (all four operations) in the Math Fact Worksheets (blackline masters available via ConnectED). Use these with the understanding that you should keep moving through the lessons, but you should do what you can to provide additional practice for students who need it.
2. The *CMC Level E* Practice Software includes a math fact strand that is organized into sets of facts that follow the instructional sequence in the lessons. It is designed to facilitate continuous review and reinforcement of the math facts as they are introduced and practiced. The Practice Software is accessed via ConnectED.

Fractions (Lessons 1–130)

CMC Level E assumes that students who start on Lesson 1 know something about fractions; however, students who have not been through the earlier levels of *CMC* are often weak in fraction-related skills. So the fraction instruction in the first thirty lessons provides a thorough review.

Students first learn that the bottom number of fractions tells how many parts are in each unit, and the top number tells how many parts are used. When the top number is bigger than the bottom number, the fraction is more than one unit. When the numbers are the same size, the fraction equals one; when the bottom number is bigger than the top number, the fraction is less than one unit.

Students write fractions for pictures. They write the number of parts in each unit as the bottom number. They write the number of parts that are shaded as the top number. Students apply this strategy to writing fractions for number lines that have shaded parts.

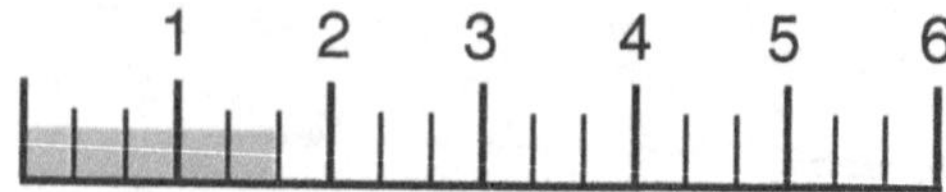

This fraction has three parts in each unit, and five parts are shaded. So the fraction is 5/3.

Students learn to create fractions that equal whole numbers by applying the formula that the whole number tells how many times bigger the top number is than the bottom number. If a fraction equals 5, the top number is 5 times the bottom number. If a fraction equals 19, the top number is 19 times the bottom number.

Finally, in the first thirty lessons, students learn conventions for adding and subtracting fractions. They learn that the denominators of all the fractions are the same and that you add or subtract to find the numerator of the answer.

If the bottom numbers are not the same, students can't work the problem.

Mastery of these skills is imperative because students will build on these skills throughout the program.

Basic Fraction Analysis

On Lesson 1, students learn that the bottom number of the fraction tells the number of parts in each unit. The top number tells how many parts are shaded.

Here's the exercise from Lesson 4:

(Teacher reference:)

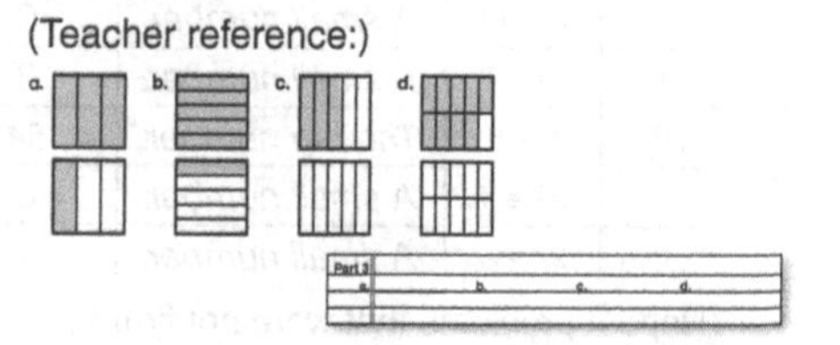

Part 3 shows pictures of fractions. Each fraction shows 2 units.

- Touch the picture for fraction A. ✔
- Look at the parts in each unit. ✔
- How many parts are in each unit? **(Signal.)** *3.*
- So what's the bottom number of fraction A? **(Signal.)** *3.*
- Look at the parts that are shaded. ✔
- How many parts are shaded? **(Signal.)** *4.*
- So what's the top number of the fraction? **(Signal.)** *4.*

b. Touch the picture for fraction B. ✔
- Count the parts that are in each unit to yourself. ✔
- How many parts are in each unit? **(Signal.)** *5.*
- So what's the bottom number of the fraction? **(Signal.)** *5.*
- Look at the parts that are shaded. ✔
- How many parts are shaded? **(Signal.)** *6.*
- So what's the top number of the fraction? **(Signal.)** *6.*

c. Touch the picture for fraction C. ✔
- Count the parts that are in each unit to yourself. ✔
- How many parts are in each unit? **(Signal.)** *5.*
- So what's the bottom number of the fraction? **(Signal.)** *5.*
- Look at the parts that are shaded. ✔
- How many parts are shaded? **(Signal.)** *3.*
- So what's the top number of the fraction? **(Signal.)** *3.*

d. Touch the picture for fraction D. ✔
- Raise your hand when you can tell me the bottom number. ✔
- What's the bottom number of fraction D? **(Signal.)** *10.*
  Get ready to tell me the top number.
- What's the top number of fraction D? **(Signal.)** *9.*
  (Repeat steps a through d that were not firm.)

e. Write part 3 on your lined paper and write the letters A through D below. Make the fraction bar for A and write the fraction. Make the fraction bar straight across. Put your pencil down when you've written fraction A.
(Observe students and give feedback.)

- Everybody, what fraction did you write for picture A? (Signal.) *4 thirds.*

(Display:) [4:8A]

Part 3				
	a. $\frac{4}{3}$	b.	c.	d.

Here's what you should have for fraction A.

f. Make the fraction bars and write the fractions for the rest of the pictures. Remember to make the line for your fractions straight across. (Observe students and give feedback.)

g. Check your work.

- Read the fraction for B. (Signal.) *6 fifths.*
- Read the fraction for C. (Signal.) *3 fifths.*
- Read the fraction for D. (Signal.) *9 tenths.*

(Display:) [4:8B]

Part 3				
	a. $\frac{4}{3}$	b. $\frac{6}{5}$	c. $\frac{3}{5}$	d. $\frac{9}{10}$

Here are the fractions you should have for part 3.

Lesson 4, Exercise 8

Teaching Note: On the preceding lessons, students did more detailed work on unit and part. Students are firm on these words, which means that you should be able to go fast through the exercise. If students make a mistake on any questions, correct the mistake and continue through step D. Repeat any steps that were not firm.

In steps E and F, students write fractions for the pictures. If they are firm on the earlier steps in the exercise, they should have no trouble.

On Lesson 5, students learn the rule that if the top number of the fraction is more than the bottom number, the fraction is more than one unit. Here's the first part of the exercise from Lesson 5:

a. (Display:) [5:8A]

$\frac{3}{4}$ $\frac{4}{3}$ $\frac{7}{3}$ $\frac{3}{7}$

$\frac{1}{5}$ $\frac{3}{10}$ $\frac{10}{3}$

You're going to read each fraction and tell me what its picture would look like.

- Does the bottom number or the top number tell about the parts in each unit? (Signal.) *The bottom number.*
- Which number tells about the parts that are shaded? (Signal.) *The top number.*

b. (Point to $\frac{3}{4}$.) Read this fraction. (Signal.) *3 fourths.*

- How many parts are in each unit? (Signal.) *4.*
- How many parts are shaded? (Signal.) *3.*

c. (Repeat the following tasks for the remaining fractions:)

(Point to ___.) Read this fraction.		How many parts are in each unit?	How many parts are shaded?
$\frac{4}{3}$	*4 thirds*	*3*	*4*
$\frac{7}{3}$	*7 thirds*	*3*	*7*
$\frac{3}{7}$	*3 sevenths*	*7*	*3*
$\frac{1}{5}$	*1 fifth*	*5*	*1*
$\frac{3}{10}$	*3 tenths*	*10*	*3*
$\frac{10}{3}$	*10 thirds*	*3*	*10*

(Repeat fractions that were not firm.)

d. Some of these fractions are more than one unit. Here's the rule: If the top number is bigger than the bottom number, the fraction is more than one whole unit.

If the top number is smaller than the bottom number, the fraction is less than one.

- (Point to $\frac{3}{4}$.) Read this fraction. (Signal.) *3 fourths.*
- How many parts are in each unit? (Signal.) *4.*
- How many parts are shaded? (Signal.) *3.*
- Are more than 4 parts or less than 4 parts shaded? (Signal.) *Less than 4 parts.*
- So is 3/4 more than 1 unit or less than 1 unit? (Signal.) *Less than 1 unit.*

e. (Point to $\frac{4}{3}$.) Read this fraction. (Signal.) *4/3.*

- How many parts are in each unit? (Signal.) *3.*
- Are more than 3 parts or less than 3 parts shaded? (Signal.) *More than 3 parts.*
- So is 4/3 more than 1 unit or less than 1 unit? (Signal.) *More than 1 unit.*

f. (Point to $\frac{7}{3}$.) Read this fraction. (Signal.) *7/3.*

- How many parts are in each unit? (Signal.) *3.*
- Are more than 3 parts or less than 3 parts shaded? (Signal.) *More than 3 parts.*
- So is 7/3 more than 1 unit or less than 1 unit? (Signal.) *More than 1 unit.*

g. (Point to $\frac{3}{7}$.) Read this fraction. (Signal.) *3/7.*

- How many parts are in each unit? (Signal.) *7.*
- Are more than 7 parts or less than 7 parts shaded? (Signal.) *Less than 7 parts.*
- So is 3/7 more than 1 unit or less than 1 unit? (Signal.) *Less than 1 unit.*

from Lesson 5, Exercise 8

Teaching Note: Starting with step E, you follow the same routine with each fraction. You direct students to read the fraction, indicate how many parts are in each unit, ask if more or less than that number of parts are shaded, and then ask if the fraction is more or less than one.

On Lesson 8, students compare fractions to 1 by making the appropriate signs. (<, >, =)

The rule they apply was taught before Lesson 8: Write the bigger end of the sign close to the side that has more. If the sides are the same size, the sign is =.

(Teacher reference:) **R Part H**

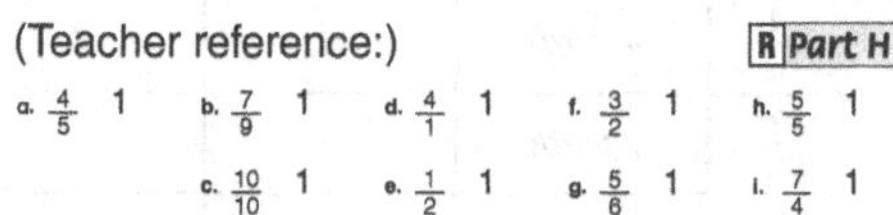

Some of these problems have fractions that are more than one unit, less than one unit, or equal to one unit.

For each problem, you're going to complete the sign to show if the fraction or 1 is more or if the sides are equal.

- Problem A: What's the fraction? **(Signal.)** *4 fifths.*
- Is the top number more than, less than, or equal to the bottom number? **(Signal.)** *Less than.*
- Is 4/5 more than, less than, or equal to 1? **(Signal.)** *Less than (1).*
- So will you make the sign so the end next to 4/5 is bigger or smaller? **(Signal.)** *Smaller.*
- Make the sign. ✔

(Display:) [8:4B]

a. $\frac{4}{5} < 1$

Here's what you should have for A. 4/5 is less than 1.

b. Problem B: What's the fraction? **(Signal.)** *7 ninths.*

- Is the top number more than, less than, or equal to the bottom number? **(Signal.)** *Less than.*
- Is 7/9 more than, less than, or equal to 1? **(Signal.)** *Less than (1).*
- So will you make the sign so the end next to 7/9 is bigger or smaller? **(Signal.)** *Smaller.*
- Make the sign. ✔

(Display:) [8:4C]

b. $\frac{7}{9} < 1$

Here's what you should have for B. 7/9 is less than 1.

from Lesson 8, Exercise 4

First students read fractions and identify if each is more, less, or equal to 1. For the Workbook exercise that follows, students read each fraction, determine if it is more than, less than, or equal to 1, and write the sign.

Teaching Note: If students are firm on steps B through D and making the signs, they should have no trouble with the Workbook exercise. If students have trouble with an item, however, use the same series of questions you presented on the preceding lessons.

For example: Students say that 8/10 is more than 1.

Ask: "How many parts are in each unit?"

"Are there more than ten parts or less than ten parts shaded?"

"So is 8/10 more, less or equal to 1?"

Practice this routine before you present the exercise on Lesson 8.

Part of the early work with fractions focuses on the relationship between diagrams and corresponding fractions. Students practice writing fractions based on pictures and practice shading parts to make pictures conform to fractions.

Here's part of the Exercise from Lesson 10:

a. Find part 2 in your workbook. ✔

(Teacher reference:)

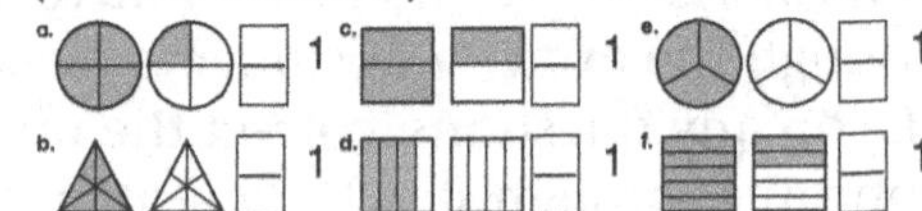

You're going to write the fraction for each picture. Then you'll make the sign to show if the fraction is more than 1, equal to 1, or less than 1.

- Touch picture A. ✔ You're going to write the fraction for this picture in the box next to it.
- Raise your hand when you know the bottom number. ✔
- What's the bottom number? **(Signal.)** *4.*
- Touch where you'll write 4. ✔

b. Everybody, what's the top number of fraction A? **(Signal.)** *5.*

- Touch where you'll write 5. ✔
- Write the fraction for picture A. ✔

(Display:) [10:3J]

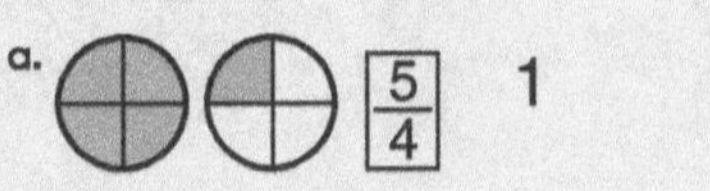

Here's what you should have so far.

c. Everybody, what's the fraction for A? **(Signal.)** *5 fourths.*

- Is 5/4 more than, less than, or equal to 1? **(Signal.)** *More than (1).*
- So will you make the sign with the bigger or smaller end next to 5 fourths? **(Signal.)** *Bigger.*

(Add to show:) [10:3K]

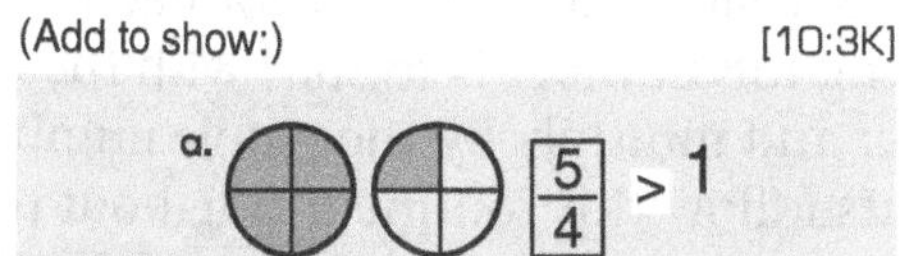

- Make the sign to show that 5/4 is more than 1. **(Observe students and give feedback.)**

d. Write fractions for the rest of the pictures in part 2. Put your pencil down when you've written all the fractions.
(Observe students and give feedback.)

e. Check your work. You'll read the fractions.

- Fraction B. **(Signal.)** *6 sixths.*
- **(Repeat for:)** C, *3/2;* D, *3/4;* E, *3/3;* F, *7/5.*

(Display:) [10:3L]

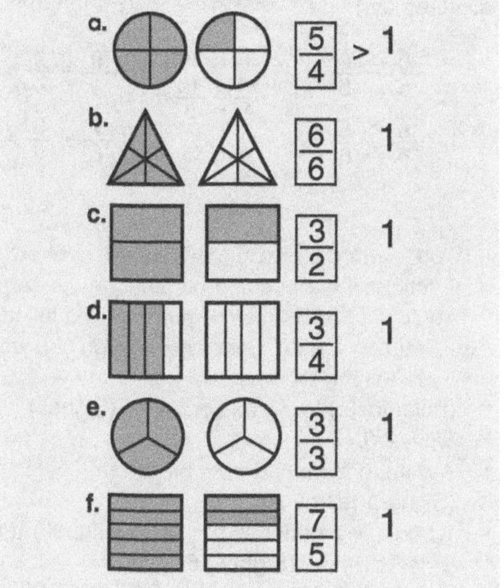

Here is what you should have for part 2. Make sure you have the correct fractions.

f. Go back to fraction B. ✔

- Is the fraction for B more than, less than, or equal to 1? **(Signal.)** *Equal to (1).*
- Make the sign to show that 6/6 is equal to 1. ✔

g. Make the sign to show if fraction C is more than, less than, or equal to 1.
(Observe students and give feedback.)

- Did you write an equation for C? **(Signal.)** *No.*
- Is the bigger or smaller end next to 3 halves? **(Signal.)** *Bigger.*

h. Make signs for the rest of the fractions.
(Observe students and give feedback.)

i. Check your work.

- Fraction D: Did you write an equation? **(Signal.)** *No.*
- Is the bigger or smaller end next to 3 fourths? **(Signal.)** *Smaller.*

j. Fraction E: Did you write an equation? **(Signal.)** *Yes.*

- Say the equation. **(Signal.)** *3/3 = 1.*

k. Fraction F: Did you write an equation? **(Signal.)** *No.*

- Is the bigger or smaller end next to 7 fifths? **(Signal.)** *Bigger.*

(Add to show:) [10:3M]

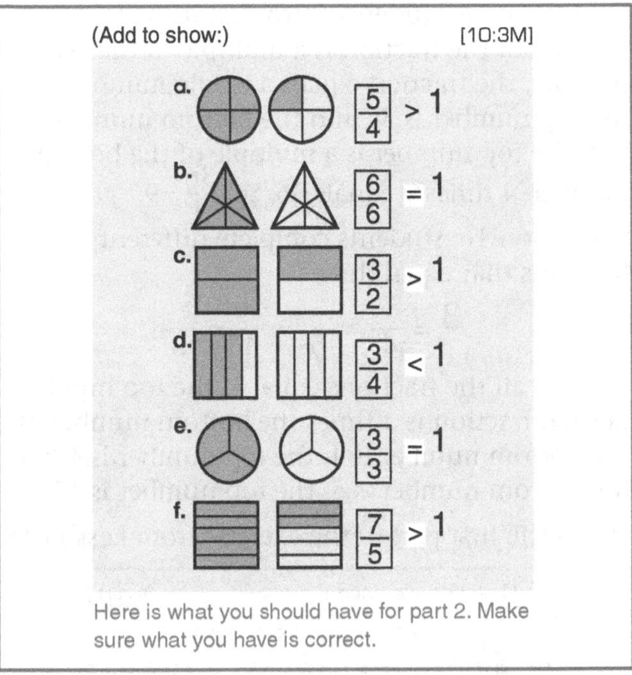

Here is what you should have for part 2. Make sure what you have is correct.

In steps A through E, students write fractions for each picture. In steps F through K, students write signs to show if the fraction is more, less, or equal to 1.

Teaching Note: In steps A and B, you direct students to touch where they will write the numbers for fraction A. These tasks were included because some students were confused about where to write the numbers. Make sure that lower performers are touching the correct place for each number. On Lesson 13, students write fractions for shaded parts shown on number lines.

For example:

Students first identify the number of parts in each unit (five). Then students indicate the number of parts that are shaded (nine). Some fractions equal whole numbers.

On Lesson 16, students learn that if the top number of the fraction is a multiple of the bottom number, the fraction equals a whole number. If the top number is 36 and the bottom number is 4, the top number is a multiple of the bottom number. 4 times 9 equals 36. So, $\frac{36}{4} = 9$.

On Lesson 16, students complete different fractions that equal three.

$$3 = \frac{\quad}{4} = \frac{\quad}{5} = \frac{\quad}{1} = \frac{\quad}{9}$$

Because all the fractions equal 3, the top number of each fraction is 3 times the bottom number. If the bottom number is 4, the top number is 12; if the bottom number is 5, the top number is 15.

Here's the first part of the exercise from Lesson 18:

(Teacher reference:) **R Part J**

a. $3 = \frac{\quad}{10} = \frac{\quad}{5} = \frac{\quad}{9} = \frac{\quad}{2}$ b. $2 = \frac{\quad}{6} = \frac{\quad}{4} = \frac{\quad}{8} = \frac{\quad}{1}$

- All the fractions for problem A are equal to a whole number. What whole number? (Signal.) *3.*
- So the top number of these fractions is how many times the bottom number? (Signal.) *3.*

b. Touch the bottom number of the first fraction. ✔
- What's the bottom number? (Signal.) *10.*
- Say the problem for finding the top number. (Signal.) *3 × 10.*
- What's the top number of that fraction? (Signal.) *30.*
  Yes, 30 tenths equals 3.

c. What's the bottom number of the next fraction? (Signal.) *5.*
- Say the problem and the answer for the top number. (Signal.) *3 × 5 = 15.*
  Yes, 15 fifths equals 3.

d. What's the bottom number of the next fraction? (Signal.) *9.*
- Say the problem and the answer for the top number. (Signal.) *3 × 9 = 27.*
  Yes, 27 ninths equals 3.

e. What's the bottom number of the next fraction? (Signal.) *2.*
- Say the problem and the answer for the top number. (Signal.) *3 × 2 = 6.*
  Yes, 6 halves equals 3.

f. Complete all the fractions for problem A. Put your pencils down when you've completed the equation for A.
(Observe students and give feedback.)

g. Check your work for equation A.
- You'll read the whole equation. Get ready. (Signal.) *3 = 30 tenths = 15 fifths = 27 ninths = 6 halves.*
  (Display:) [18:4A]

  a. $3 = \frac{30}{10} = \frac{15}{5} = \frac{27}{9} = \frac{6}{2}$

  Here's what you should have.

from Lesson 18, Exercise 4

The steps that students follow for fractions with a missing top number is to start with the bottom number and multiply by the whole number. Students follow this routine throughout their work with converting whole numbers into fractions.

Adding and subtracting fractions starts on Lesson 21. For the first two days, all the problems add.

Here's the exercise from Lesson 25. Students read each problem and indicate whether the bottom numbers are the same and if they can work the problem. Students say the answers to the problems they can work.

a. (Display:) [25:4A]

$\frac{8}{5} - \frac{2}{5} =$ $\frac{14}{4} + \frac{15}{5} =$ $\frac{9}{1} - \frac{9}{4} =$

$\frac{6}{5} - \frac{5}{6} =$ $\frac{7}{6} + \frac{3}{6} =$

$\frac{14}{1} - \frac{10}{1} =$

You learned that you can add and subtract fractions if the bottom numbers are the same. You'll read each fraction problem and tell me if you can work it. For problems you can work, you'll tell me the answer.
- (Point to $\frac{8}{5}$.) Read this problem. (Signal.) *8/5 – 2/5.*
- Are the bottom numbers the same? (Signal.) *Yes.*
- So can you work this problem? (Signal.) *Yes.*
- What is 8 – 2? (Signal.) *6.*
- So what's 8 **fifths** – 2 **fifths**? (Signal.) *6/5.*

b. (Point to $\frac{6}{5}$.) Read this problem. (Signal.) *6/5 – 5/6.*
- Are the bottom numbers the same? (Signal.) *No.*
- So can you work this problem? (Signal.) *No.*

c. (Point to $\frac{14}{4}$.) Read this problem. (Signal.) *14/4 + 15/5.*
- Are the bottom numbers the same? (Signal.) *No.*
- So can you work this problem? (Signal.) *No.*

d. (Point to $\frac{14}{1}$.) Read this problem. (Signal.) *14 over 1 – 10 over 1.*
- Are the bottom numbers the same? (Signal.) *Yes.*
- So can you work this problem? (Signal.) *Yes.*
- What is 14 – 10? (Signal.) *4.*
- So what's 14 over 1 – 10 over 1? (Signal.) *4 over 1.*

e. (Point to $\frac{9}{1}$.) Read this problem. (Signal.) *9 over 1 – 9/4.*
- Are the bottom numbers the same? (Signal.) *No.*
- So can you work this problem? (Signal.) *No.*

f. (Point to $\frac{7}{6}$.) Read this problem. (Signal.) *7/6 + 3/6.*
- Are the bottom numbers the same? (Signal.) *Yes.*
- So can you work this problem? (Signal.) *Yes.*
- What's 7 + 3? (Signal.) *10.*
- So what's 7 **sixths** + 3 **sixths**? (Signal.) *10/6.*
  (Repeat problems that were not firm.)

from Lesson 25, Exercise 4

In the Textbook practice, all the problems can be worked. Students work them and write the answers.

FRACTIONS LESSONS 31–130

Students who have completed *CMC Level D* enter the program at Lesson 31, which is the first lesson that presents fraction operations and discriminations that were not introduced in *CMC Level D*. The work that begins on Lesson 31 prepares students to identify fractions that equal whole numbers.

Here's the exercise from Lesson 33:

a. Find part 4 in your workbook. ✔
(Teacher reference:) **R Part K**

a. $\frac{26}{5} - \frac{5}{3} =$ c. $\frac{6}{10} - \frac{6}{10} =$ e. $\frac{12}{7} + \frac{5}{7} =$

b. $\frac{18}{3} + \frac{2}{3} =$ d. $\frac{8}{4} - \frac{8}{3} =$ f. $\frac{10}{3} + \frac{3}{8} =$

You can work some of these problems. You can't work others. You'll read each problem and tell me if you can work it.
- Read problem A. **(Signal.)** *26/5 – 5/3.*
- Are the bottom numbers the same? **(Signal.)** *No.*
- So can you work it? **(Signal.)** *No.*

b. Read problem B. **(Signal.)** *18/3 + 2/3.*
- Are the bottom numbers the same? **(Signal.)** *Yes.*
- So can you work it? **(Signal.)** *Yes.*

(If students are 100%, skip to step d.)

c. **(Repeat the following tasks for problems C through F:)**

Read problem __.		Are the bottom numbers the same?	So can you work it?
C	*6/10 – 6/10*	*Yes*	*Yes*
D	*8/4 – 8/3*	*No*	*No*
E	*12/7 + 5/7*	*Yes*	*Yes*
F	*10/3 + 3/8*	*No*	*No*

(Repeat problems that were not firm.)

d. Listen: Cross out all the problems you can't work. Put your pencil down when you've done that much.
(Observe students and give feedback.)

e. Check your work.
- What's the letter of the first problem you crossed out? **(Signal.)** *A.*
- What's the letter of the next problem you crossed out? **(Signal.)** *D.*
- What's the letter of the next problem you crossed out? **(Signal.)** *F.*

(Display:) [33:4A]

a. ~~$\frac{26}{5} - \frac{5}{3}$~~ = c. $\frac{6}{10} - \frac{6}{10} =$ e. $\frac{12}{7} + \frac{5}{7} =$

b. $\frac{18}{3} + \frac{2}{3} =$ d. ~~$\frac{8}{4} - \frac{8}{3}$~~ = f. ~~$\frac{10}{3} + \frac{3}{8}$~~ =

Here's what you should have for part 4 so far.

f. You can work the other problems in part 4.
- Do it. Put your pencil down when you've worked the remaining problems in part 4.
(Observe students and give feedback.)

g. Check your work. You'll read the problem and the answer for the problems you worked.
- Problem B. **(Signal.)** *18/3 + 2/3 = 20/3.*
- Problem C. **(Signal.)** *6/10 – 6/10 = 0/10.*
- Problem E. **(Signal.)** *12/7 + 5/7 = 17/7.*

(Add to show:) [33:4B]

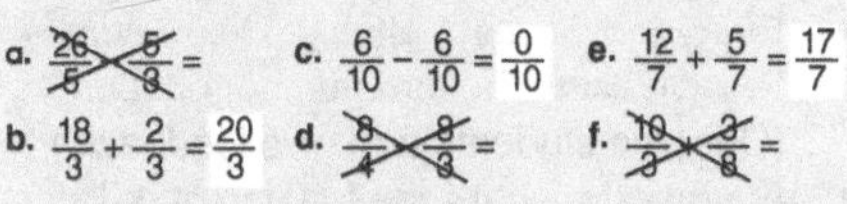

a. 26/5 − 5/3 =
c. 6/10 − 6/10 = 0/10
e. 12/7 + 5/7 = 17/7
b. 18/3 + 2/3 = 20/3
d. 8/4 − 8/3 =
f. 10/3 + 3/8 =

Here's what you should have for part 4.

Lesson 33, Exercise 4

Steps A through C provide a verbal review. Students indicate whether the bottom numbers of the fractions are the same and, therefore, whether students can work the problem as written.

In steps D through F, they cross out problems they can't work as written and work the problems they can work as written. Students indicate the answers to the problems they worked.

On Lesson 37, students rewrite whole numbers as fractions with specified denominators. For instance, "Rewrite 6 as a fraction with a bottom number of 5."

Here's the exercise from Lesson 37:

(Teacher reference:)

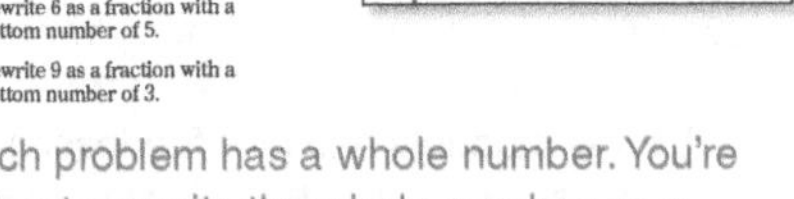

Each problem has a whole number. You're going to rewrite the whole number as a fraction. The directions in each problem tell you about the bottom number of the fraction.
- Write part 3 on your lined paper with the letters A through D below.
(Observe students and give feedback.)

b. I'll read problem A. Rewrite 7 as a fraction with a bottom number of 2.
- What's the whole number? **(Signal.)** *7.*

c. **(Display:)** [37:7A]

Part 3
a. $\frac{}{2}$ 7̸

- Do it. Make a fraction bar after A. Then write 7 and rewrite it as a fraction with a bottom number of 2.
(Observe students and make sure they cross out 7 when they write it.)
- What fraction did you write for 7? **(Signal.)** *14/2.*

(Add to show:) [37:7B]

Part 3

a. $\frac{14}{2}$ 7

Here's what you should have.

d. Touch B and read the directions. Get ready. (Signal.) *Rewrite 4 as a fraction with a bottom number of 10.*

- What's the whole number? (Signal.) *4.*
- Everybody, write 4 after B. Then rewrite 4 as a fraction with a bottom number of 10. (Observe students and give feedback.)
- Everybody, what's the fraction of 4 with a bottom number of 10? (Signal.) *40/10.*

(Display:) [37:7C]

b. $\frac{40}{10}$ 4

Here's what you should have.

e. Touch C and read the directions. Get ready. (Signal.) *Rewrite 6 as a fraction with a bottom number of 5.*

- What's the whole number? (Signal.) *6.*
- Everybody, write 6 after C. Then rewrite 6 as a fraction with a bottom number of 5. (Observe students and give feedback.)
- Everybody, what's the fraction of 6 with a bottom number of 5? (Signal.) *30/5.*

(Display:) [37:7D]

c. $\frac{30}{5}$ 6

Here's what you should have.

f. Touch D and read the directions. Get ready. (Signal.) *Rewrite 9 as a fraction with a bottom number of 3.*

- What's the whole number? (Signal.) *9.*
- Everybody, write 9 after D. Then rewrite 9 as a fraction with a bottom number of 3. (Observe students and give feedback.)
- Everybody, what's the fraction of 9 with a bottom number of 3? (Signal.) *27/3.*

(Display:) [37:7E]

d. $\frac{27}{3}$ 9

Here's what you should have.

Lesson 37, Exercise 7

Teaching Note: Students who started *CMC Level E* at Lesson 1 should be well-practiced on converting whole numbers into fractions with a specified denominator. See Lesson 18 on page 85.

If students are confused, remind them of the procedure: "Start with the bottom number of the fraction and multiply by the whole number."

Problem D may present difficulties because the bottom number of the fraction is 3 and the whole number is a multiple of 3–9. So some students may simply divide 9 by 3. Remind them: "Start with the bottom number of the fraction and multiply by the whole number."

Mixed Numbers

On Lesson 38, students learn to read and write mixed numbers for fractions on a number line. The procedure they learn for reading mixed numbers is to read the whole number part, say *and,* and read the fraction part. 3 1/4 is read "three and one fourth." The procedure for writing mixed numbers is to count the number of *units* that are shaded. That tells the whole number part of the mixed number. Then write the fraction for the remaining shaded parts.

Here's part of the exercise from Lesson 38:

a. (Display:) [38:8A]

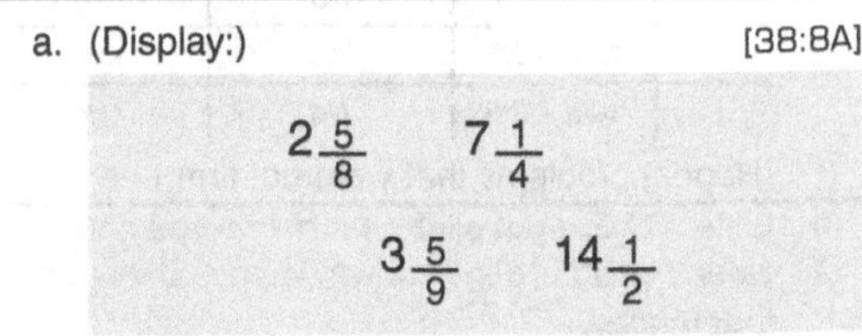

These are mixed numbers. When you read mixed numbers, you just say the number and the fraction.

- (Point to **2.**) This mixed number is 2 and 5 eighths.
- Say the mixed number. (Signal.) *2 and 5/8.* (Repeat until firm.)

b. (Point to **7.**) Say the mixed number. (Signal.) *7 and 1/4.*

- (Point to **3.**) Say the mixed number. (Signal.) *3 and 5/9.*
- (Point to **14.**) Say the mixed number. (Signal.) *14 and 1/2.* (Repeat until firm.)

Remember, mixed numbers have a whole number and a fraction.

c. (Display:) [38:8B]

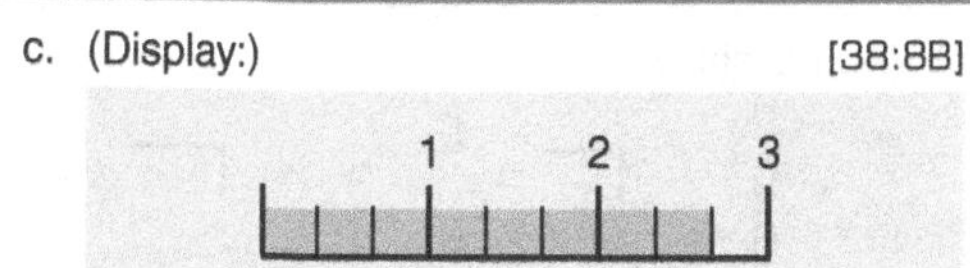

- Raise your hand when you know the fraction for this number line. ✔
- Everybody, what fraction? (Signal.) *8/3.*

(Add to show:) [38:8C]

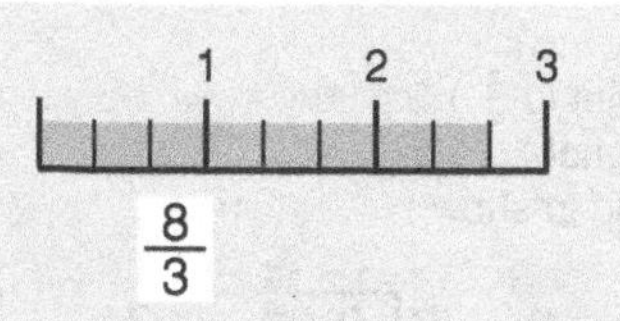

d. I'll show you the mixed number that equals 8/3.
- (Point to **2.**) How many whole units are shaded? (Signal.) *2.*
- So 2 is the first part of the mixed number. What's the first part of the mixed number? (Signal.) *2.*

(Add to show:) [38:8D]

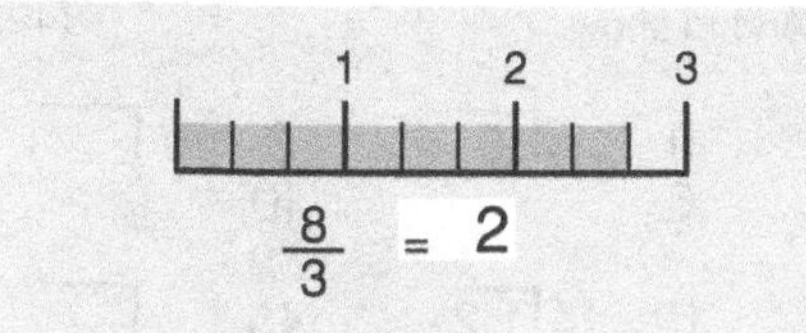

e. My turn to figure out the fraction part. I count the parts **after** 2 units. (Touch and count.) 1/3, 2/3.
- What's the fraction part of the mixed number? (Signal.) *2/3.*

(Add to show:) [38:8E]

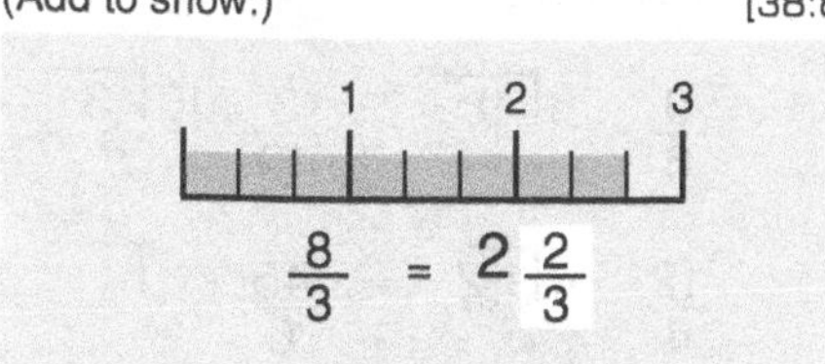

Here's the mixed number for 8/3: 2 and 2/3.
- Read the mixed number for 8/3. (Signal.) *2 and 2/3.*

from Lesson 38, Exercise 8

In steps A and B, the teacher models reading mixed numbers. In step C, students figure out the fraction for the shaded part of the number line (8/3).

In step D, students identify the number of whole units that are shaded (2).

In step E, the teacher models how to figure out the fraction part of the mixed number 2 2/3.

Steps F through H present another number line and test students on identifying the fraction and mixed number for the shaded part (7/4).

On Lesson 41, students work problems that add a whole number and a fraction. Students convert the whole number into a fraction.

Here's the first part of the exercise:

a. (Display:) [41:7A]

$$5 + \frac{1}{4} \qquad 2 + \frac{5}{8}$$
$$3 + \frac{7}{2}$$

You learned that you can add or subtract fractions if the bottom numbers are the same.
- (Point to $5 + \frac{1}{4}$.) Read this problem. (Signal.) *5 + 1/4.*
- Are the bottom numbers the same? (Signal.) *No.*
- So can you work that problem? (Signal.) *No.*

b. (Point to $2 + \frac{5}{8}$.) Read this problem. (Signal.) *2 + 5/8.*
- (Point to $3 + \frac{7}{2}$.) Read this problem. (Signal.) *3 + 7/2.*

c. Listen: You can't add whole numbers and fractions. You have to rewrite the whole number as a fraction. The fraction for the whole number must have the same bottom number as the other fraction.
- (Point to $5 + \frac{1}{4}$.) What's the bottom number of the fraction in this problem? (Signal.) *4.*
- So you have to rewrite 5 as a fraction with what bottom number? (Signal.) *4.*

d. (Point to $2 + \frac{5}{8}$.) What's the bottom number of the fraction in this problem? (Signal.) *8.*
- So you have to rewrite 2 as a fraction with what bottom number? (Signal.) *8.*

e. (Point to $3 + \frac{7}{2}$.) What's the bottom number of the fraction in this problem? (Signal.) *2.*
- So you have to rewrite 3 as a fraction with what bottom number? (Signal.) *2.*

(Repeat steps c through e until firm.)

f. (Display:) [41:7B]

$$5 + \frac{1}{4}$$

- To work this problem, I rewrite 5 as a fraction with what bottom number? (Signal.) *4.*

(Add to show:) [41:7C]

$$\frac{}{4}\!\!\not{5} + \frac{1}{4}$$

- Raise your hand when you know the top number of the fraction for 5. ✔
- What's the top number of the fraction? (Signal.) *20.*

(Add to show:) [41:7D]

$$\frac{20}{4}\!\!\not{5} + \frac{1}{4}$$

g. What's the fraction that equals 5? (Signal.) *20/4.*
- Read the fraction problem. (Signal.) *20/4 + 1/4.*
- What's the answer? (Signal.) *21/4.*

(Add to show:) [41:7E]

$$\frac{20}{4}\!\!\not{5} + \frac{1}{4} = \frac{21}{4}$$

- What does 5 + 1/4 equal? (Signal.) *21/4.*

from Lesson 41, Exercise 7

Teaching Note: Make sure that students respond very well to steps C through E. Repeat this part several times if necessary. If students are facile at answering the series of questions presented in this segment, everything that follows—through working mixed-number problems—is easy. The only difference is that mixed numbers don't have a plus sign between the whole number and the fraction. All the conventions are the same, however.

On Lesson 42, students learn a tool concept: You can read fractions as division problems and work them as division problems.

Students have performed division operations earlier when they worked with fractions that equal whole numbers. To figure out what whole number a fraction like 12/4 equals, students indicate that the top number is 3 times 4. This is an important concept for higher math operations. The problem:

$$\frac{12}{4} \qquad 12 \div 4 \qquad 4\overline{)12}$$

Here's part of the exercise from Lesson 42:

a. (Display:) [42:6A]

$$\frac{9}{6} \qquad \frac{3}{10}$$
$$\frac{12}{4} \qquad \frac{45}{6}$$

- (Point to $\frac{9}{6}$.) Read this fraction. (Signal.) *9/6.*
- (Point to $\frac{12}{4}$.) Read this fraction. (Signal.) *12/4.*
- (Point to $\frac{3}{10}$.) Read this fraction. (Signal.) *3/10.*
- (Point to $\frac{45}{6}$.) Read this fraction. (Signal.) *45/6.*

b. Here's a rule about fractions: You can write any fraction as a division problem.

- (Point to $\frac{9}{6}$.) I'll say the division problem for 9 sixths. 9 divided by 6.
- Your turn: Say the division problem. (Signal.) *9 ÷ 6.*
- (Point to $\frac{12}{4}$.) Say the division problem for 12/4. (Signal.) *12 ÷ 4.*
- (Point to $\frac{3}{10}$.) Say the division problem for 3/10. (Signal.) *3 ÷ 10.*
- (Point to $\frac{45}{6}$.) Say the division problem for 45/6. (Signal.) *45 ÷ 6.*

(Repeat step b until firm.)

c. (Add to show:) [42:6B]

$$\frac{9}{6}\ \overline{)} \qquad \frac{3}{10}\ \overline{)}$$
$$\frac{12}{4}\ \overline{)} \qquad \frac{45}{6}\ \overline{)}$$

This time, you'll say the division problem, and I'll write it.

- (Point to $\frac{9}{6}$.) Say the division problem for 9/6. (Signal.) *9 ÷ 6.*

(Add to show:) [42:6C]

$$\frac{9}{6}\ 6\overline{)9} \qquad \frac{3}{10}\ \overline{)}$$
$$\frac{12}{4}\ \overline{)} \qquad \frac{45}{6}\ \overline{)}$$

d. (Point to $\frac{12}{4}$.) Say the division problem for 12/4. (Signal.) *12 ÷ 4.*
(Add to show:) [42:6D]

$$\frac{9}{6}\ 6\overline{)9} \qquad \frac{3}{10}\ \overline{)}$$
$$\frac{12}{4}\ 4\overline{)12} \qquad \frac{45}{6}\ \overline{)}$$

e. (Point to $\frac{3}{10}$.) Say the division problem for 3/10. (Signal.) *3 ÷ 10.*
(Add to show:) [42:6E]

$$\frac{9}{6}\ 6\overline{)9} \qquad \frac{3}{10}\ 10\overline{)3}$$
$$\frac{12}{4}\ 4\overline{)12} \qquad \frac{45}{6}\ \overline{)}$$

f. (Point to $\frac{45}{6}$.) Say the division problem for 45/6. (Signal.) *45 ÷ 6.*
(Add to show:) [42:6F]

$$\frac{9}{6}\ 6\overline{)9} \qquad \frac{3}{10}\ 10\overline{)3}$$
$$\frac{12}{4}\ 4\overline{)12} \qquad \frac{45}{6}\ 6\overline{)45}$$

from Lesson 42, Exercise 6

Teaching Note: The procedure is simple. Start with the top number and say the division problem. The validity of the operation is shown by fractions that equal whole numbers. 20/2 is 20 divided by 2, which is 10.

On Lesson 49 students write addition problems from mixed numbers and write the fractions they equal.

Here's part of the exercise from Lesson 49:

(Teacher reference:)

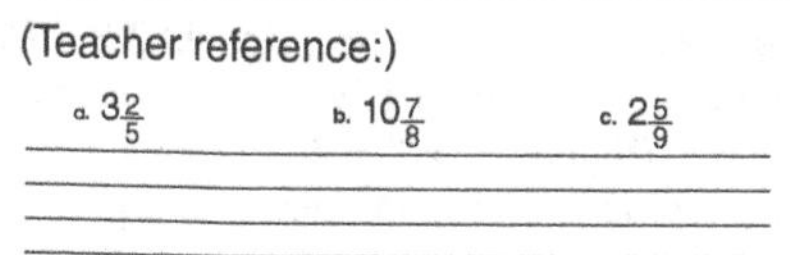

These are mixed numbers. You'll write the addition for each mixed number on the lines below it.

- Read mixed number A. **(Signal.)** *3 and 2/5.*
- Say the addition for 3 and 2/5. **(Signal.)** *3 plus 2/5.*

(Display:) [49:3B]

a. $3\frac{2}{5}$

$3 + \frac{2}{5}$

Here's the addition for A.

b. Write the addition problem for mixed number A. **(Observe students and give feedback.)**

c. Read mixed number B. **(Signal.)** *10 and 7/8.*

- Say the addition problem for 10 and 7/8. **(Signal.)** *10 plus 7/8.*
- Write the addition problem for 10 and 7/8 on the lines below it. **(Observe students and give feedback.)**

(Display:) [49:3C]

b. $10\frac{7}{8}$

$10 + \frac{7}{8}$

Here's the addition for B.

d. Read mixed number C. **(Signal.)** *2 and 5/9.*

- Write the addition problem for 2 and 5/9 on the lines below it. **(Observe students and give feedback.)**
- Read the addition for 2 and 5/9. **(Signal.)** *2 plus 5/9.*

(Display:) [49:3D]

c. $2\frac{5}{9}$

$2 + \frac{5}{9}$

Here's the addition for C.

e. Go back to problem A and read the addition for 3 and 2/5 again. **(Signal.)** *3 plus 2/5.*

- Rewrite the whole number as a fraction. Then complete the equation. **(Observe students and give feedback.)**

f. Check your work for problem A.

- What fraction did you write for 3? **(Signal.)** *15/5.*
- Read the fraction problem and the answer. **(Signal.)** *15/5 + 2/5 = 17/5.*
- What fraction does the mixed number 3 and 2/5 equal? **(Signal.)** *17/5.*

(Display:) [49:3E]

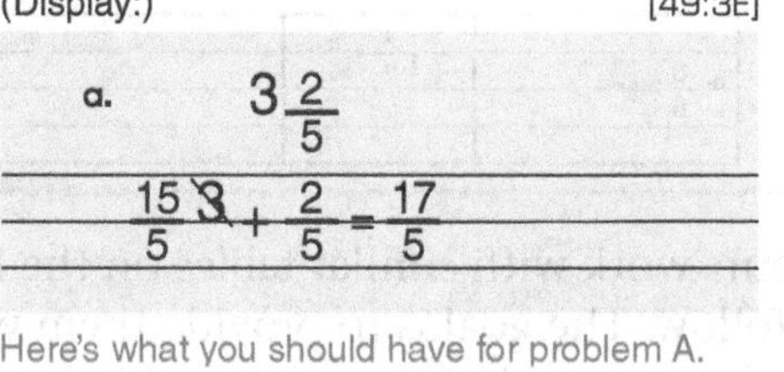

Here's what you should have for problem A.

from Lesson 49, Exercise 3

Students learn to write mixed numbers as addition problems. 3 2/5 is 3 + 2/5.

Students apply what they have learned about converting the whole number into a fraction and adding.

Starting on Lesson 61, students complete tables that have columns for fraction equations, related division problems and answers, and mixed numbers or whole numbers. Here's the table students work with on Lesson 62:

	Fraction	Division	Mixed Number
a.			$4\frac{7}{8}$
b.	$\frac{8}{6}$		
c.			$3\frac{3}{7}$
d.		$9\overline{)24}$	

For the top row, students first figure out the fraction for the mixed number and write it (39/8).

In the same box they complete the fraction equation (39/8 = 4 7/8).

Finally, they write the division problem for 39/8 and write the mixed number it equals.

For the second row, students write the division problem for the fraction and figure out the mixed-number answer (8/6=1 2/6). With this information they complete the fraction equation and write the mixed number for the row:

b. $\frac{8}{6} = 1\frac{2}{6}$	$6\overline{)8}$ 1 2, 6	$1\frac{2}{6}$

Students work with similar tables on the lessons that follow. The goal is to provide them with enough practice that the relationship between division fractions and mixed numbers or whole numbers becomes fairly automatic.

On Lesson 64, students start to work with tables that require calculations that students probably can't do mentally.

Here's the table from Workbook Lesson 64:

Fraction	Division	Mixed Number	
a.	$\overline{)}$	$24\frac{1}{4}$	
b. $\frac{103}{9}$	$\overline{)}$		
c.	$5\overline{)537}$		
d.	$\overline{)}$	$215\frac{2}{3}$	

To figure out the fraction for 24 1/4, students work the problem 24 × 4 and then add one.

Fraction Multiplication

Starting on Lesson 54, students work basic problems that multiply a fraction by a fraction. They learn about two differences between adding and multiplying fractions:

- The denominators do not have to be the same for multiplication.
- Also, you work both the problem for the top numbers and the problem for the bottom numbers when multiplying.

Here's the Workbook practice from Lesson 55:

(Teacher reference:) **R Part M**

a. $\frac{7}{8} \times \frac{1}{9} =$ b. $\frac{10}{7} \times \frac{6}{5} =$ c. $\frac{2}{3} \times \frac{8}{7} =$ d. $\frac{5}{2} \times \frac{6}{9} =$

These problems multiply a fraction by a fraction.

- Touch and read problem A. **(Signal.)** *7/8 × 1/9.*
- Say the problem and the answer for the top numbers. **(Signal.)** *7 × 1 = 7.*
- Say the problem and the answer for the denominators. **(Signal.)** *8 × 9 = 72.*

b. Touch and read problem B. **(Signal.)** *10/7 × 6/5.*

- Say the problem and the answer for the top numbers. **(Signal.)** *10 × 6 = 60.*
- Say the problem and the answer for the denominators. **(Signal.)** *7 × 5 = 35.*

(If students are 100%, skip to step e.)

c. Touch and read problem C. **(Signal.)** *2/3 × 8/7.*

- Say the problem and the answer for the top numbers. **(Signal.)** *2 × 8 = 16.*
- Say the problem and the answer for the denominators. **(Signal.)** *3 × 7 = 21.*

d. Touch and read problem D. **(Signal.)** *5/2 × 6/9.*

- Say the problem and the answer for the top numbers. **(Signal.)** *5 × 6 = 30.*
- Say the problem and the answer for the denominators. **(Signal.)** *2 × 9 = 18.*

(Repeat problems that were not firm.)

e. Go back to problem A and work it.

(Observe students and give feedback.)

- Problem A: 7/8 × 1/9. What's the answer? **(Signal.)** *7/72.*

f. Work problem B.

(Observe students and give feedback.)

- Read problem B and the answer. **(Signal.)** *10/7 × 6/5 = 60/35.*

(Display:) [55:3K]

a. $\frac{7}{8} \times \frac{1}{9} = \frac{7}{72}$ b. $\frac{10}{7} \times \frac{6}{5} = \frac{60}{35}$

Here are equations A and B.

g. Work problems C and D.

(Observe students and give feedback.)

h. Check your work. You'll read the equations for problems C and D.

- Problem C. Say the equation with **over** in the answer. Get ready. **(Signal.)** *2/3 × 8/7 = 16* ***over*** *21.*
- Problem D. Get ready. **(Signal.)** *5/2 × 6/9 = 30/18.*

(Display:) [55:3L]

c. $\frac{2}{3} \times \frac{8}{7} = \frac{16}{21}$ d. $\frac{5}{2} \times \frac{6}{9} = \frac{30}{18}$

Here are equations C and D.

Remember, when you multiply fractions, the denominators don't have to be the same. You multiply the top numbers and multiply the denominators.

from Lesson 55, Exercise 3

In steps A through D, students say the multiplication and the answer for the top numbers and **denominators** of three problems. (Students no longer refer to the "bottom number" of the fraction.) In steps E through H, students work the problems and check their answers.

On Lesson 59, students work a mixed set of problems. Some multiply fractions; some add or subtract fractions. Students answer a pair of questions for each problem:

"Does this problem add or subtract?"

"Do the denominators have to be the same?"

A variation of the mixed problem set appears first on Lesson 62. Some of the problems can't be worked the way they are written.

Here's the Workbook activity:

Part 5

a. $\frac{7}{2} + \frac{7}{5}$

b. $\frac{5}{4} - \frac{3}{4}$

c. $\frac{6}{9} \times \frac{5}{6}$

d. $\frac{2}{7} + \frac{5}{7}$

e. $\frac{5}{4} - \frac{3}{9}$

f. $\frac{5}{4} \times \frac{3}{9}$

On Lesson 65, students work problems that multiply three fractions (1/3 × 2/4 × 5/1).

On Lesson 67, students work problems that multiply whole numbers by fractions. Students learn to make a simple fraction for the whole number (7 × 3/2). Students change 7 into 7/1, then multiply the top numbers and multiply the denominators.

Students continue to work problems of this type on the following lessons.

Here's the activity from Lesson 68:

a. I'll say a simple fraction that equals 53: 53 over 1.

- Your turn: Say a simple fraction that equals 13. (Signal.) *13 over 1.*
- Say a simple fraction that equals B. (Signal.) *B over 1.*
- Say a simple fraction that equals 199. (Signal.) *199 over 1.*
- Say a simple fraction that equals K. (Signal.) *K over 1.*

(Repeat until firm.)

(If students were 100% on the last lesson, skip to **Textbook Practice.**)

b. (Display:) [68:8A]

$$\frac{8}{3} \times 5$$

- Read this problem. (Signal.) *8/3 × 5.*
- What's the whole number in this problem? (Signal.) *5.*
- Say the simple fraction that equals 5. (Signal.) *5 over 1.*

(Add to show:) [68:8B]

$$\frac{8}{3} \times \frac{5}{1}$$

c. Say the problem for the numerators. (Signal.) *8 × 5.*

- What's the answer? (Signal.) *40.*

(Add to show:) [68:8C]

$$\frac{8}{3} \times \frac{5}{1} = 40$$

- Say the problem for the denominators. (Signal.) *3 × 1.*
- What's the answer? (Signal.) *3.*

(Add to show:) [68:8D]

$$\frac{8}{3} \times \frac{5}{1} = \frac{40}{3}$$

- Read the equation. (Signal.) *8/3 × 5 over 1 = 40/3.*
- What does 8/3 × 5 equal? (Signal.) *40/3.*

TEXTBOOK PRACTICE

a. Find part 4 in your textbook. ✔
(Teacher reference:)

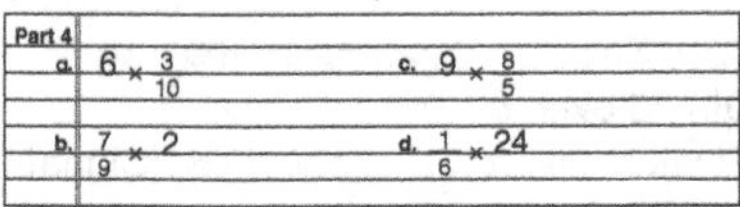

Part 4
a. $6 \times \frac{3}{10}$ c. $9 \times \frac{8}{5}$
b. $\frac{7}{9} \times 2$ d. $\frac{1}{6} \times 24$

These problems multiply a whole number and a fraction.

- Copy part 4 on your lined paper.
(Observe students and give feedback.)

b. Read problem A on your lined paper. (Signal.) *6 × 3/10.*

- What fraction will you change the whole number into? (Signal.) *6 over 1.*
- Work problem A.
(Observe students and give feedback.)
- Read the equation for A. Get ready. (Signal.) *6 over 1 × 3/10 = 18/10.*
- What does 6 × 3/10 equal? (Signal.) *18/10.*
(Display:) [68:8E]

Part 4
a. $\frac{6}{1} \times \frac{3}{10} = \frac{18}{10}$

Here's what you should have for A.

c. Read problem B. (Signal.) *7/9 × 2.*

- What fraction will you change the whole number into? (Signal.) *2 over 1.*
- Work problem B.
(Observe students and give feedback.)
- Read the equation for B. (Signal.) *7/9 × 2 over 1 = 14/9.*
- What does 7/9 × 2 equal? (Signal.) *14/9.*

d. Read problem C. (Signal.) *9 × 8/5.*

- What fraction will you change the whole number into? (Signal.) *9 over 1.*
- Work problem C.
(Observe students and give feedback.)
- Read the equation for C. (Signal.) *9 over 1 × 8/5 = 72/5.*
- What does 9 × 8/5 equal? (Signal.) *72/5.*

e. Read problem D. (Signal.) *1/6 × 24.*

- What fraction will you change the whole number into? (Signal.) *24 over 1.*
- Work problem D.
(Observe students and give feedback.)
- Read the equation for D. (Signal.) *1/6 × 24 over 1 = 24/6.*
- What does 1/6 × 24 equal? (Signal.) *24/6.*

Lesson 68, Exercise 8

On Lesson 72, students work a mixed set of problems. Some problems multiply; some add. For addition problems, students change the whole number into a fraction with the same denominator as the other fraction in the problem. If the problem multiplies, students write the whole number as a fraction with a denominator of one.

Here's the first part of the Textbook activity:

(Teacher reference:)

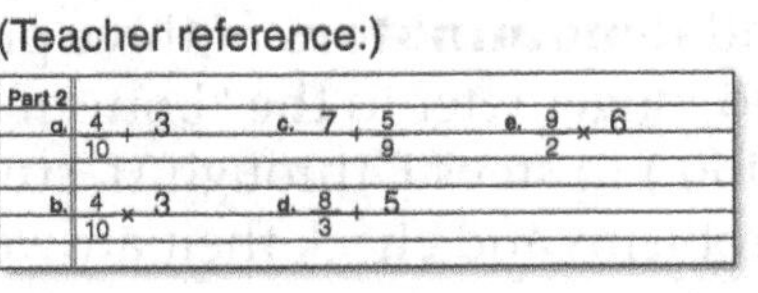

Part 2
a. $\frac{4}{10} + 3$ c. $7 + \frac{5}{9}$ e. $\frac{9}{2} \times 6$
b. $\frac{4}{10} \times 3$ d. $\frac{8}{3} + 5$

Some of these problems add, and some multiply. For each whole number, you'll tell me the denominator of the fraction. Remember, if the problem adds, the denominators must be the same. If the problem multiplies, the denominator for the whole number is 1.

b. Touch and read problem A. (Signal.) *4/10 + 3.*

- Does the problem add? (Signal.) *Yes.*
- So tell me the denominator of the fraction you'll write for 3. (Signal.) *10.*

c. Touch and read problem B. (Signal.) *4/10 × 3.*

- Does the problem add? (Signal.) *No.*
- So tell me the denominator of the fraction you'll write for 3. (Signal.) *1.*

(Repeat steps b and c until firm.)

d. (Repeat the following tasks for problems C through E:)

Read problem __.		Does the problem add?	So tell me the denominator of the fraction you'll write for __.	
C	*7 + 5/9*	*Yes*	7	*9*
D	*8/3 + 5*	*Yes*	5	*3*
E	*9/2 × 6*	*No*	6	*1*

(Repeat problems that were not firm.)

e. Copy part 2. Then write the fraction for the whole number in problem A and complete the equation.
(Observe students and give feedback.)

- Everybody, read the fraction equation for problem A. (Signal.) *4/10 spell 30/10 = 34/10.*
- So what fraction does 4/10 plus 3 equal? (Signal.) *34/10.*
(Display:) [72:6F]

Part 2
a. $\frac{4}{10} + 3\frac{30}{10} = \frac{34}{10}$

Here's what you should have for A.

from Lesson 72, Exercise 6

Teaching Note: In steps B through D, students answer a series of questions about each problem. Make sure their responses are firm. If students miss one or more questions about a problem, repeat all the questions for the item.

EQUIVALENT FRACTIONS

Work on equivalent fractions starts on Lesson 68 after students are well practiced in multiplying fractions by fractions and whole numbers by fractions. If the multiplication involves a fraction and a whole number, the whole number is rewritten as the number with a denominator of 1. Fraction multiplication is used extensively in *CMC Level E* to create equivalent fractions and ratios. The basic rule that students learn about equivalent fractions is that if a fraction is multiplied by 1, the resulting fraction is equal to the starting fraction. Therefore, if a fraction is multiplied by *any fraction that equals one,* the resulting fraction is equal to the starting fraction.

This relationship implies the test for determining if fractions are equal. 3/4 and 12/16 may be compared by identifying the fraction needed to convert 3/4 into 12/16. The fraction is 4/4; therefore, the fractions 3/4 and 12/16 are equal. If the second fraction multiplied is not equal to one, the fractions being compared are not equal. 3/4 does not equal 15/16.

Conversely, to create fractions that are equivalent, students multiply by a fraction that equals one.

For example: $\frac{3}{7} \times \frac{\quad}{\quad} = \frac{84}{\quad}$

Students work the problem for the numerators: 3 × __ = 84.

The missing number is 28. So the fraction that equals 1 is 28/28.

The problem for the denominators is 7 × 28.

The equivalent fraction is 84/196.

On Lesson 68, students learn about multiplying by fractions of one.

a. Listen: 4 × 1. What's the answer? **(Signal.)** *4.*
- New problem: 412 × 1. What's the answer? **(Signal.)** *412.*
- New problem: 4/12 times 1. What's the answer? **(Signal.)** *4/12.*
- What does the number you start with times 1 equal? **(Call on a student.)** *The number you start with.*
  Yes, the number you start with times 1 equals the number you start with.
- Everybody, what does the number you start with times 1 equal? **(Signal.)** *The number you start with.*
  It works for fractions, too.

b. **(Display:)** [68:3A]

$$\frac{5}{7} \times \frac{3}{3}$$

- **(Point to $\frac{5}{7}$.)** Read this problem. **(Signal.)** *5/7 × 3/3.*
- Does 3/3 equal 1? **(Signal.)** *Yes.*
- So will the answer be equal to 5/7? **(Signal.)** *Yes.*

c. **(Change to show:)** [68:3B]

$$\frac{5}{7} \times \frac{2}{3}$$

- **(Point to $\frac{5}{7}$.)** Read this problem. **(Signal.)** *5/7 × 2/3.*
- Does 2/3 equal 1? **(Signal.)** *No.*
- So will the answer be equal to 5/7? **(Signal.)** *No.*

d. **(Change to show:)** [68:3C]

$$5 \times \frac{2}{3}$$

- **(Point to 5.)** Read this problem. **(Signal.)** *5 × 2/3.*
- Does 2/3 equal 1? **(Signal.)** *No.*
- So will the answer be equal to 5? **(Signal.)** *No.*

e. **(Change to show:)** [68:3D]

$$5 \times \frac{2}{2}$$

- **(Point to 5.)** Read this problem. **(Signal.)** *5 × 2/2.*
- Does 2/2 equal 1? **(Signal.)** *Yes.*
- So will the answer be equal to 5? **(Signal.)** *Yes.*

f. (Display:) [68:3E]

$$\frac{13}{5} \times \frac{12}{12}$$

- (Point to $\frac{13}{5}$.) Read this problem. (Signal.) *13/5 × 12/12.*
- Does 12/12 equal 1? (Signal.) *Yes.*
- So will the answer be equal to 13/5? (Signal.) *Yes.*

(Repeat steps b through f that were not firm.)

g. Now I'm going to show you equations that multiply by fractions. You're going to tell me if the answer is equal to the fraction the equation starts with.

(Display:) [68:3F]

$$\frac{5}{7} \times \frac{2}{3} = \frac{10}{21}$$

- (Point to $\frac{5}{7}$.) Look at the fractions. Read this equation. (Signal.) *5/7 × 2/3 = 10 over 21.*
- Is 5/7 multiplied by 1? (Signal.) *No.*
- So is 5/7 equal to 10/21? (Signal.) *No.*

h. (Display:) [68:3G]

$$5 \times \frac{2}{3} = \frac{10}{3}$$

- (Point to **5.**) Read this equation. (Signal.) *5 × 2/3 = 10/3.*
- Is 5 multiplied by a fraction that equals 1? (Signal.) *No.*
- So is 5 equal to 10/3? (Signal.) *No.*

i. (Display:) [68:3H]

$$5 \times \frac{2}{2} = \frac{10}{2}$$

- (Point to **5.**) Read this equation. (Signal.) *5 × 2/2 = 10/2.*
- Is 5 multiplied by a fraction that equals 1? (Signal.) *Yes.*
- So is 5 equal to 10/2? (Signal.) *Yes.*
- What fraction does 5 equal? (Signal.) *10/2.*

j. (Display:) [68:3I]

$$\frac{13}{5} \times \frac{12}{12} = \frac{156}{60}$$

- (Point to $\frac{13}{5}$.) Read this equation. (Signal.) *13/5 × 12/12 = 156/60.*
- Is 13/5 multiplied by 1? (Signal.) *Yes.*
- What fraction does 13/5 equal? (Signal.) *156/60.*

(Repeat steps g through j that were not firm.)

Lesson 68, Exercise 3

On Lesson 71, students analyze fraction-multiplication equations to determine whether the starting value and the answer are equal. If the problem multiplies by a fraction that equals one, the starting value and the answer are equal.

Here's part of the exercise:

(Teacher reference:) **R Part G**

a. $8 \times \frac{9}{10} = \frac{72}{10}$ ___ ___

c. $\frac{7}{5} \times \frac{15}{12} = \frac{105}{60}$ ___ ___

e. $\frac{5}{7} \times \frac{14}{16} = \frac{70}{112}$ ___ ___

b. $\frac{3}{4} \times \frac{13}{13} = \frac{39}{52}$ ___ ___

d. $2 \times \frac{27}{27} = \frac{54}{27}$ ___ ___

f. $\frac{8}{11} \times \frac{9}{7} = \frac{72}{77}$ ___ ___

Below each equation, you're going to write what the equation starts with and the answer. Then you're going to complete the statement to show if they are equal or not equal.

- Read equation A. (Signal.) *8 × 9/10 = 72/10.*
- What does the equation start with? (Signal.) *8.*
- What's the answer? (Signal.) *72/10.*
- Is 8 multiplied by a fraction that equals 1? (Signal.) *No.*
- So is 8 equal to 72/10? (Signal.) *No.*

(If students have been 100% for the last two lessons, skip to step g.)

b. Read equation B. (Signal.) *3/4 × 13/13 = 39 over 52.*
- What does the equation start with? (Signal.) *3/4.*
- What's the answer? (Signal.) *39 over 52.*
- Is 3/4 multiplied by a fraction that equals 1? (Signal.) *Yes.*
- So is 3/4 equal to 39 over 52? (Signal.) *Yes.*

c. Read equation C. (Signal.) *7/5 × 15/12 = 105/60.*
- What does the equation start with? (Signal.) *7/5.*
- What's the answer? (Signal.) *105/60.*
- Is 7/5 multiplied by a fraction that equals 1? (Signal.) *No.*
- So is 7/5 equal to 105/60? (Signal.) *No.*

d. Read equation D. (Signal.) *2 × 27/27 = 54/27.*
- What does the equation start with? (Signal.) *2.*
- What's the answer? (Signal.) *54/27.*
- Is 2 multiplied by a fraction that equals 1? (Signal.) *Yes.*
- So is 2 equal to 54/27? (Signal.) *Yes.*

e. Read equation E. (Signal.) *5/7 × 14/16 = 70 over 112.*
- What does the equation start with? (Signal.) *5/7.*
- What's the answer? (Signal.) *70 over 112.*
- Is 5/7 multiplied by a fraction that equals 1? (Signal.) *No.*
- So is 5/7 equal to 70 over 112? (Signal.) *No.*

f. Read equation F. (Signal.) *8/11 × 9/7 = 72/77.*
- What does the equation start with? (Signal.) *8/11.*
- What's the answer? (Signal.) *72/77.*
- Is 8/11 multiplied by a fraction that equals 1? (Signal.) *No.*
- So is 8/11 equal to 72/77? (Signal.) *No.*

(Repeat equations that were not firm.)

g. Let's do those again. Equation A is $8 \times 9/10 = 72/10$.

- What does equation A start with? (Signal.) *8.*
- What's the answer? (Signal) *72/10.*
- Raise your hand when you can tell me if 8 and 72/10 are equal or not equal. ✔
- Everybody, is 8 equal or not equal to 72/10? (Signal.) *Not equal.*

(Display:) [71:3A]

a. $8 \times \frac{9}{10} = \frac{72}{10}$

$\frac{8}{\quad} \neq \frac{72}{10}$

Here's equation A with the statement below.

- Complete the statement below equation A. (Observe students and give feedback.)
- Read the statement for A. Get ready. (Signal.) *8 is not equal to 72/10.*

h. Equation B is $3/4 \times 13/13 = 39$ over 52.

- What does equation B start with? (Signal.) *3/4.*
- What's the answer? (Signal.) *39 over 52.*
- Raise your hand when you can tell me if 3/4 and 39 over 52 are equal or not equal. ✔
- Everybody, is 3/4 equal or not equal to 39 over 52? (Signal.) *Equal.*

(Display:) [71:3B]

b. $\frac{3}{4} \times \frac{13}{13} = \frac{39}{52}$

$\frac{3}{4} = \frac{39}{52}$

- Here's the statement. Complete the statement below equation B. (Observe students and give feedback.)
- Read the statement for B. Get ready. (Signal.) *3/4 = 39 over 52.*

i. Equation C is $7/5 \times 15/12 = 105/60$.

- Below equation C, write what the equation starts with and the answer. Then complete the statement to show if they are equal or not equal. Put your pencil down when you're finished. (Observe students and give feedback.)

from Lesson 71, Exercise 3

Teaching Note: The exercise presents a series of questions about each equation. These questions are not difficult, but they are important. So even if it seems "redundant" to ask the same set of questions for each equation, the series provides students with a basic strategy for understanding the role of each number in the equation.

Starting on Lesson 73, students work fraction-multiplication problems that have two missing values.

For example: $\frac{\quad}{4} \times \frac{3}{5} = \frac{15}{\quad}$

Students say the problem for the numerators as "What number times 3 equals 15?"

The problem for the denominators is $4 \times 5 =$ what number?

The resulting equation is $\frac{5}{4} \times \frac{3}{5} = \frac{15}{20}$.

Students apply what they know about multiplying fractions by fractions to conclude that 5/4 is not equal to 15/20.

A simple extension of this analysis occurs on Lesson 77. The new problem type does not show the middle fraction or one part of the answer. The missing fraction equals one.

For example: $\frac{3}{4} \times \frac{\quad}{\quad} = \frac{\quad}{72}$

Students have to figure out the middle fraction. They can't work the problem for the numerators because the numerator is not shown for the last fraction. So students first work the problem $4 \times \_ = 72$. The answer is 18, so the fraction that equals one is 18/18.

$$\frac{3}{4} \times \frac{18}{18} = \frac{\quad}{72}$$

Now students work the problem 3×18 to complete the last fraction. The resulting equation is $\frac{3}{4} \times \frac{18}{18} = \frac{54}{72}$.

Because the equation multiplies by 1, students conclude that $3/4 = 54/72$.

Here's the first display of the exercise from Lesson 78, Exercise 9. It presents items of this type.

$\frac{5}{3} \times \frac{\quad}{\quad} = \frac{\quad}{12}$ $\frac{4}{7} \times \frac{\quad}{\quad} = \frac{20}{\quad}$

$\frac{10}{9} \times \frac{\quad}{\quad} = \frac{\quad}{54}$

On later lessons, students work problems of this type that have larger numbers, which require students to work a division problem and a multiplication problem.

Here's the problem set from Textbook Lesson 84, Part 4:

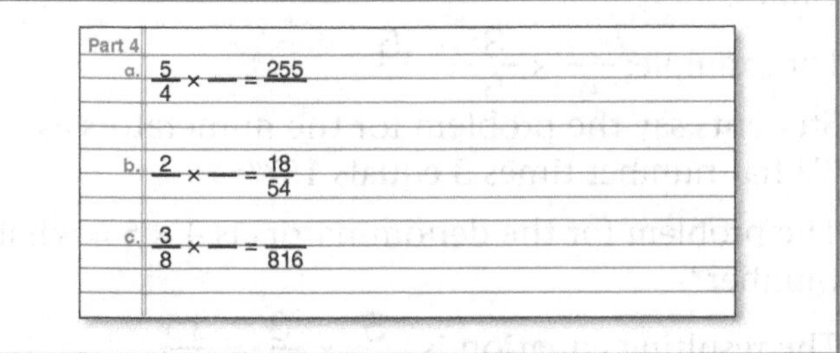

Part 4

a. $\frac{5}{4} \times \frac{}{} = \frac{255}{}$

b. $\frac{2}{} \times \frac{}{} = \frac{18}{54}$

c. $\frac{3}{8} \times \frac{}{} = \frac{}{816}$

For problem A, students first work the division problem for the numerators: 255/5.

They write the fraction that equals one—51/51.

Students next work the problem 51 × 4. (Students follow the convention of writing the multi-digit number first, not 4 × 51.)

Students conclude that 5/4 = 255/204.

This problem type occurs later when students work ratio word problems. (See Page 161) For most problems, they set up two equivalent fractions, one of which has a missing numerator or denominator. They complete the fraction that equals 1 and then multiply or divide to complete the fraction with a missing number.

Exercises that "compare" fractions begin on Lesson 95. The program introduces two problem types. One consists of a pair of fractions that have the same denominator; the other presents fractions that students compare to 1. For all problems, students write the sign <, >, or =.

Here's the exercise from Lesson 96:

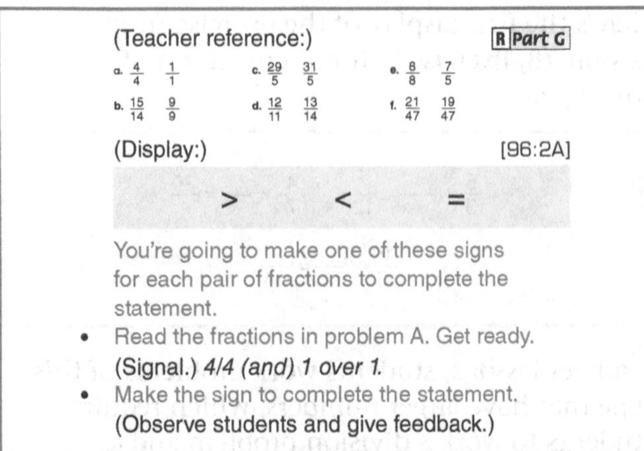

(Teacher reference:) **R Part G**

a. $\frac{4}{4}$ $\frac{1}{1}$ c. $\frac{29}{5}$ $\frac{31}{5}$ e. $\frac{8}{8}$ $\frac{7}{5}$

b. $\frac{15}{14}$ $\frac{9}{9}$ d. $\frac{12}{11}$ $\frac{13}{14}$ f. $\frac{21}{47}$ $\frac{19}{47}$

(Display:) [96:2A]

> < =

You're going to make one of these signs for each pair of fractions to complete the statement.

- Read the fractions in problem A. Get ready. **(Signal.)** *4/4 (and) 1 over 1.*
- Make the sign to complete the statement. **(Observe students and give feedback.)**

b. Check your work.
- Read the statement for A. Get ready. **(Signal.)** *4/4 = 1 over 1.*
- How do you know 4/4 and 1 over 1 are equal? **(Call on a student.)** *Both of the fractions equal one.*

c. Problem B. Read the fractions. Get ready. **(Signal.)** *15/14 (and) 9/9.*
- Make the sign to complete the statement. **(Observe students and give feedback.)**

d. Check your work for problem B.
- Tell me the fraction that's more. Get ready. **(Signal.)** *15/14.*
- How do you know 15/14 is more than 9/9? **(Call on a student. Idea:)** *15/14 is more than 1. 9/9 equals 1.*

e. Problem C. Read the fractions. Get ready. **(Signal.)** *29/5 (and) 31/5.*
- Make the sign to complete the statement. **(Observe students and give feedback.)**

f. Check your work for problem C.
- Tell me the fraction that's more. Get ready. **(Signal.)** *31/5.*
- How do you know 31/5 is more than 29/5? **(Call on a student. Idea:)** *The denominators are the same. 31 is more than 29.*

g. Make the signs for the rest of the fraction pairs. **(Observe students and give feedback.)**

h. Check your work.
- Problem D. Are the fractions equal? **(Signal.)** *No.*
- Which fraction is more? **(Signal.)** *12/11.*

i. Problem E. Are the fractions equal? **(Signal.)** *No.*
- Which fraction is more? **(Signal.)** *7/5.*

j. Problem F. Are the fractions equal? **(Signal.)** *No.*
- Which fraction is more? **(Signal.)** *21/47.*

(Display:) [96:2B]

a. $\frac{4}{4} = \frac{1}{1}$ c. $\frac{29}{5} < \frac{31}{5}$ e. $\frac{8}{8} < \frac{7}{5}$

b. $\frac{15}{14} > \frac{9}{9}$ d. $\frac{12}{11} > \frac{13}{14}$ f. $\frac{21}{47} > \frac{19}{47}$

Here's what you should have for part 2.

Lesson 96, Exercise 2

On Lesson 98, students learn how to compare fractions that have the same numerators. Students refer to the denominators. The fraction with the *smaller* denominator has *larger* parts. So the fraction with the smaller denominator is more than the other fraction.

Here's the first part of the exercise:

a. (Display:) [98:2A]

$\frac{7}{3} \quad \frac{7}{5}$

- Read the fractions. Get ready. **(Signal.)** *7/3 (and) 7/5.*
- These fractions have the same numerator. What's the numerator? **(Signal.)** *7.*
  The denominators are different.

b. Think about which is bigger—thirds or fifths.

- Which is bigger? **(Signal.)** *Thirds.*
  (Display:) [98:2B]

Yes, here's a picture of thirds and fifths. You can see that thirds are bigger than fifths.

- So which is bigger—7/3 or 7/5? **(Signal.)** *7/3.*
  (Display:) [98:2C]

$\frac{7}{3} > \frac{7}{5}$

c. (Display:) [98:2D]

$\frac{3}{9} \quad \frac{3}{8}$

- Read the fractions. Get ready. **(Signal.)** *3/9 (and) 3/8.*
  The numerators are the same so the fraction that has the denominator with the bigger parts is bigger.
- Which has bigger parts—ninths or eighths? **(Signal.)** *Eighths.*
- Eighths has bigger parts, so which fraction is more—3/9 or 3/8? **(Signal.)** *3/8.*
  (Add to show:) [98:2E]

$\frac{3}{9} < \frac{3}{8}$

d. (Display:) [98:2F]

$\frac{1}{20} \quad \frac{1}{10}$

- Read the fractions. **(Signal.)** *1/20 (and) 1/10.*
- Are the numerators the same? **(Signal.)** *Yes.*
- Which has bigger parts—twentieths or tenths? **(Signal.)** *Tenths.*
- So which fraction is more—1/20 or 1/10? **(Signal.)** *1/10.*
  (Add to show:) [98:2G]

$\frac{1}{20} < \frac{1}{10}$

from Lesson 98, Exercise 2

On Lesson 102, students compare fractions that can't be readily analyzed by inspection. To figure out which is bigger, students work each fraction as a division problem.

For instance, on Lesson 102, students compare 47/9 and 19/4. They work each fraction as a division problem. The results are:

$$47 \div 9 = 5\frac{2}{9}$$

$$19 \div 4 = 4\frac{3}{4}$$

On Lesson 104, students use all four strategies to compare fractions.

If denominators are the same, the fraction with the larger numerator is more. If numerators are the same, the fraction with the larger parts is more.

If one fraction is less than 1 and the other fraction is more than 1, the larger fraction is more.

If inspection does not reveal which fraction is more, division discloses which fraction is more. Note that not all problem types are presented because students haven't learned how to find the lowest common denominator. So division problems that require this calculation do not appear in problem sets.

Here's the first part of the exercise from Lesson 106:

(Teacher reference:) **R Part C**

a.	$\frac{16}{15}$	$\frac{27}{27}$	e.	$\frac{23}{19}$	$\frac{21}{19}$
b.	$\frac{30}{6}$	$\frac{35}{5}$	f.	$\frac{53}{54}$	$\frac{4}{3}$
c.	$\frac{12}{7}$	$\frac{12}{9}$	g.	$\frac{9}{2}$	$\frac{35}{9}$
d.	$\frac{16}{2}$	$\frac{40}{5}$	h.	$\frac{37}{37}$	$\frac{5}{5}$

You can compare some of the pairs of fractions in part 1 by inspection. You cannot compare the other pairs of fractions by inspection. You'll tell me if you can compare the fractions in each problem by inspection or not.

b. Read the fractions for problem A. Get ready. **(Signal.)** *16/15 (and) 27/27.*

- Are the numerators or denominators the same? **(Signal.)** *No.*
- Tell me yes or no if either of the fractions is less than or equal to one. Get ready. **(Signal.)** *Yes.*
- So can you compare the fractions by inspection? **(Signal.)** *Yes.*
  Later, you'll write the sign to complete the statement for problem A.

c. Read the fractions for problem B. Get ready. (Signal.) *30/6 (and) 35/5.*

- Are the numerators or denominators the same? (Signal.) *No.*
- Tell me if either of the fractions is less than or equal to one. Get ready. (Signal.) *No.*
- So can you compare the fractions by inspection? (Signal.) *No.*
  Later, you'll figure out what the fractions equal to complete the statement for B.

d. Read the fractions for problem C. Get ready. (Signal.) *12/7 (and) 12/9.*

- Are the numerators or denominators the same? (Signal.) *Yes.*
- So can you compare the fractions by inspection? (Signal.) *Yes.*
  (If students are firm, skip to step h.)

e. Read the fractions for problem D. Get ready. (Signal.) *16/2 (and) 40/5.*

- Are the numerators or denominators the same? (Signal.) *No.*
- Tell me if either of the fractions is less than or equal to one. Get ready. (Signal.) *No.*
- So can you compare the fractions by inspection? (Signal.) *No.*
  Later, you'll figure out what the fractions equal to complete the statement for D.

from Lesson 106, Exercise 1

Teaching Note: The questions are ordered to provide students with a systematic strategy for testing problems. They first ask, "Are the numerators or denominators the same?"

If the answer is no, they ask the next question, "Is one of the fractions less than or equal to one?"

If the answer is no, they know that they have to divide to figure out which fraction is more.

Later in the program, students expand what they have learned to compare fractions, decimals, percents, and whole numbers.

Whole Numbers (Lessons 5–120)

This track begins on Lesson 5 and continues intermittently through Lesson 120.

In this track, students first review reading and writing four-digit numbers. Then they learn to write numbers with up to six digits. They also write statements that compare numbers.

Here's the first review exercise:

(Teacher reference:)

Part 2	
	2020
a.	
b.	
c.	
d.	
e.	

You're going to write thousands numbers.

- Listen: You write three digits after the thousands digit. How many digits do you write after the thousands digit? **(Signal.)** *3.*

b. The number on the top line is already written.

- What is the number on the top line? **(Signal.)** *2020.*
- Copy part 2. Write 2020 and the letters A through E on the lines below part 2.
(Observe students and give feedback.)

c. You'll write 6 thousand 14 for A.

- What number? **(Signal.)** *6014.*
- What's the part after the thousands digit? **(Signal.)** *14.*
- How many digits do you write after the thousands digit? **(Signal.)** *3.*
- Say the three digits you'll write for 14. **(Call on a student.)** *Zero, 1, 4.*
- Yes, you'll write zero, 1, 4 after the thousands digit. Everybody, say the digits you'll write after the thousands digit. **(Signal.)** *Zero, 1, 4.*
- Write 6014.
(Observe students and give feedback.)
(Display:) [5:7B]

Part 2	
	2 0 2 0
a.	6 0 1 4

Here's what you should have for A. The digits for 6014 are 6, zero, 1, 4.

- What are the digits of 6014? **(Signal.)** *6, zero, 1, 4.*

d. You'll write 4 thousand 5 for B.

- What number? **(Signal.)** *4005.*
- Say the three digits you'll write after the thousands digit. Get ready. **(Signal.)** *Zero, zero, 5.*
- Write 4005.
(Observe students and give feedback.)
- What are the digits of 4 thousand 5? **(Signal.)** *4, zero, zero, 5.*

e. You'll write 4 thousand 5 hundred for C.

- What number? **(Signal.)** *4500.*
- Say the three digits you'll write after the thousands digit. Get ready. **(Signal.)** *5, zero, zero.*
- Write 4500.
(Observe students and give feedback.)
- What are the digits of 4 thousand 500? **(Signal.)** *4, 5, zero, zero.*

f. You'll write 4 thousand 50 for D.

- What number? **(Signal.)** *4050.*
- Say the three digits you'll write after the thousands digit. Get ready. **(Signal.)** *Zero, 5, zero.*
- Write 4050.
(Observe students and give feedback.)
- What are the digits of 4 thousand 50? **(Signal.)** *4, zero, 5, zero.*

g. You'll write 4 thousand 1 hundred 59 for E.

- What number? **(Signal.)** *4159.*
- Write 4159.
(Observe students and give feedback.)
- What are the digits of 4 thousand 1 hundred 59? **(Signal.)** *4, 1, 5, 9.*

from Lesson 5, Exercise 7

On the following lessons, students work similar exercises. On Lesson 14, students read and write thousands numbers that have five digits. When writing these numbers, students follow the rule of making a comma after the thousands digit. On Lesson 16, students work with six-digit numbers.

Here's the exercise from Lesson 16:

a. You're going to write thousands numbers with 2 or 3 digits for the thousands part. You'll write the digits for thousands. Then you'll write a comma. Then you'll write the three digits that come after the thousands digit.

- What symbol do you write after the digits for thousands? (Signal.) *A comma.*
- How many digits do you write after the comma? (Signal.) *3.*

(Display:) [16:8A]

Part A	
a.	c.
b.	d.

- Write part A on your lined paper and write the letters A through D below. (Observe students and give feedback.)

b. You'll write **29 thousand 470** for A. What number? (Signal.) *29 thousand 470.*

- What do you write before the comma? (Signal.) *29.*
- What symbol do you write after the 29? (Signal.) *A comma.*
- What do you write after the comma? (Signal.) *470.*

(Repeat until firm.)

- Write 29,470. (Observe students and give feedback.)
- Number A is 29,470. Read the symbols you wrote. (Signal.) *2, 9, comma, 4, 7, zero.* (Add **29,470** to show.) [16:8B] Here's what you should have written for A.

c. You'll write **356 thousand 10** for B. What number? (Signal.) *356 thousand 10.*

- What do you write before the comma? (Signal.) *356.*
- What do you write after the comma? (Signal.) *10.*

(Repeat until firm.)

- Write 356,010. (Observe students and give feedback.)
- Number B is 356,010. (Add **356,010** to show.) [16:8C] Here's what you should have written for B.

d. You'll write **490 thousand 9** for C. What number? (Signal.) *490 thousand 9.*

- What do you write before the comma? (Signal.) *490.*
- What do you write after the comma? (Signal.) *9.*

(Repeat until firm.)

- Write 490,009. (Observe students and give feedback.)
- Number C is 490,009. (Add **490,009** to show.) [16:8D] Here's what you should have written for C.

e. You'll write **521 thousand 6** for D. What number? (Signal.) *521 thousand 6.*

- What do you write before the comma? (Signal.) *521.*
- What do you write after the comma? (Signal.) *6.*

(Repeat until firm.)

- Write 521,006. (Observe students and give feedback.)
- Number D is 521,006. (Add to show:) [16:8E]

Part A		
a.	29,470	c. 490,009
b.	356,010	d. 521,006

Here's what you should have written for D.

Lesson 16, Exercise 8

Teaching Note: The biggest problem that students have is writing missing zeros after the comma. If students make mistakes, remind them, "You need three digits after the comma." Explain that if the number does not name all three digits, you have to put zeros in the empty columns. For 24 thousand 5, you put 5 in the ones column and then put zeros in the hundreds and tens columns.

24,005

Students do not write commas in numbers that have fewer than five whole-number digits.

On Lesson 114, students write statements that compare more than two whole numbers. The comparison shows which number is more. To compare 350, 305, and 355 students write:

$$305 < 350 < 355$$

Here's part of the exercise from Lesson 117:

(Teacher reference:)

a. 1526, 1652, 1256, 1562
b. 8370, 8307, 8703, 8073
c. 5194, 5491, 5419, 5941

Part 3
a.
b.
c.

Each problem shows 4 numbers. For problems A and B, you're going to write the statement with the numbers from smallest to largest.

- Write part 3 on your lined paper with the letters A, B, and C below.
  (Observe students and give feedback.)

b. Read the numbers for problem A. Get ready. **(Signal.)** *1526, 1652, 1256, 1562.*

- For problem A, you'll write the statement with the numbers from smallest to largest. For problem A, will you write the largest or smallest number first? **(Signal.)** *(The) smallest (number).*
- Which is smaller—1526 or 1652? **(Signal.)** *1526.*
- Which is smaller—1526 or 1256? **(Signal.)** *1256.*
- Which is smaller—1256 or 1562? **(Signal.)** *1256.*
- So which is the smallest number? **(Signal.)** *1256.*

(Display:) [117:5A]

Part 3	
a.	1256 <
b.	
c.	

Here's the smallest number and the first sign of the statement for problem A.

(Repeat step b until firm.)

(If students are firm on writing comparison statements for numbers, skip to step d.)

c. Look at the other three numbers. Which is smallest—1526, 1652, or 1562? **(Signal.)** *1526.*

- Look at the other two numbers. Which is smaller—1652 or 1562? **(Signal.)** *1562.*
- Which is the largest number? **(Signal.)** *1652.*

(Repeat step c until firm.)

d. For problem A, write the statement with the numbers from smallest to largest.
(Observe students and give feedback.)

e. Check your work for problem A.

(Add to show:) [117:5B]

Part 3	
a.	1256 < 1526 < 1562 < 1652
b.	
c.	

Here's what you should have for part 3 so far.

- Touch the smallest number in the statement for problem A. ✔
- Start with the smallest number and read the statement for A. Get ready. **(Signal.)** *1256 is less than 1526 is less than 1562 is less than 1652.*

(If students are firm on writing comparison statements for numbers, skip to step g.)

f. Read the numbers for problem B. Get ready. **(Signal.)** *8370, 8307, 8703, 8073.*

- Raise your hand when you know which is the smallest number. ✔
- Which is the smallest number in problem B? **(Signal.)** *8073.*
  Yes, 8073 is the smallest number.
- Raise your hand when you know which is the next-smallest number. ✔
- Which is the next-smallest number in problem B? **(Signal.)** *8307.*
- The last two numbers are 8370 and 8703. Which is smaller? **(Signal.)** *8370.*
- So which number is the largest number in problem B? **(Signal.)** *8703.*

g. For problem B, write the statements with the numbers from smallest to largest.
(Observe students and give feedback.)

h. Check your work.

- Start with the smallest number and read the statement. Get ready. **(Signal.)** *8073 is less than 8307 is less than 8370 is less than 8703.*

(Add to show:) [117:5C]

Part 3	
a.	1256 < 1526 < 1562 < 1652
b.	8073 < 8307 < 8370 < 8703
c.	

Here are the statements for problems A and B.

from Lesson 117, Exercise 5

Teaching Note: If students have trouble identifying the numbers in any problem, tell them to find the first number for the statement they will write. For example: 1526, 1652, 1256, 1562. The smallest number is 1256. Students arrange the other numbers in order of size and connect them using < signs.

Decimal and Percent (Lessons 27–130)

DECIMAL

Here's an outline of the major concepts that students learn in this track. Students will:

- Review decimal operations by working addition and subtraction problems involving dollar-and-cents amounts. (Lesson 27)
- Learn to read and write decimals for tenths and hundredths. (Lesson 31)
- Add and subtract decimal values. (Lesson 36)
- Complete tables that show equivalent fractions. (Lesson 43)
- Read and write percents. (Lesson 76)
- Complete tables that show equivalent fractions and percents. (Lesson 86)
- Use ratio equations to answer questions about fractions and corresponding percents. (Lesson 94)
- Apply multiple strategies to figure out corresponding fractions, decimals and percents. (Lesson 101)

On Lesson 27, students review reading dollar-and-cents amounts.

Here's the exercise from Lesson 27:

a. (Display:) [27:2A]

$2.16
$34.05
$503.21
$96.50
$3188.01

These are dollars-and-cents amounts. They all begin with a dollar sign, and they all have a decimal point.

- **(Point to $ in $2.16.)** What sign is this? **(Signal.)** *A dollar sign.*
- **(Point to decimal point.)** The numbers in front of the decimal point tell about dollars. What do the numbers in front of the decimal point tell about? **(Signal.)** *Dollars.*
- The numbers after the decimal point tell about cents. What do the numbers after the decimal point tell about? **(Signal.)** *Cents.*

b. Tell me if the part tells about dollars or if the part tells about cents.

- **(Point to 2.)** Is this the number for dollars or for cents? **(Signal.)** *Dollars.*
- **(Point to 16.)** What does this number tell about? **(Signal.)** *Cents.*

c. **(Point to .05 in $34.05.)** What does this number tell about? **(Signal.)** *Cents.*

- **(Point to 34.)** What does this number tell about? **(Signal.)** *Dollars.*

d. Listen: When you read dollars and cents, you say **and** for the decimal point.

- What do you say for the decimal point? **(Signal.)** *And.*

e. My turn to read some of these dollars-and-cents amounts: **(Touch symbols.)** 2 dollars and 16 cents. 34 dollars and 5 cents.

- Your turn: Read all the dollars-and-cents amounts.
- **(Point to $2.16.)** Read this amount. **(Touch symbols.)** *2 dollars and 16 cents.*
- **(Point to $34.05.)** Read this amount. **(Touch.)** *34 dollars and 5 cents.*
- **(Point to $503.21.)** Read this amount. **(Touch.)** *503 dollars and 21 cents.*
- **(Point to $96.50.)** Read this amount. **(Touch.)** *96 dollars and 50 cents.*
- **(Point to $3188.01.)** This is a big amount. Read this amount. **(Touch.)** *3 thousand 188 dollars and 1 cent.*

(Repeat until firm.)

Lesson 27, Exercise 2

Teaching Note: Students are to say *dollars* for the amount before the decimal, say the word *and* for the decimal, and then say the cents amounts. These conventions should not be new, but may require some review. Expect some students to have trouble reading $3188.01. Stepping out the process will help. Direct them to say only the dollar amount. Then read the whole amount.

On Lesson 29, students review how to work addition-subtraction problems that have dollar-and-cents amounts. Students first write the dollar sign and the decimal point in the answer. Then they add or subtract as they normally would.

Here's part of the exercise from Lesson 29:

g. (Point to **$40.07.**) Read the problem. (Signal.) *40 dollars and 7 cents minus 21 dollars and 1 cent.*
- What are the first things I write in the answer? (Call on a student.) *A dollar sign and decimal point.*
(Add **$** and **.** to show.) [29:2G]

h. Read the problem for the one-cent column. (Signal.) *7 – 1.*
- What's the answer? (Signal.) *6.*
(Add **6** to show.) [29:2H]

i. Read the problem for the ten-cent column. (Signal.) *Zero – zero.*
- What's the answer? (Signal.) *Zero.*
(Add **0** to show.) [29:2I]

j. Read the problem for the one-dollar column. (Signal.) *Zero – 1.*
- Can you work that problem? (Signal.) *No.*
(Add to show:) [29:2J]

```
    1 1          3
  $35.21     $4̶¹0.07
  +64.95     -21.01
 $100.16     $   .06
```

- Say the new problem in the one-dollar column. (Signal.) *10 – 1.*
- What's the answer? (Signal.) *9.*
(Add to show:) [29:2K]

```
    1 1          3
  $35.21     $4̶¹0.07
  +64.95     -21.01
 $100.16     $  9.06
```

k. Read the new problem for the ten-dollar column. (Signal.) *3 – 2.*
- What's the answer? (Signal.) *1.*
(Add to show:) [29:2L]

```
    1 1          3
  $35.21     $4̶¹0.07
  +64.95     -21.01
 $100.16     $19.06
```

- Read the whole problem and the answer. (Signal.) *40 dollars and 7 cents – 21 dollars and 1 cent = 19 dollars and 6 cents.*

WORKBOOK PRACTICE

a. Find part 3 in your workbook. ✔
(Teacher reference:)

```
a. $16.37     c. $19.87
   - 4.44        -12.59

b. $45.08
   +16.07
```

- Touch and read problem A. Get ready. (Signal.) *$16.37 – $4.44.*
- Work problem A. Then stop. Remember the dollar sign and the decimal point in your answer. (Observe students and give feedback.)

b. Check your work.
- Read problem A and the answer. (Signal.) *$16.37 – $4.44 = $11 and 93 cents.*
- Touch the dollar sign in your answer. ✔
- Touch the decimal point in your answer. ✔
(Display:) [29:2M]

```
        5
a. $1 6̶ ¹3 7
   -  4.4 4
   $1 1.9 3
```

Here's what you should have for problem A.

c. Work the rest of the problems in part 3. (Observe students and give feedback.)

from Lesson 29, Exercise 2

Teaching Note: The directions students are to follow refer to the one-cent column, the ten-cent column, the one-dollar column, and the ten-dollar column.

If students have trouble following the directions, review touching the columns. Start with, "What's the top number in the one-cent column?" and present parallel questions about the ten-cent column, the one-dollar column, and the ten-dollar column.

Work on decimals that do not involve dollar amounts begins on Lesson 31. Students review hundredths.

In the first part of the exercise (not shown) students read pairs of numbers—the first is a dollar-and-cents amount; the second is a parallel decimal value without a dollar sign.

For example:

$56.02

56.02

56.02 is read as 56 and 2 hundredths.

In the same exercise, students also write hundredths numbers.

Here's that part of the exercise:

h. You're going to write hundredths numbers. (Display:) [31:5B]

Part A	
a.	c.
b.	d.

- Write part A on your lined paper. Then write the letters A through D below.
- For hundredths numbers, how many digits do you write after the decimal point? (Signal.) *2.* Yes, you write two digits after the decimal point for each number.

i. The decimal number for A is 17 and 92 hundredths.
- What decimal number? (Signal.) *17 and 92 hundredths.*
- What number will you write before the decimal point? (Signal.) *17.*
- What symbol will you write after the number 17? (Signal.) *A decimal point.*
- What number will you write after the decimal point for 17 and 92 hundredths? (Signal.) *92.*
- Write 17 and 92 hundredths. (Observe students and give feedback.) (Add to show:) [31:5C]

Part A	
a. 17.92	c.
b.	d.

Here's decimal number A.

j. Decimal number B is 217 and 5 hundredths.
- What decimal number? (Signal.) *217 and 5 hundredths.*
- What number will you write before the decimal point? (Signal.) *217.*
- What symbol will you write after the number 217? (Signal.) *A decimal point.* The digits after the decimal point work just like cents. The digits zero, 5 go after the decimal point.
- What digits will you write after the decimal point? (Signal.) *Zero, 5.*
- Write 217 and 5 hundredths. (Observe students and give feedback.) (Add to show:) [31:5D]

Part A	
a. 17.92	c.
b. 217.05	d.

Here's decimal number B.

from Lesson 31, Exercise 5

The same steps are used with the remaining numbers:

c. 54.08

d. 304.60

Students work addition and subtraction problems that have hundredths numbers. Students are also introduced to tenths numbers in Lesson 36.

Students learn that tenths have one digit after the decimal point and hundredths have two digits after the decimal point.

Here's part of the exercise from Lesson 38:

a. For decimal numbers that are tenths and hundredths, you learned how many digits there are after the decimal point.

- Tenths have how many digits after the decimal point? (Signal.) *1.*
- Hundredths have how many digits after the decimal point? (Signal.) *2.*

(Repeat until firm.)

b. (Display:) [38:5A]

39.06	39.6	7.13
390.6	3.96	71.3

These are decimal numbers. Some of these numbers are tenths. The rest of the numbers are hundredths. For each number, you'll tell me tenths or hundredths. Then you'll read the number.

- (Point to **39.06.**) How many digits are there after the decimal point? (Signal.) *2.*
- So is this number tenths or hundredths? (Signal.) *Hundredths.*
- What number is this? (Signal.) *39 and 6 hundredths.*

c. (Point to **390.6.**) How many digits are there after the decimal point? (Signal.) *1.*

- So is this number tenths or hundredths? (Signal.) *Tenths.*
- What number is this? (Signal.) *390 and 6 tenths.*

d. (Point to **39.6.**) How many digits are there after the decimal point? (Signal.) *1.*

- So is this number tenths or hundredths? (Signal.) *Tenths.*
- What number is this? (Signal.) *39 and 6 tenths.*

e. (Point to **3.96.**) How many digits are there after the decimal point? (Signal.) *2.*

- So is this number tenths or hundredths? (Signal.) *Hundredths.*
- What number is this? (Signal.) *3 and 96 hundredths.*

(If students were perfect on steps b through e, skip to step h.)

f. (Point to **7.13.**) How many digits are there after the decimal point? (Signal.) *2.*

- So is this number tenths or hundredths? (Signal.) *Hundredths.*
- What number is this? (Signal.) *7 and 13 hundredths.*

g. (Point to **71.3.**) How many digits are there after the decimal point? (Signal.) *1.*

- So is this number tenths or hundredths? (Signal.) *Tenths.*
- What number is this? (Signal.) *71 and 3 tenths.*

h. Now you're going to write decimal numbers.
(Display:) [38:5B]

Part A	
a.	**d.**
b.	**e.**
c.	

- Write part A on your lined paper. Then write A through E on the lines below. Put your pencil down when you're ready to write the decimal number for A.

(Observe students and give feedback.)

i. Listen: Decimal number A is 50 and 7 tenths. What number? (Signal.) *50 and 7 tenths.*

- The number is tenths, so how many digits will you write after the decimal point? (Signal.) *1.*
- What number do you write before the decimal point for 50 and 7 tenths? (Signal.) *50.*
- What symbol do you write after 50? (Signal.) *(A) decimal point.*
- What do you write after the decimal point? (Signal.) *7.*
- Write 50 and 7 tenths for A.

(Observe students and give feedback.)

- Check your work. Say the symbols for A. Get ready. (Signal.) *5, zero, decimal point, 7.*

(Display:) [38:5C]

Part A	
a.	50.7

Here's what you should have written for 50 and 7 tenths.

j. Decimal number B is 50 and 7 hundredths. What number? (Signal.) *50 and 7 hundredths.*

- The number is hundredths, so how many digits will you write after the decimal point? (Signal.) *2.*
- What number do you write before the decimal point for 50 and 7 hundredths? (Signal.) *50.*
- What symbol do you write after 50? (Signal.) *(A) decimal point.*
- How many digits do you write after the decimal point? (Signal.) *2.*
- Tell me the digits you write after the decimal point for 50 and 7 hundredths. (Signal.) *Zero, 7.*
- Write 50 and 7 hundredths for B.

(Observe students and give feedback.)

- Check your work. Say the symbols for B. Get ready. (Signal.) *5, zero, decimal point, zero, 7.*

(Display:) [38:5D]

b.	50.07

Here's what you should have written for 50 and 7 hundredths.

from Lesson 38, Exercise 5

Teaching Note: The pair of questions you ask in steps B through G prompts students to discriminate between tenths and hundredths.

How many digits are there after the decimal point?

So is the number tenths or hundredths?

This is a good pair of questions to present later, when students make mistakes in reading a decimal value.

Steps I and J direct students to write minimally different decimal values for tenths and hundredths. The value 50.07 is particularly difficult for students to write because it is irregular. They may think that .70 is correct because it has the digit 7 and ends in the hundredths place. The value 7 hundredths (.07) has the digit 7 in the hundredths place with a zero in the tenths place. Students often make the mistake of omitting the zero.

If students write 50.70 for B, ask them to read the value after the decimal point (seventy hundredths.) Say, "That's not seven hundredths. To write 7 hundredths, you write zero seven." If students write 50.7 for B, follow the same procedure. Ask them to read the number. Then tell them how to write seven hundredths.

On Lesson 38, students work problems with tenths numbers.

Here's the set of problems:

a. $\begin{array}{r} 36.3 \\ -\ \ .9 \\ \hline \end{array}$ b. $\begin{array}{r} .52 \\ 20.08 \\ +\ 7.40 \\ \hline \end{array}$ c. $\begin{array}{r} 48.3 \\ -47.4 \\ \hline \end{array}$

On Lesson 43, students write equations that show fractions less than one with denominators of 10 or 100 and the equivalent decimal value. For instance:

$$\frac{6}{10} = .6$$

Students read each fraction and then write the decimal value it equals. Note that when they read the fraction, they say the decimal value they will write.

Here's the set of problems from Lesson 43:

a. $\frac{6}{10} =$ c. $\frac{6}{100} =$ e. $\frac{0}{100} =$

b. $\frac{16}{100} =$ d. $\frac{3}{10} =$

On Lesson 45, students complete a table that has two columns—one for tenths and hundredths fractions; the other for corresponding decimal values.

When students read the values shown, they say the fraction or the decimal they will write.

Here's the table from Workbook Lesson 45:

Fractions	Decimals
a.	.09
b. $\frac{5}{10}$	
c. $\frac{4}{100}$	
d.	.03
e.	.00
f. $\frac{14}{100}$	

Starting on Lesson 46, students complete tables that have decimals and percents that are more than one. The strategy students learn is to write the same digits for the decimal value and the numerator of the fraction. To write the fraction for the decimal value 3.95, students write 395. They write the denominator of 100 (because the decimal value shows two places after the decimal point) 395/100.

To write the decimal value for the fraction 27 5/10, students write the digits 275 and write the decimal point to show one decimal place (because the denominator of the fraction is tenths) 27.5.

Here's part of the exercise from Lesson 46:

(Teacher reference:) R Part L

Fractions	Decimals
a.	8.09
b. $\frac{275}{10}$	
c. $\frac{614}{100}$	
d.	70.3
e.	42.50
f. $\frac{38}{10}$	

Each row shows a decimal or a fraction that is more than 1. You're going to tell me what you'll write to complete each row so it has a fraction and a decimal that are equal.

Here's the rule for completing these rows: For the top number of the fraction and for the decimal, the digits are the same.

- What do you know about the digits? (Signal.) *They are the same.*

b. Read the decimal for row A. Get ready. (Signal.) *8 and 9 hundredths.*

- What are the digits of 8 and 9 hundredths? (Signal.) *8, zero, 9.*
- So what are the digits of the top number of the fraction? (Signal.) *8, zero, 9.*

(Display:) [46:2A]

Fractions	Decimals
a. 809	8.09

- What's the bottom number for the fraction that equals 8 and 9 **hundredths?** (Signal.) *100.*

(Add to show:) [46:2B]

Fractions	Decimals
a. $\frac{809}{100}$	8.09

- Here's the fraction for 8 and 9 hundredths. What fraction? (Signal.) *809 hundredths.*
- Complete row A. ✔

c. Read the fraction for row B. (Signal.) *275 tenths.*

- What are the digits of the top number? (Signal.) *2, 7, 5.*

2, 7, 5 are the digits for the decimal.

(Display:) [46:2C]

b. 275 / 10	27 5

- How many digits are after the decimal point for 275 tenths? (Signal.) *1.*

(Add to show:) [46:2D]

b. $\frac{275}{10}$	27.5

- Here's the decimal for 275 tenths. What decimal? (Signal.) *27 and 5 tenths.*
- Complete row B. ✔

d. Read the fraction for row C. (Signal.) *614 hundredths.*

- What are the digits of the top number? (Signal.) *6, 1, 4.*
- So what are the digits for the decimal? (Signal.) *6, 1, 4.*
- How many digits will you write after the decimal point for 614 hundredths? (Signal.) *2.*
- Complete row C.

(Observe students and give feedback.)

e. Check your work for row C.

- Read the symbols you wrote for the decimal. Get ready. (Signal.) *6, decimal point, 1, 4.*
- What decimal equals 614 hundredths? (Signal.) *6 and 14 hundredths.*

(Display:) [46:2E]

c. $\frac{614}{100}$	6.14

Here's what you should have for row C.

from Lesson 46, Exercise 2

Teaching Note: For fourteen lessons, students work separately on tables that have only fraction and decimal values less than 1 and tables that have only values more than 1 or equal to 1. This practice is designed to make them fluent in applying strategies for values that are more than 1 and strategies for values less than 1.

On Lesson 60, the table has some values more than 1 and some less than 1.

Here's part of the exercise from Lesson 60:

a. For values less than 1 or more than 1, you've written fractions and the decimals that are equal.
- Tell me the denominators of the fractions you've worked with. Get ready. **(Signal.)** *10 (and) 100.*
- The digits for values more than one must be the same. What do you know about the digits for values more than one? **(Call on a student.)** *They must be the same.*

b. Find part 6 in your textbook. ✔
(Teacher reference:)

Part 6	Fraction	Decimal
a.		23.04
b.	$\frac{7}{100}$	
c.		.3
d.	$\frac{219}{10}$	
e.		.08

This table shows fractions and decimals that are more than 1 and less than 1. You're going to complete each row so it shows the fraction and the decimal it equals.
- Read the value in row A. **(Signal.)** *23 and 4 hundredths.*
- Is that value more than one? **(Signal.)** *Yes.*
- So do the digits for the top number of the fraction have to be the same? **(Signal.)** *Yes.*
- Say the digits for the top number of the fraction. **(Signal.)** *2, 3, 0, 4.*
(Repeat until firm.)

c. Read the value in row B. **(Signal.)** *7 hundredths.*
- Is that value more than one? **(Signal.)** *No.*
- So do the digits in the decimal have to be the same? **(Signal.)** *No.*
- Tell me the digits for the decimal for row B. Just the digits. Get ready. **(Signal.)** *Zero, 7.*

d. Read the value for row C. **(Signal.)** *3 tenths.*
- Is that value more than one? **(Signal.)** *No.*
- So do the digits for the top number of the fraction have to be the same? **(Signal.)** *No.*
- Tell me the digits for the top number of the fraction. Get ready. **(Signal.)** *3.*

e. Read the value in row D. **(Signal.)** *219/10.*
- Is that value more than one? **(Signal.)** *Yes.*
- So do the digits in the decimal have to be the same? **(Signal.)** *Yes.*
- Tell me the digits for the decimal. Get ready. **(Signal.)** *2, 1, 9.*

f. Read the value in row E. **(Signal.)** *8 hundredths.*
- Is that value more than one? **(Signal.)** *No.*
- So do the digits for the top number of the fraction have to be the same? **(Signal.)** *No.*
- Tell me the digits for the top number of the fraction. **(Signal.)** *8.*
(Repeat steps c through f until firm.)

from Lesson 60, Exercise 9

PERCENTS

On Lesson 76, students learn the relationship between decimals and percents. For decimals that are less than 1, the decimal number translates into a corresponding percent number. In the exercise on Lesson 76, students first indicate percents for decimal numbers, then they write equations that show percents and equivalent decimal values.

For example: .76 translates into 76%, .12 translates into 12%.

Note that these relationships hold only if the decimal value is less than 1.

On Lesson 77, students are introduced to percents that are more than 100%. The digits in the percent number are the same digits written for the hundredths decimal number. The decimal point shows hundredths.

For example: 256% = 2.56
17% = .17
3% = .03

Because all decimal numbers for percents will be hundredths values, students write at least two decimal places for percents. If the percent has more than two places, the decimal point goes before the last two digits.

Here's part of the exercise from Lesson 77:

d. **(Display:)** [77:2F]

130%

- Read this value. **(Signal.)** *130%.*
- What decimal does 130% equal? **(Signal.)** *130 hundredths.*
- Say the digits for 130 hundredths. Get ready. **(Signal.)** *1, 3, 0.*
- How many digits go after the decimal point? **(Signal.)** *two.*
(Add to show:) [77:2G]

130% = 1.30

- Read the equation. Get ready. **(Signal.)** *130 percent equals 1 and 30 hundredths.*
Yes, 130% equals 1 and 30 hundredths.

e. **(Display:)** [77:2H]

256%

- Read this value. **(Signal.)** *256%.*
- How many hundredths is 256%? **(Signal.)** *256.*
- Say the **symbols** for 256 hundredths. Get ready. **(Signal.)** *2, decimal point, 5, 6.*
(Add to show:) [77:2I]

256% = 2.56

- Read the equation. Get ready. (Signal.) *256 percent equals 2 and 56 hundredths.*

f. (Display:) [77:2J]

304%

- Read this value. (Signal.) *304%.*
- Say the **symbols** for the decimal number that equals 304%. Get ready. (Signal.) *3, decimal point, 0, 4.*

(Add to show:) [77:2K]

304% = 3.04

- Read the equation. Get ready. (Signal.) *304 percent equals 3 and 4 hundredths.*

g. (Display:) [77:2L]

300%

- Read this value. (Signal.) *300%.*
- Say the symbols for the decimal number that equals 300%. Get ready. (Signal.) *3, decimal point, 0, 0.*

(Add to show:) [77:2M]

300% = 3.00

- Read the equation. Get ready. (Signal.) *300 percent equals 3 and zero hundredths.*

h. So 300% equals 3. What does 300% equal? (Signal.) *3.*
- What does 100% equal? (Signal.) *1.*
- What does 800% equal? (Signal.) *8.*
- What does 500% equal? (Signal.) *5.*

(Repeat step h until firm.)

Yes, 500% of a number is 5 times that number.

WORKBOOK PRACTICE

a. Find part 2 in your workbook. ✔

(Teacher reference:)

a. 20% c. 5% e. 1%

b. 320% d. 400% f. 708%

You'll read each percent. Then you'll write the decimal number for each percent.

- Read the percent for A. Get ready. (Signal.) *20%.*
- Say the **symbols** you'll write for the decimal number that equals 20%. (Signal.) *Decimal point, 2, 0.*

b. Read the percent for B. (Signal.) *320%.*
- Say the symbols you'll write for the decimal number that equals 320%. (Signal.) *3, decimal point, 2, 0.*

from Lesson 77, Exercise 2

Teaching Note: Students should find this work easy because the relationships of decimals to percents are parallel to the relationships of decimals to fractions. The biggest challenge is going from percents to hundredths decimals. If students are firm in their understanding that the decimals are hundredths, however, they know that there must be two decimal places after the decimal point.

Starting on Lesson 79, students complete tables that show percents and corresponding hundredths decimals.

Here's the table from Workbook Lesson 79:

%	Decimal
a.	6.03
b. 7%	
c. 216%	
d.	.04
e. 500%	

On Lesson 86, students convert decimals that show tenths into hundredths by writing a zero:

.7 = .70

2.5 = 2.50

4.0 = 4.00

The hundredths values convert directly into percents:

.70 = 70%

2.50 = 250%

4.00 = 400%

So to convert .7 into percent, students first convert .7 to hundredths: .70, which converts into 70%. If students have trouble with this discrimination, remind them that percents are hundredths, so you have to change tenths decimals into hundredths.

On Lesson 87, students complete tables that show equivalent tenths decimals, hundredths decimals, and percents.

Here's the exercise from Lesson 88:

(Teacher reference:)

Part 6	Tenths	Hundredths	%
a.			690%
b.	.8		
c.	14.6		
d.			40%
e.	270.0		

This table shows decimals and percents. You'll copy this table, and for each row, you'll write the tenths number, the hundredths number, and the percent that are equivalent.

- Read the value that's shown in row A. Get ready. **(Signal.)** ***690%.***
- Say the hundredths decimal that equals 690%. Get ready. **(Signal.)** ***6 and 90/100.***

b. Read the value that's shown in row B. Get ready. **(Signal.)** ***8/10.***

- Say the hundredths decimal that equals 8 tenths. Get ready. **(Signal.)** ***80/100.***

c. **(Repeat the following tasks for rows C through E:)**

	Read the value shown in row ___. Get ready.	Say the hundredths decimal for ___. Get ready.	
C	*14 and 6 tenths*	14 and 6 tenths	*14 and 60 hundredths*
D	*40%*	40%	*40 hundredths*
E	*270 and 0 tenths*	270 and 0 tenths	*270 and 0 hundredths*

(Repeat rows that were not firm.)

d. Copy part 6 and write the hundredths decimal for each row. Do not complete the rest of the table. Just write the hundredths decimal for each row.
(Observe students and give feedback.)

e. Check your work.
(Display:) [88:7A]

Part 6	Tenths	Hundredths	%
a.		6.90	690%
b.	.8	.80	
c.	14.6	14.60	
d.		.40	40%
e.	270.0	270.00	

Here's what you should have.

f. Read the hundredths number you wrote in row A. Get ready. **(Signal.)** ***6 and 90 hundredths.***

- Say the tenths decimal that equals 6 and 90 hundredths. Get ready. **(Signal.)** ***6 and 9 tenths.***

g. Read the hundredths number you wrote in row B. **(Signal.)** ***80 hundredths.***

- Say the percent that equals 80 hundredths. Get ready. **(Signal.)** ***80%.***

(If students have been 100% on the last two lessons, skip to step k.)

h. Read the hundredths number you wrote in row C. Get ready. **(Signal.)** ***14 and 60/100.***

- Say the percent that equals 14 and 60/100. Get ready. **(Signal.)** ***1460%.***

i. Read the hundredths number you wrote in row D. Get ready. **(Signal.)** ***40/100.***

- Say the tenths decimal that equals 40/100. Get ready. **(Signal.)** ***4/10.***

j. Read the hundredths number you wrote in row E. **(Signal.)** ***270 and zero/100.***

- Say the percent that equals 270 and zero/100. Get ready. **(Signal.)** ***27,000%.***

(Repeat problems that were not firm.)

k. Complete the table. Put your pencil down when you've written the tenths number and the percent for each row.
(Observe students and give feedback.)

l. Check your work.

- Row A. Tell me the tenths number that 690% equals. Get ready. **(Signal.)** ***6 and 9/10.***
- Row B. Tell me the percent that 8/10 equals. Get ready. **(Signal.)** ***80%.***
- Row C. Tell me the percent that 14 and 6/10 equals. Get ready. **(Signal.)** ***1460%.***
- Row D. Tell me the tenths number that 40% equals. Get ready. **(Signal.)** ***4/10.***
- Row E. Tell me the percent that 270 and zero/10 equals. Get ready. **(Signal.)** ***27,000%.***

(Add to show:) [88:7B]

Part 6	Tenths	Hundredths	%
a.	6.9	6.90	690%
b.	.8	.80	80%
c.	14.6	14.60	1460%
d.	.4	.40	40%
e.	270.0	270.00	27,000%

Here's the table you should have for part 6.

Lesson 88, Exercise 7

The strategy that students follow is to convert the value shown in each row to hundredths and then write the missing value.

Students work with variations of the three-column tables through the end of *CMC Level E.* Related items give students information about whole numbers, decimals, or percents and ask about one of the other categories.

Here's a problem set from Textbook Lesson 92:

Questions

a. What tenths decimal equals 7010%?
b. What percent does 6 equal?
c. What percent does 15.3 equal?
d. What hundredths decimal does 153% equal?
e. What whole number does 1000% equal?
f. What percent does 40 equal?

Part 2		
a.		d.
b.		e.
c.		f.

Starting on Lesson 95, students work problems of the following form:

$$\frac{7}{20} \times \frac{}{} = \frac{}{100} \text{ OR } \frac{9}{} \times \frac{}{} = \frac{36}{100}$$

$$\frac{7}{20} = \_\_\_\% \text{ OR } \frac{9}{} = \_\_\_\%$$

After completing the fraction-multiplication equation, students complete the equation below. They translate the completed hundredths fraction into percent.

Here's the Textbook activity from Lesson 96:

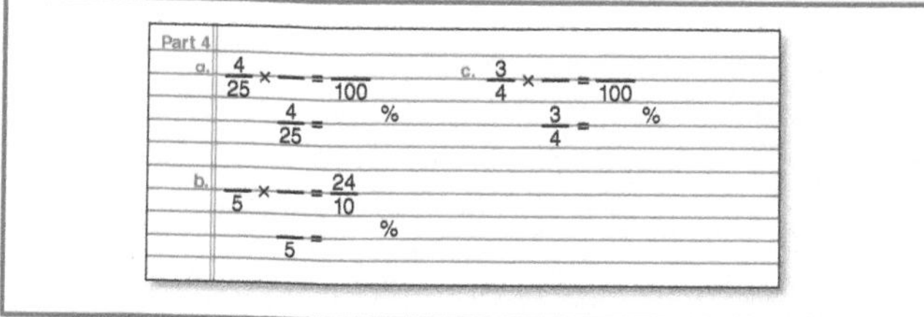

Students work problems of this type through Lesson 99 with less structure.

On Lesson 101, the problems are of this form:

What percent is ____ (fraction with denominator of 2, 4, 5, 20, 25, or 50)?

Here's part of the exercise from Lesson 101:

(Teacher reference:)

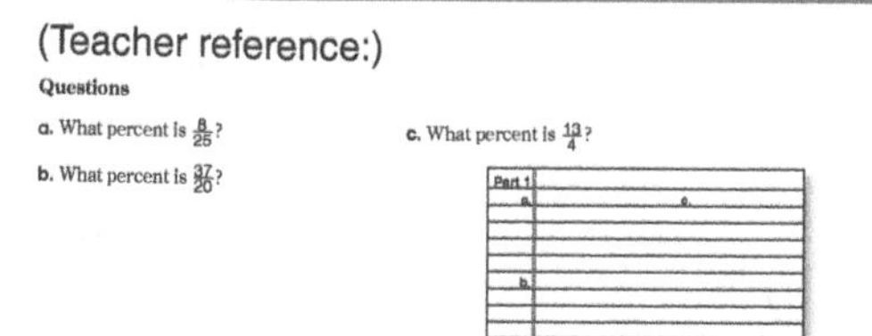

You've completed fraction-multiplication equations and written statements to show the percent fractions equal. For each of the questions in part 1, you'll set up a fraction-multiplication equation to figure out the percent each fraction equals.

- Write part 1 on your lined paper with the letters A, B, and C below.
(Observe students and give feedback.)

b. Read question A. (Call on a student.) *What percent is 8/25?*

- To work this problem, we set up a fraction-multiplication equation to find a fraction that equals 8/25. What's the first fraction in that equation? (Signal.) *8/25.*
- The denominator of the fraction equal to 8/25 is 100. What's the denominator? (Signal.) *100.*

(Display:) [101:3A]

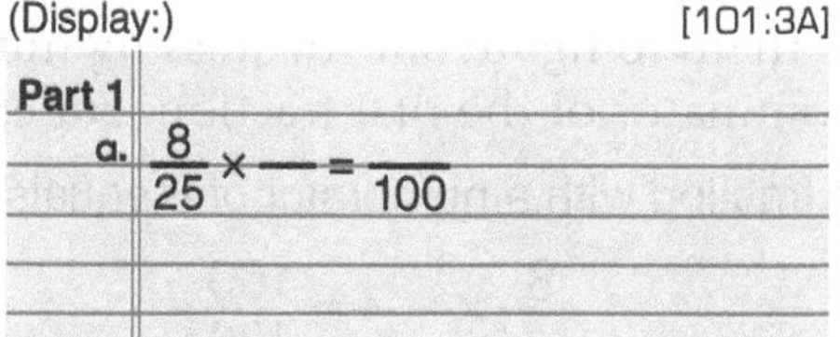

Here's the problem A setup.

- Set up problem A and complete the fraction-multiplication equation. Then write the statement below to show the percent 8/25 equals.
(Observe students and give feedback.)

c. Check your work for problem A.

- Read the fraction-multiplication equation. Get ready. (Signal.) *8/25 x 4/4 = 32/100.*
- Read the statement below. (Signal.) *8/25 = 32%.*

(Add to show:) [101:3B]

Part 1

a. $\frac{8}{25} \times \frac{4}{4} = \frac{32}{100}$

$\frac{8}{25} = 32\%$

Here's what you should have for problem A.

d. Read question B. (Call on a student.) *What percent is 37/20?*
(If students were 100%, skip to step e.)

- To work this problem, we set up a fraction-multiplication equation to find a fraction that equals 37/20. What's the first fraction in the equation? (Signal.) *37/20.*
- What's the denominator of the fraction equal to 37/20? (Signal.) *100.*

(Display:) [101:3C]

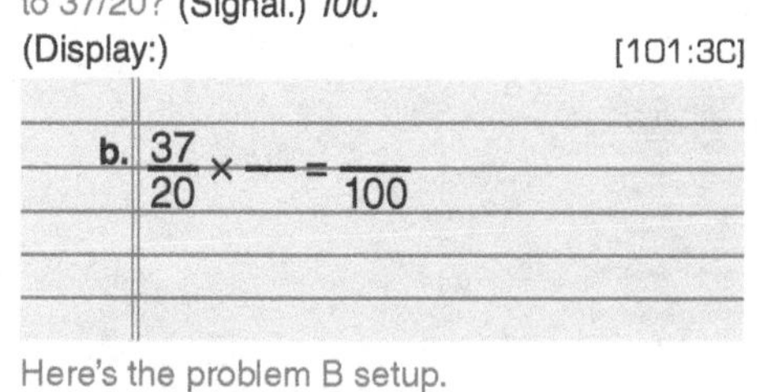

Here's the problem B setup.

e. Set up problem B and complete the fraction-multiplication equation. Then write the statement below to show the percent 37/20 equals.
(Observe students and give feedback.)

f. Check your work for problem B.

- Read the fraction-multiplication equation. Get ready. (Signal.) *37/20 x 5/5 = 185/100.*
- Read the statement below. (Signal.) *37/20 = 185%.*

(Add to show:) [101:3D]

b. $\frac{37}{20} \times \frac{5}{5} = \frac{185}{100}$

$\frac{37}{20} = 185\%$

Here's what you should have for problem B.

from Lesson 101, Exercise 3

On Lesson 107, students work problems that require them to figure out the missing numerator or denominator of the first fraction. For example:

What fraction with a numerator of 8 equals 160%?

$$\frac{8}{\quad} \times \frac{\quad}{\quad} = \frac{160}{100}$$

By Lesson 109, students have learned to answer the full range of questions involving fractions and related whole numbers, decimals, and percents.

Here's the set of problems from Lesson 109:

Questions

a. What fraction with a denominator of 25 does 436% equal?

b. What tenths decimal does 590% equal?

c. What percent does $\frac{6}{5}$ equal?

d. What hundredths decimal does $\frac{6}{5}$ equal?

e. What tenths decimal does $\frac{6}{5}$ equal?

Part 4
a.
d.
b.
e.
c.

Geometry (Lessons 1–130)

Work on geometry starts on Lesson 1 and continues intermittently through the end of the program. Students will:

- Review of names for two-dimensional shapes (hexagon, pentagon, quadrilateral, and triangle).
- Review area and perimeter.
- Learn to identify angles that are multiples of 90 degrees.
- Write the number of degrees for different angles.
- Add or subtract to find missing angles shown in diagrams.
- Compute supplementary and vertical angles.
- Measure and draw angles using protractors.
- Write X and Y values for points on a coordinate grid.
- Plot values for X and Y on a coordinate grid.
- Complete function tables and plot the points described by the table.
- Given a line on a coordinate grid, complete an X-Y table.
- Identify acute, obtuse, and right angles (and triangles).
- Use nomenclature to refer to parallel, parallelograms, perpendicular, intersect, lines, line segments, and rays.

On Lesson 1, students review names of two-dimensional shapes.

The teacher first reviews names of hexagons, pentagons, quadrilaterals, and triangles. The teacher then presents figures, directs students to count the number of sides a figure has, and then tells the name of the figure. Students then answer the question: What are figures with ___ sides?

PERIMETER-AREA

On Lesson 3, students begin a review of perimeter and area.

The rule for perimeter is: The perimeter is the distance all the way around the outside of the figure.

Students work perimeter problems involving different shapes (rectangles, triangles, hexagons. . .). Initially students write the answers as a number followed by the word *units*.

On Lesson 8, the review of area begins. Students learn that the number of squares inside a figure is the area of the figure. To find the area of a rectangle, students multiply the length times the width. The unit in the answer is squares.

On Lesson 12, students review the length units: inches, feet, miles, and yards. Students write these names (or abbreviations) in the answer.

On the next lesson, students work area problems that have these units. (They answer as square inches, square feet . . .)

Work with metric units begins on Lesson 17.

Here's an exercise from Lesson 25 that helps students consolidate what they have learned about area and perimeter and the units for each type of measurement:

a. I'm going to say units for area or perimeter. You'll say the units and then tell me if they're for area or perimeter.
- Listen: Centimeters. Say **centimeters.** (Signal.) *Centimeters.*
- Is that a unit for perimeter or area? (Signal.) *Perimeter.*

b. Say **square centimeters.** (Signal.) *Square centimeters.*
- Is that a unit for perimeter or area? (Signal.) *Area.*

c. (Repeat the following tasks for the specified units:)

Say ___.	Is that a unit for perimeter or area?
Meters	*Perimeter*
Miles	*Perimeter*
Square inches	*Area*
Yards	*Perimeter*
Square feet	*Area*
Square miles	*Area*

(Repeat units that were not firm.)

Lesson 25, Exercise 8

Teaching Note: Make sure that students are very firm on this exercise. Too often students will know how to compute area or perimeter but will not know the unit name for each type of measurement.

After the structured work on area and perimeter are completed on Lesson 27, students work problems, some of which have large numbers, as part of their Independent Work.

Here's the exercise from Lesson 29:

(Teacher reference:) **R Test 4**

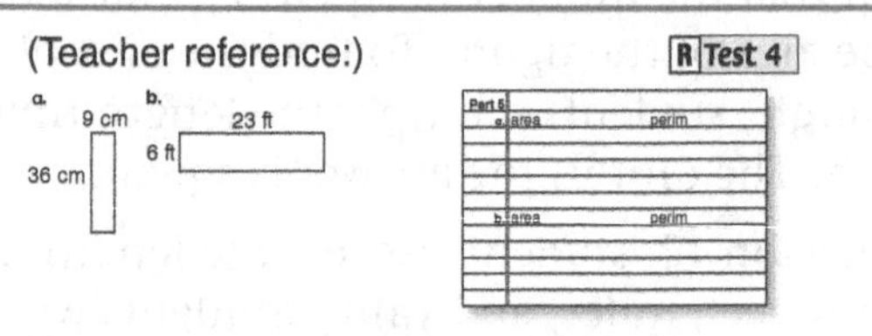

You'll write column problems to figure out the area and perimeter for each shape.

- Look at shape A. Start with the number that has more than one digit and say the area problem. Get ready. **(Signal.)** *36 × 9.*
- Start with the top number and say the problem for the perimeter. Get ready. **(Signal.)** *9 + 36 + 9 + 36.*

(Display:) [29:9A]

Part 5		
a.	area 36 × 9	perim 9 36 9 +36

You'll write the column problems and work them for the shapes in part 5 as part of your independent work. Remember, write the number with more than one digit first in column multiplication problems.

Lesson 29, Exercise 9

Finding the Missing Length

On Lesson 90, students learn to solve problems that give the area and ask about the length or width of a rectangle.

a. (Display:) [90:7A]

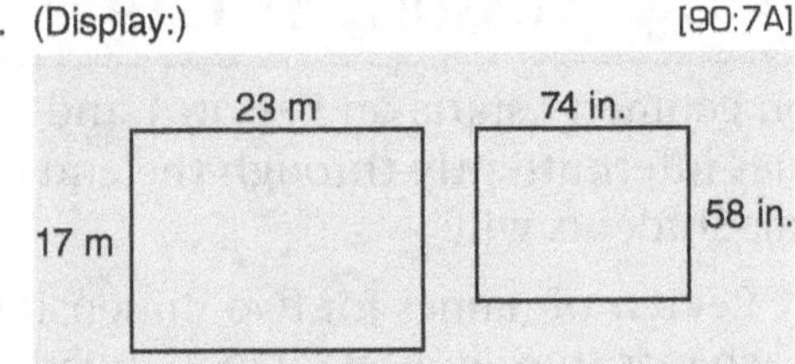

You've learned how to find the area of rectangles. For both rectangles, you'll tell me the number problem for finding the area, and you'll tell me the units in the answer.

- **(Point to 17 m.)** Tell me the length of the top and the length of a side. Get ready. **(Signal.)** *23 meters (and) 17 meters.*
- Say the number problem for finding the area. Get ready. **(Signal.)** *23 × 17.*
- What are the units in the answer? **(Signal.)** *Square meters.*

b. **(Point to 58 in.)** Tell me the length of the top and the length of a side. Get ready. **(Signal.)** *74 inches (and) 58 inches.*

- Say the problem for finding the area. Get ready. **(Signal.)** *74 × 58.*
- What are the units in the answer? **(Signal.)** *Square inches.*

c. (Display:) [90:7B]

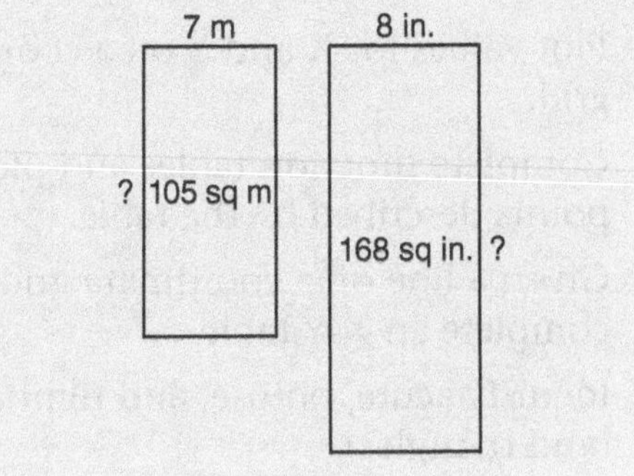

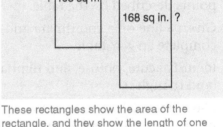

These rectangles show the area of the rectangle, and they show the length of one side. You divide the area by the side to find the missing length.

- **(Point to 7 m.)** Tell me the length of the side that's shown. Get ready. **(Signal.)** *7 meters.*
- Tell me the area of the rectangle. Get ready. **(Signal.)** *105 square meters.*
- The problem for finding the missing length is 105 ÷ 7. Say the problem for finding the missing length. **(Signal.)** *105 ÷ 7.*
- What are the units for the missing length? **(Signal.)** *Meters.*

d. **(Point to 8 in.)** Tell me the length of the side that's shown. Get ready. **(Signal.)** *8 inches.*

- Tell me the area of the rectangle. Get ready. **(Signal.)** *168 square inches.*
- Say the problem for finding the missing length. Get ready. **(Signal.)** *168 ÷ 8.*
- What are the units for the missing length? **(Signal.)** *Inches.*

from Lesson 90, Exercise 7

Teaching Note: Expect students to have some trouble with the unit name. It may not be obvious to students that dividing square units by units gives units—not square units. Remind students that when you solve for length, you are solving for units, not square units.

On Lesson 92, students work mixed problem types. For some problems, students solve for the area of a rectangle. For others they solve for the height or side.

Here's a set of problems from Textbook Lesson 93:

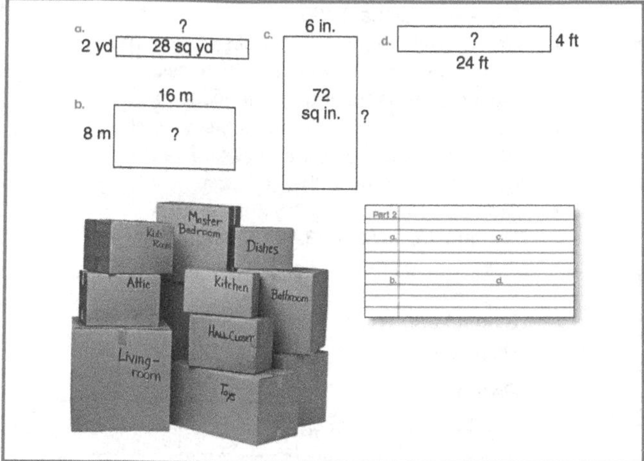

On Lesson 94, students work problems that have mixed-number answers.

For example:

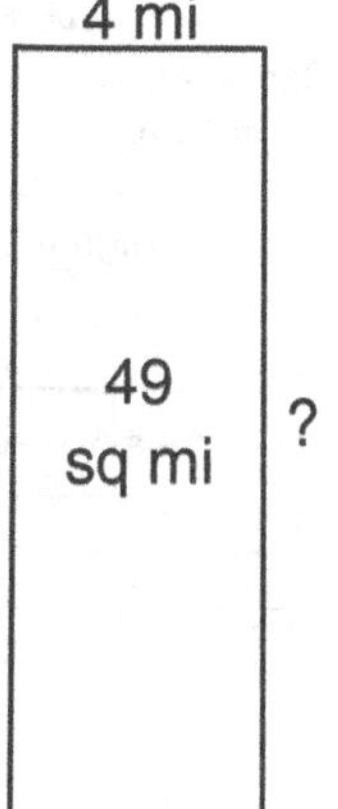

Students figure out the height, which is 12 1/4 miles. (See Division, page 137, for the method students use to work the problem.)

On Lesson 97, students work problems that require finding the area or the length of a side and finding the perimeter of the rectangle.

Here are the problems from Lesson 100:

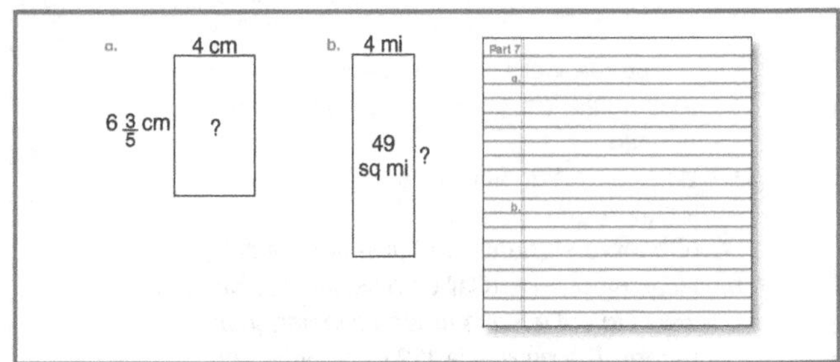

The problem students work for A is 6 3/5 × 4. (See Mixed-Number Computation, page 153, for the method students use to work the problem.)

On Lesson 124, students work word problems that refer to area, length, and perimeter.

For some problems, students multiply a fraction by a whole number; for others, students figure out a missing length and the perimeter of the figure.

Here's the exercise from Lesson 126:

(Teacher reference:)

a. A farmer wants to put a fence around a rectangular garden. The garden is 132 yards wide and 58 yards long. How many yards of fencing does the farmer need to put a fence around that garden?

b. Jim has enough paint to cover 100 square feet of a wall. The wall he wants to paint is 8 feet tall. If he has just enough paint to cover the wall, how long is the wall?

These are word problems that tell about rectangles. You'll figure out the answer to the questions.

- Write part 4 on your lined paper with the letters A and B below.
  (Observe students and give feedback.)

b. Read problem A. **(Call on a student.)** *A farmer wants to put a fence around a rectangular garden. The garden is 132 yards wide and 58 yards long. How many yards of fencing does the farmer need to put a fence around that garden?*

- Tell me the length of the longest side of the rectangle. Get ready. **(Signal.)** *132 yards.*
- What's the length of the other side? **(Signal.)** *58 yards.*
- Make a sketch of the rectangle on your lined paper with the sides labeled. Make sure you label the longest side of your sketch with the biggest length.
  (Observe students and give feedback.)
  (Display:) [126:6A]

Part 4
a.
132 yd
58 yd

Here's a sketch of the garden.

- The question asks: How many yards of fencing does the farmer need to put a fence around that garden? Does that question ask about the area or perimeter? **(Signal.)** *Perimeter.*
- Start with 132 and say the problem for finding the perimeter. Get ready. **(Signal.)** *132 + 58 + 132 + 58.*

c. Read problem B. **(Call on a student.)** *Jim has enough paint to cover 100 square feet of a wall. The wall he wants to paint is 8 feet tall. If he has just enough paint to cover the wall, how long is the wall?*

- The length of one of the sides of the wall is given. What's that length? **(Signal.)** *8 feet.*
- Does the problem give another length or the area of the rectangle? **(Signal.)** *The area.*
- What's the area of the rectangle? **(Signal.)** *100 square feet.*
- Make a sketch of the rectangle on your lined paper with a side labeled and the area labeled.
  (Observe students and give feedback.)
  (Display:) [126:6B]

Here's a sketch of the wall.

- The question asks: How long is the wall? Say the problem for finding the length of the wall. Get ready. **(Signal.)** *100 ÷ 8.*

d. Write the problem for each word problem below each rectangle and figure out the answer.
(Observe students and give feedback.)

e. Check your work. For each problem, you'll read the problem and the answer. Then you'll tell me the answer to the question.

- Read the problem and the answer for A. **(Signal.)** *132 + 58 + 132 + 58 = 380 (yards).*
- How many yards of fencing does the farmer need to put a fence around that garden? **(Signal.)** *380 yards.*
  (Display:) [126:6C]

Part 4
a.
132 yd
58 yd

```
   1 2
   1 3 2
     5 8
   1 3 2
 +   5 8
   3 8 0 yd
```

Here's what you should have for problem A.

f. Read the problem and the answer for B. **(Signal.)** *100 ÷ 8 = 12 and 4/8 (feet).*

- If Jim has just enough paint to finish the wall, how long is the wall? **(Signal.)** *12 and 4/8 feet.*
  (Display:) [126:6D]

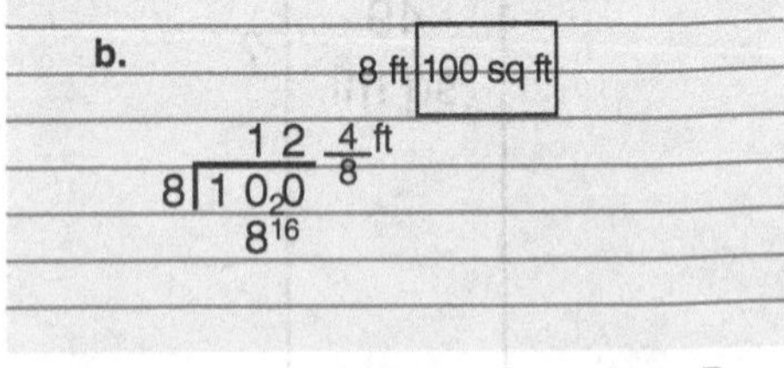

Here's what you should have for problem B.

Lesson 126, Exercise 6

ANGLES

Work on angles begins on Lesson 62 and continues through the end of *CMC Level E.*

The initial work focuses on angles. Angles are formed where two lines meet. Angles are shown with part of a circle and are expressed in degrees. An arc that spans 1/4 circle shows a 90-degree angle. Angles that are more than 90 degrees have arcs that are more than 1/4 circle. Multiples of 90-degree angles are 180 degrees, 270 degrees, and 360 degrees.

Here's the last part of the exercise from Lesson 62:

l. **(Display:)** [62:2G]

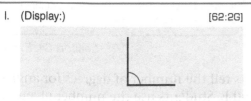

- How many degrees is this angle? **(Signal.)** *90.*

m. Listen: 90 degrees is 1/4 of a circle. If you turn 90 degrees, you turn 1/4 of a circle.

- How many degrees is 1/4 of a circle? **(Signal.)** *90.*
- If you turn and face to your right or left, how far do you turn? **(Signal.)** ***90 degrees.***

n. Let's add another 1/4 of a circle to 90 degrees. **(Change to show:)** [62:2H]

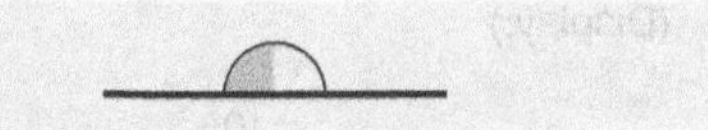

This angle is 90 plus 90 degrees. That's 180 degrees.
(Add to show:) [62:2I]

You can see that 180 degrees is 2/4 of a circle. That's half a circle.

- How many degrees is 2/4 of a circle? **(Signal.)** *180.*
- If you turn around and face the opposite way, how far do you turn? **(Signal.)** ***180 degrees.***

o. Let's add another 1/4 of a circle to 180 degrees. **(Change to show:)** [62:2J]

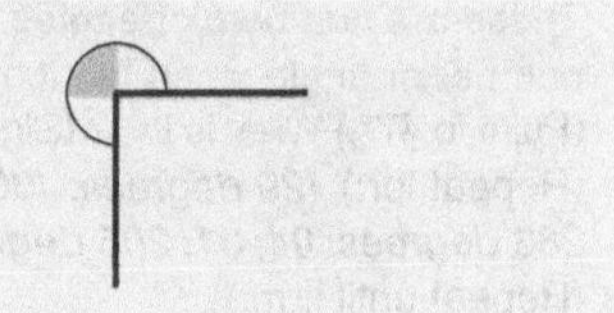

This angle is 90 plus 90 plus 90 degrees. That's 270 degrees.

(Add to show:) [62:2K]

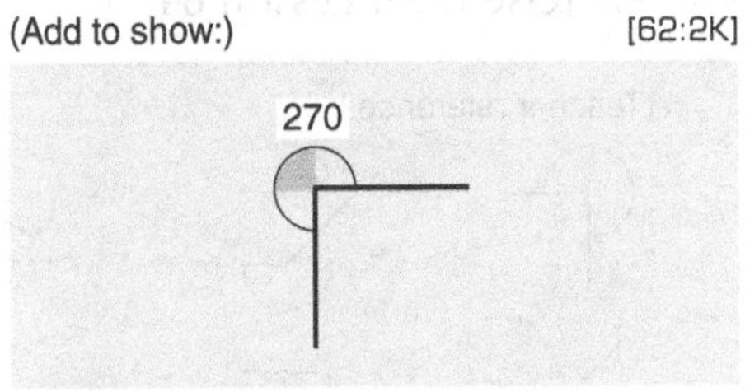

You can see that 270 is 3/4 of a circle.

- How many degrees is 3/4 of a circle? **(Signal.)** *270.*

p. Let's add another 1/4 of a circle to 270 degrees. **(Change to show:)** [62:2L]

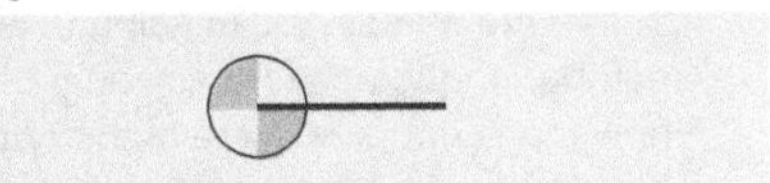

This angle is 90 plus 90 plus 90 plus 90. That's 360 degrees.
(Add to show:) [62:2M]

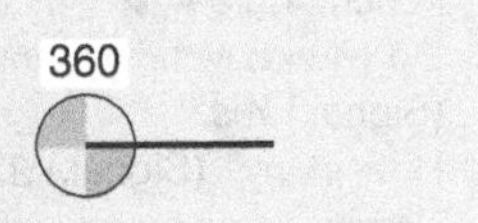

You can see that 360 degrees is 4/4 of a circle. That's the whole circle.

- How many degrees is 4/4 of a circle? **(Signal.)** *360.*
- If you turn all the way around, how far do you turn? **(Signal.)** ***360 degrees.***

q. Let's do those again. **(Display:)** [62:2N]

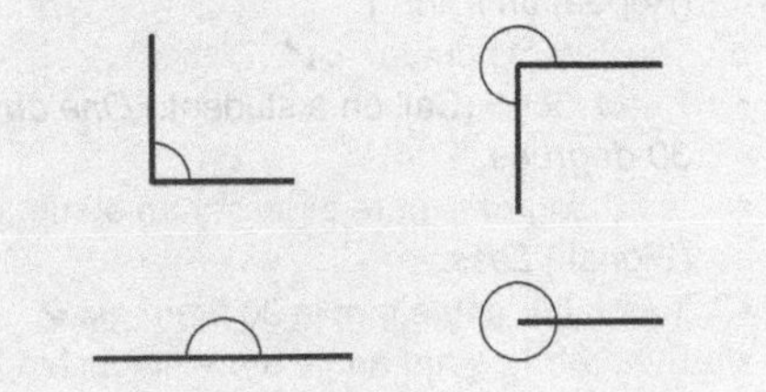

- **(Point to ∟.)** How many degrees are in 1/4 of a circle? **(Signal.)** *90.*
- **(Point to ⌒.)** How many degrees are in 2/4 of a circle? **(Signal.)** *180.*
- **(Point to ⌐.)** How many degrees are in 3/4 of a circle? **(Signal.)** *270.*
- **(Point to ⊖.)** How many degrees are in 4/4 of a circle? **(Signal.)** *360.*
- **(Repeat until firm.)**

from Lesson 62, Exercise 2

On the following lessons, students use information about multiples of 90-degree angles to figure out other angles.

Here's an exercise from Lesson 64:

(Teacher reference:)

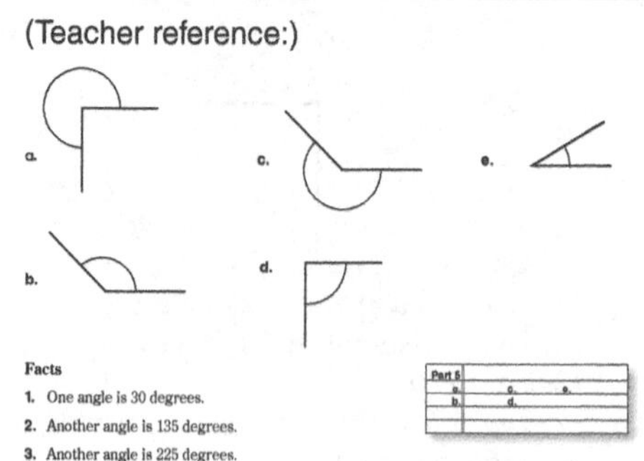

a.
b.
c.
d.
e.
Facts
1. One angle is 30 degrees.
2. Another angle is 135 degrees.
3. Another angle is 225 degrees.

These are angles. You're going to write the number of degrees for each angle. You know how many degrees some of the angles are. You can figure out the rest of the angles from the facts. You're going to tell me the number of degrees for each angle. You'll write the numbers as part of your independent work.

- Touch angle A. ✔
- Do you know how many degrees angle A is? (Signal.) *Yes.*
- How many? (Signal.) *270 (degrees).*
- Do you know how many degrees angle B is? (Signal.) *No.*
- Do you know how many degrees angle C is? (Signal.) *No.*
- Do you know how many degrees angle D is? (Signal.) *Yes.*
- How many? (Signal.) *90 (degrees).*
- Do you know how many degrees angle E is? (Signal.) *No.*

(Repeat until firm.)

b. Touch fact number 1. ✔
- Read fact 1. (Call on a student.) *One angle is 30 degrees.*
- Is 30 degrees more or less than 90 degrees? (Signal.) *Less.*
- Touch the angle that is 30 degrees. ✔
- Everybody, what angle are you touching? (Signal.) *(Angle) E.*

c. Read fact 2. (Call on a student.) *Another angle is 135 degrees.*
- Is 135 degrees more or less than 90 degrees? (Signal.) *More.*
- Is 135 degrees more or less than 180 degrees? (Signal.) *Less.*
- So that angle is larger than 1/4 of a circle, but smaller than 2/4 of a circle. Touch the angle that is 135 degrees. ✔
- Everybody, what angle are you touching? (Signal.) *(Angle) B.*

d. Read fact 3. (Call on a student.) *Another angle is 225 degrees.*
- Is 225 degrees more or less than 90 degrees? (Signal.) *More.*
- Is 225 degrees more or less than 180 degrees? (Signal.) *More.*
- Is 225 degrees more or less than 270 degrees? (Signal.) *Less.*
- So that angle is larger than 2/4 of a circle, but smaller than 3/4 of a circle. Touch the angle that is 225 degrees. ✔
- Everybody, what angle are you touching? (Signal.) *(Angle) C.*

(Repeat angles that were not firm.)

As part of your independent work, you'll write part 5 on your lined paper with the letters A through E below. Then you'll write the number of degrees for each angle.

Lesson 64, Exercise 9

The facts tell the number of degrees for angles B, C, and E. Students use the number given for an angle to compare the angle to angles that are multiples of 90 degrees.

The degree symbol is introduced on Lesson 65.

Here's the first part of the exercise from Lesson 66:

a. Last time we worked with angles, and you learned the degree symbol.

(Display:) [66:6A]

106°

This shows 106 degrees.
- Everybody, what shape is the degree symbol? (Signal.) *(A) circle.*
- Is the degree symbol a big or little circle? (Signal.) *Little.*
- Does the degree symbol go before or after an angle number? (Signal.) *After.*
- Does the degree symbol go near the top or bottom of an angle number? (Signal.) *Top.*

(Repeat until firm.)

b. (Display:) [66:6B]

47°	100	283°	31
129°	315°	94	205°

These are numbers or angles. You'll read each number or angle when I touch it.
- (Point to **47°**.) What is this? (Signal.) *47 degrees.*
- (Repeat for:) *129 degrees; 100; 315 degrees; 283 degrees; 94; 31; 205 degrees.*

(Repeat until firm.)

from Lesson 66, Exercise 6

After students complete this part, they write the angle number and degree symbol for eight angles.

Calculating Missing Angles

Starting with Lesson 69, students work with problems that show degrees for two nested angles. Students figure out the third angle.

Here's the first part of the exercise from Lesson 69:

a. (Display:) [69:6A]

This angle is divided into two parts. You're going to figure out the degrees for one of the parts.

- Does the whole angle show fourths of a circle? (Signal.) *Yes.*
- How many degrees is the whole angle? (Signal.) *90 (degrees).*
- (Point to **23°.**) This shows the number of degrees for one part. How many degrees? (Signal.) *23 (degrees).*

b. Listen: The whole angle is 90 degrees. One part is 23 degrees. You have to figure out the degrees for the other part.

- Raise your hand when you can say the problem you'll work. ✔
- Everybody, what problem? (Signal.) *90 – 23.*

c. (Display:) [69:6B]

Part A

$$\begin{array}{r} 90 \\ -23 \\ \hline \end{array}$$

- Write part A on your lined paper. Write the problem, 90 – 23, and work it. (Observe students and give feedback.)
- Everybody, what's 90 – 23? (Signal.) *67.*
- So what's the missing part? (Signal.) *67 degrees.* (Display:) [69:6C]

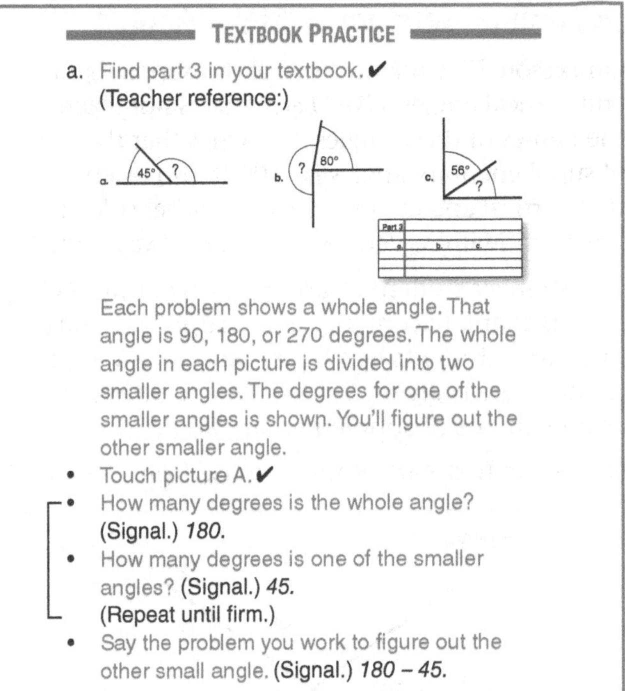

TEXTBOOK PRACTICE

a. Find part 3 in your textbook. ✔ (Teacher reference:)

Each problem shows a whole angle. That angle is 90, 180, or 270 degrees. The whole angle in each picture is divided into two smaller angles. The degrees for one of the smaller angles is shown. You'll figure out the other smaller angle.

- Touch picture A. ✔
- How many degrees is the whole angle? (Signal.) *180.*
- How many degrees is one of the smaller angles? (Signal.) *45.* (Repeat until firm.)
- Say the problem you work to figure out the other small angle. (Signal.) *180 – 45.*

from Lesson 69, Exercise 6

Note that these problems are referenced to multiples of 90 degrees. For B, the whole angle is 270 degrees, one component of the angle is 80 degrees. 270 minus 80 gives the size of the unknown angle.

On Lesson 71, students work a mixed set. The angle they solve for is the whole angle or a component angle.

Here's the set of problems from Textbook Lesson 71:

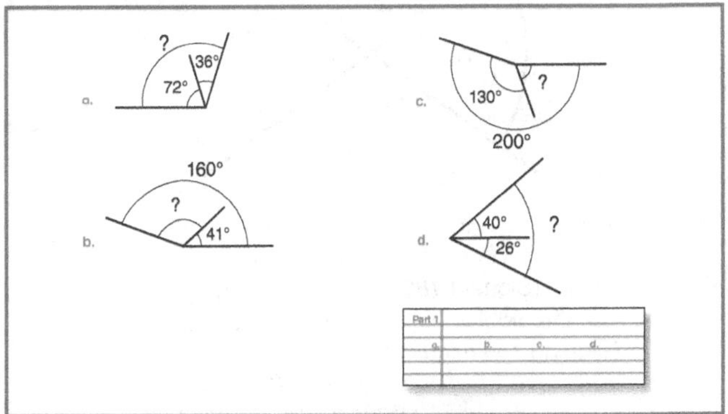

The routine students follow is to first touch the question mark and indicate if they solve for the whole angle or a smaller angle. Then they say the problem for finding the angle that has a question mark.

Supplementary and Vertical Angles

On Lesson 77, students learn about supplementary and vertical angles. *CMC Level E* does not teach the names of these angles. It teaches that the sum of supplementary angles is 180°. It also teaches that vertical angles are formed on either side of intersecting lines. Vertical angles are always equal.

The strategy students learn about two intersecting lines is that adjacent angles are supplementary. Therefore students can figure out all four angles if one angle is given. Students make different arrangements of supplementary angles.

Here's the first part of the exercise from Lesson 78:

a. (Display:) [78:2A]

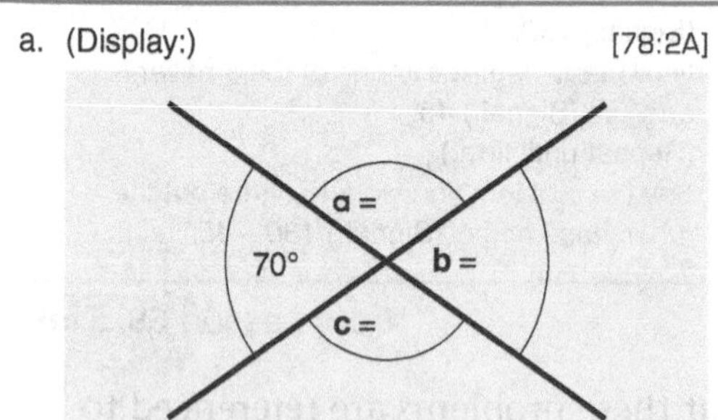

You worked this kind of problem last time. One angle is shown. You can figure out the angles that are not shown.

- How many degrees is the angle that is shown? (Signal.) *70.*
  (Add to show:) [78:2B]

  a =
  70°
  b =
  c =

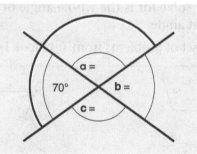

- If you add angle A to 70, you end up with an angle you know. How many degrees is that angle? (Signal.) *180.*
- Say the problem for figuring out angle A. (Signal.) *180 minus 70.*
- What's the answer? (Signal.) *110.*
- So what's angle A? (Signal.) *110 degrees.*
  (Add to show:) [78:2C]

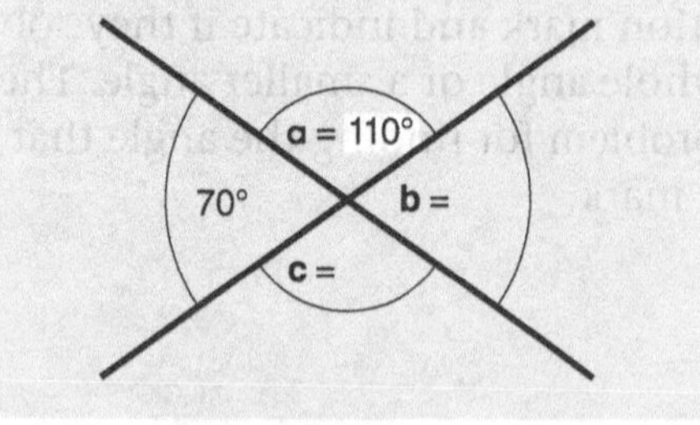

b. (Add to show:) [78:2D]

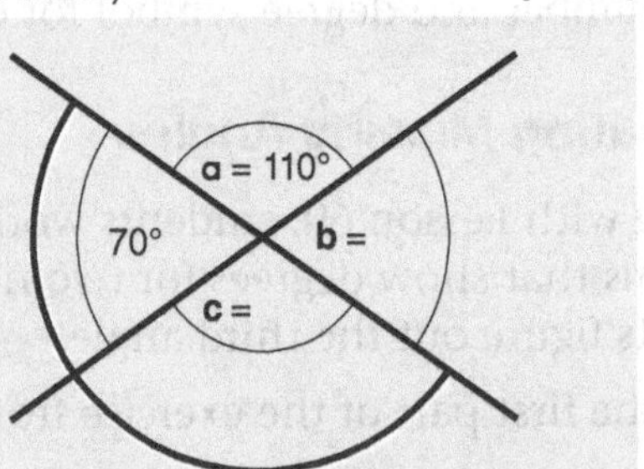

- If you add angle C to 70, you end up with an angle you know. How many degrees is that angle? (Signal.) *180.*
- Say the problem for figuring out angle C. (Signal.) *180 – 70.*
- What's the answer? (Signal.) *110.*
- So what's angle C? (Signal.) *110 degrees.*
  (Add to show:) [78:2E]

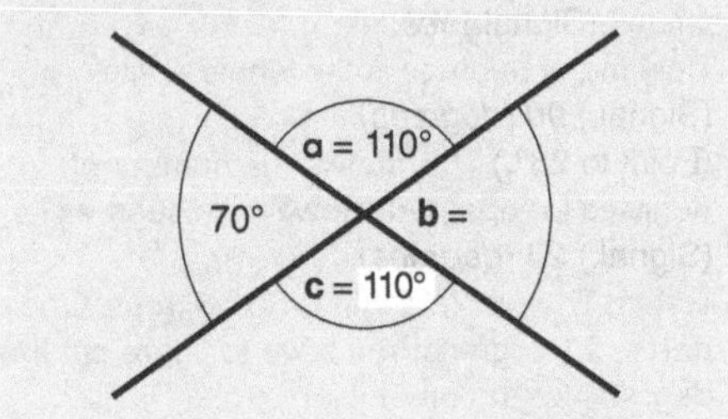

c. (Add to show:) [78:2F]

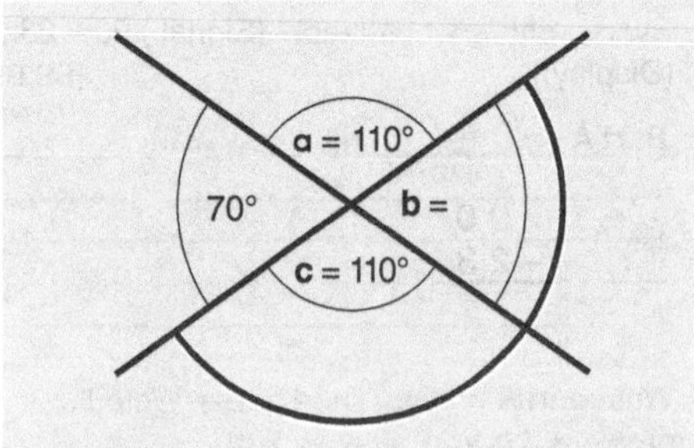

- If you add angle B to 110, you end up with an angle you know. How many degrees is that angle? (Signal.) *180.*
- Say the problem for figuring out angle B. (Signal.) *180 – 110.*
- What's the answer? (Signal.) *70.*
- So what's angle B? (Signal.) *70 degrees.*
  (Add to show:) [78:2G]

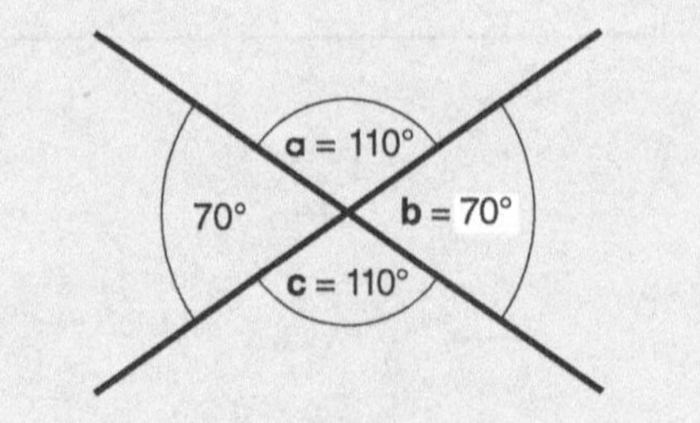

d. We end up with two pairs of angles that are across from each other.
- (Point to **c =.**) How many degrees is angle C? (Signal.) *110.*
- How many degrees is the angle that is across from C? (Signal.) *110.*
- (Point to **b =.**) How many degrees is angle B? (Signal.) *70.*
- How many degrees is the angle that is across from B? (Signal.) *70.*

from Lesson 78, Exercise 2

Right, Acute, Obtuse Angles

Right angles are introduced on Lesson 107. In this lesson, students learn 90-degree angles are also called right angles. On Lesson 108, students identify right angles and angles that are not 90°.

Lesson 108 also teaches that triangles with right angles are called right triangles.

Here's the exercise:

a. In the last lesson, you learned what a right angle is.
- Everybody, what's another name for a 90° angle? (Signal.) *A right angle.*
- What's another name for the angle of a quarter circle? (Signal.) *A right angle.*
- How many degrees is a right angle? (Signal.) *90.*

b. (Display:) [108:4A]

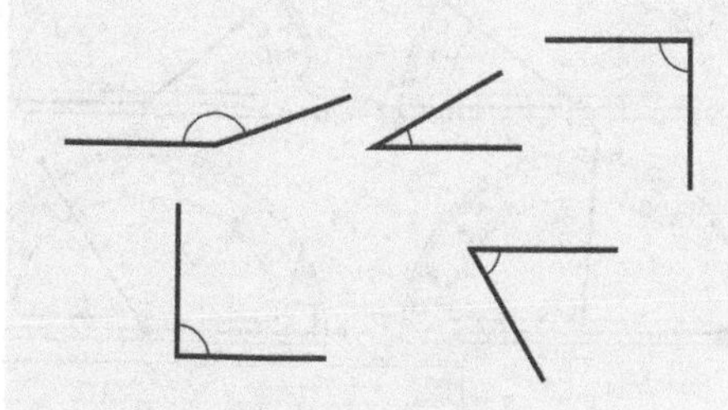

For each angle you'll tell me if it's a right angle or not right angle.
- (Point to ⌒.) Tell me right angle or not right angle. (Signal.) *Not right angle.*
- (Point to ∠.) Tell me. (Signal.) *Not right angle.*
- (Point to ⌉.) Tell me. (Signal.) *Right angle.*
- (Point to ⌊.) Tell me. (Signal.) *Right angle.*
- (Point to ⌐.) Tell me. (Signal.) *Not right angle.*

(Repeat angles that were not firm.)

c. (Display:) [108:4B]

Some of these triangles have right angles in them. Here's the rule: A triangle with a right angle is called a **right triangle**.

- What do we call a triangle that has a right angle? (Signal.) *A right triangle.*

(Repeat until firm.)

d. For each triangle, you'll tell me if it has a right angle. Then you'll tell me if it's a right triangle or not a right triangle.
- (Point to ◿.) Does this triangle have a right angle? (Signal.) *Yes.*
- So is it a right triangle? (Signal.) *Yes.*

e. (Point to △.) Does this triangle have a right angle? (Signal.) *No.*
- So is it a right triangle? (Signal.) *No.*

f. (Point to ◸.) Does this triangle have a right angle? (Signal.) *Yes.*
- So is it a right triangle? (Signal.) *Yes.*

g. (Point to ⊿.) Does this triangle have a right angle? (Signal.) *No.*
- So is it a right triangle? (Signal.) *No.*

h. (Point to ◹.) Does this triangle have a right angle? (Signal.) *Yes.*
- So is it a right triangle? (Signal.) *Yes.*

(Repeat steps d through h that were not firm.)

i. Let's do those again. This time you'll tell me right triangle or not right triangle.
- (Point to ◿.) Tell me right triangle or not right triangle. Get ready. (Signal.) *Right triangle.*
- (Point to △.) Tell me right triangle or not right triangle. Get ready. (Signal.) *Not right triangle.*
- (Point to ◸.) Tell me. (Signal.) *Right triangle.*
- (Point to ⊿.) Tell me. (Signal.) *Not right triangle.*
- (Point to ◹.) Tell me. (Signal.) *Right triangle.*

(Repeat until firm.)

Lesson 108, Exercise 4

On Lesson 112, students learn to identify that angles smaller than 90 degrees are acute angles. They also learn that acute triangles have three acute sides. Students indicate whether different triangles are right triangles, acute triangles, or neither right nor acute.

On Lesson 116, students learn to identify obtuse angles, those that are larger than 90 degrees and have an obtuse angle. Students then identify different triangles as right, acute, or obtuse.

Here are examples from Textbook Lesson 116:

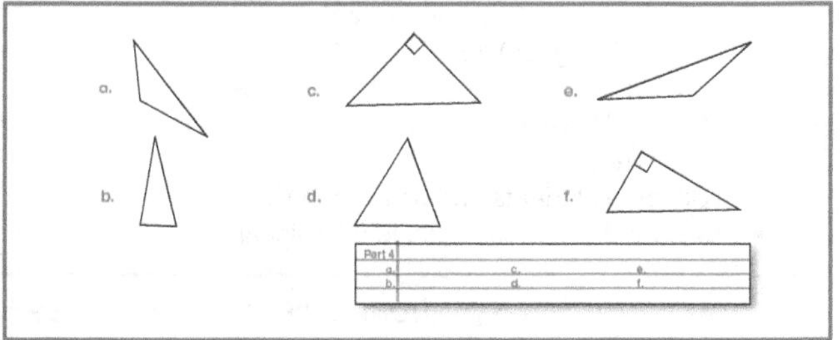

Protractors

On Lesson 121, students learn how to use a protractor to measure angles, first finding the line segment that goes through zero on the protractor. That number is either on the inside or the outside of the numbers on the semi-circle.

Students go from zero to the tens number that is closest to the line segment that marks the angle. They then count the remaining degrees from the tens number to the line segment.

Here's part of the exercise from Lesson 121:

(Teacher reference:)

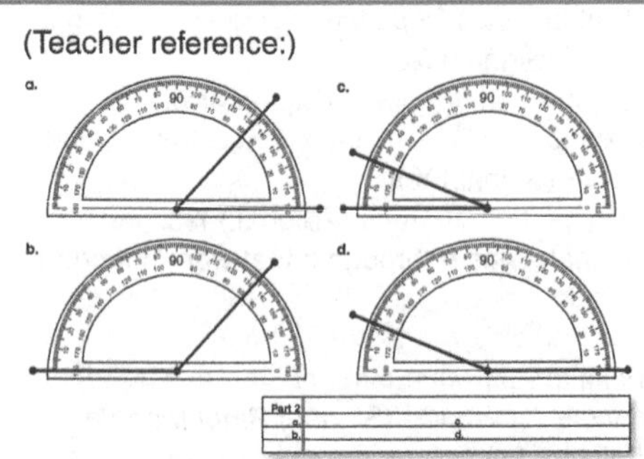

You're going to write the degrees for each angle.

- Write part 2 on your lined paper with the letters A through D below.
  (Observe students and give feedback.)

b. Look at angle A.

- Touch the line segment that goes through zero on the protractor. Is that zero for the numbers on the inside or outside? (Signal.) *The inside.*
- Look at the inside numbers and touch the tens number for the angle. ✔
- Everybody, what's the biggest tens number for angle A? (Signal.) *40.*
- Touch and count the rest of the degrees to yourself. Raise your hand when you know how big angle A is.
  (Observe students and give feedback.)
- Everybody, how big is angle A? (Signal.) *47°.*

c. Look at the line segment of angle B that goes through zero on the protractor. Is that zero for the numbers on the inside or outside?
(Signal.) *The outside.*

- Look at the outside numbers and touch the tens number for the angle. ✔
- Everybody, what's the tens number for angle B? (Signal.) *130.*
- Touch and count the rest of the degrees to yourself. Raise your hand when you know how big angle B is.
  (Observe students and give feedback.)
- Everybody, how big is angle B? (Signal.) *133°.*

from Lesson 121, Exercise 5

Teaching Note: Observe students when they touch the tens number that is closest to the angle line. Make sure they are touching the correct tens number. If they make mistakes, tell them to touch the line that goes through zero and then count the numbers for that scale to the tens number closest to the other line of the angle.

If students become confused about how far to count, direct them to touch both lines of the angle. Tell them, "You are going to figure out the number of degrees for that angle. You will not go outside that angle."

On Lesson 123, students identify which angles can't be measured as they are shown. The lines must intersect at the center marker. One of the lines must go through a zero.

Here are the examples from Textbook Lesson 123:

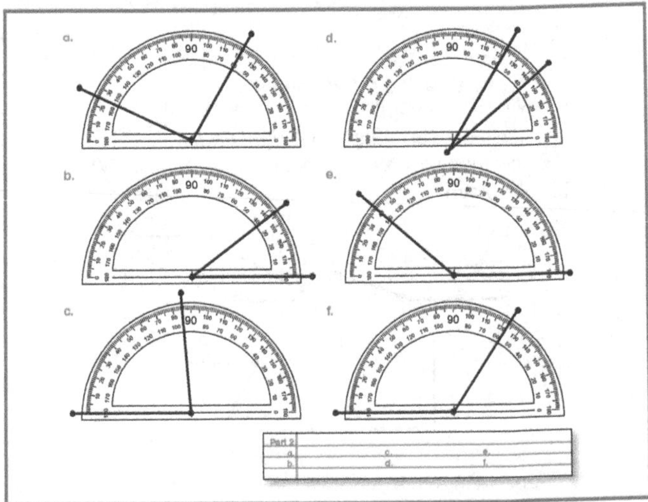

A is incorrect because no line goes through a zero.

D is incorrect because the lines do not intersect at the center mark and no line goes through zero.

Students measure angles with protractors on Textbook Lesson 124.

Here are the angles they measure:

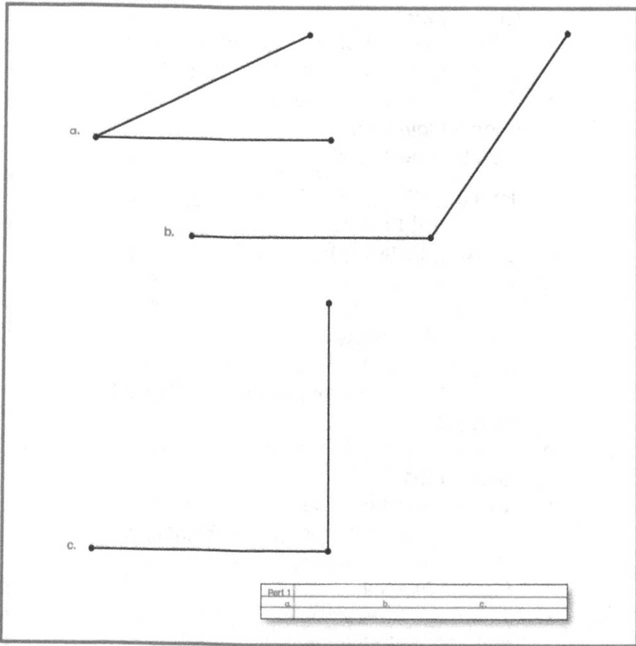

Students use variations of the same rules to measure angles that they used to figure out angles. They position the protractor so the center mark is at the intersection of the lines and one line goes through zero.

Here is the exercise from Lesson 120:

a. You've learned a lot about angles. I'm going to teach you some math names for the parts of angles.
(Display:) [120:5A]

Here's an angle. The straight sides of this angle are called line segments.

- Say **segments.** (Signal.) *Segments.*
- Say **line segments.** (Signal.) *Line segments.*
- What do you call the sides of this angle? (Signal.) *Line segments.*
(Repeat until firm.)

b. The sides aren't called lines because a line in math goes on forever.
(Change to show:) [120:5B]

Here's the angle with lines. The arrows on the ends of the lines show that they keep going.
(Change to show:) [120:5C]

Here's the angle with line segments again.

- Think of a triangle. Is a triangle made up of lines or line segments? (Signal.) *Line segments.*
- What are rectangles made up of? (Signal.) *Line segments.*

c. Line segments that touch intersect. Say **intersect.** (Signal.) *Intersect.*

- Line segments that touch do what? (Signal.) *Intersect.*
(Display:) [120:5D]

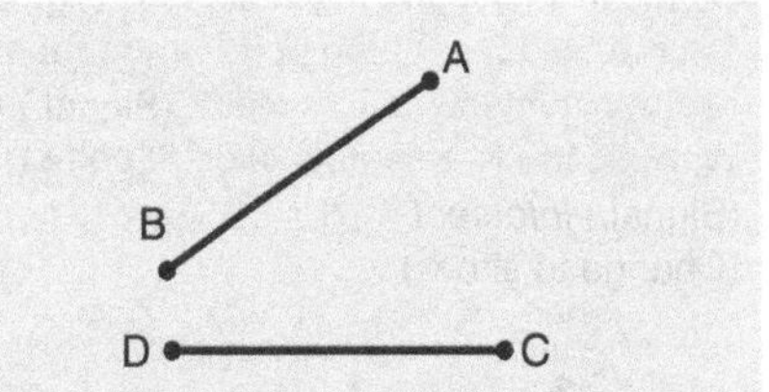

Here are two line segments. The point at the end of each line segment is labeled with a letter.

- (Point to •**B.**) What's the letter for the point at this end of the top line segment? (Signal.) *B.*
- What's the letter for the point at the other end? (Signal.) *A.*
- (Point to •**C.**) What's the letter for this point? (Signal.) *C.*
- What's the point at the other end? (Signal.) *D.*
- Do these line segments touch? (Signal.) *No.*
- So do these line segments intersect? (Signal.) *No.*

d. (Change to show:) [120:5E]

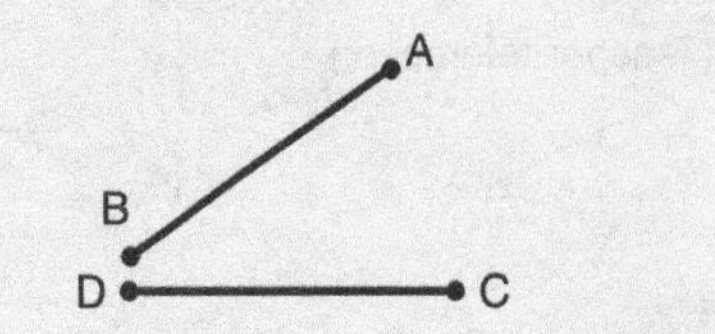

- Do these line segments touch? (Signal.) *No.*
- So do these line segments intersect? (Signal.) *No.*

e. (Change to show:) [120:5F]

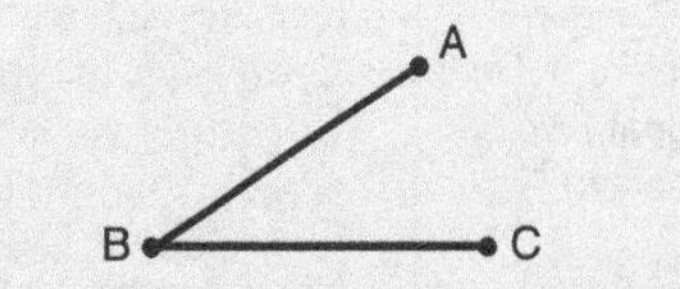

- Do these line segments touch? (Signal.) *Yes.*
- So do these line segments intersect? (Signal.) *Yes.*
- Tell me the letter of the point where the line segments intersect. Get ready. (Signal.) *B.*
- What do the line segments do at point B? (Signal.) *Intersect.*

f. (Display:) [120:5G]

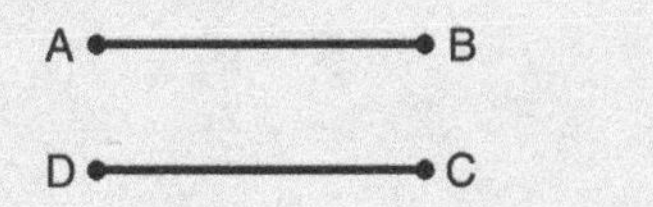

- Do these line segments intersect? (Signal.) *No.*

g. (Change to show:) [120:5H]

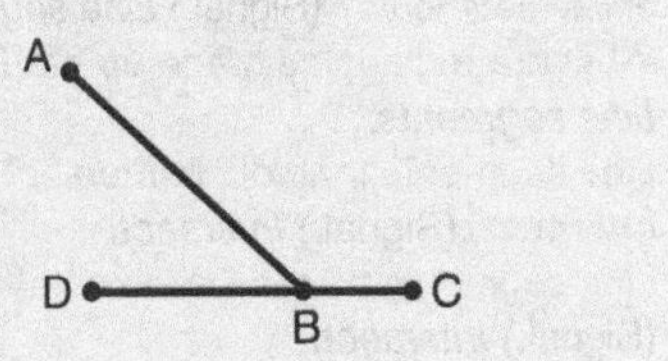

- Do these line segments intersect? (Signal.) *Yes.*
- Tell me the letter of the point where the line segments intersect. Get ready. (Signal.) *B.*
- What do the line segments do at point B? (Signal.) *Intersect.*

h. (Change to show:) [120:5I]

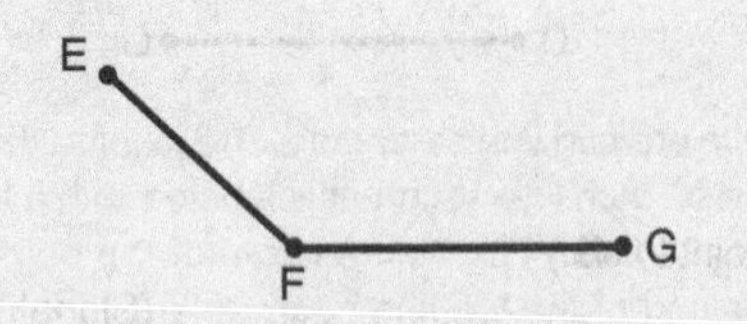

- Do these line segments intersect? (Signal.) *Yes.*
- Tell me the letter of the point where the line segments intersect. Get ready. (Signal.) *F.*
- What do the line segments do at point F? (Signal.) *Intersect.*

(Repeat steps d through h that were not firm.)

TEXTBOOK PRACTICE

a. Find part 3 in your textbook. ✔

(Teacher reference:)

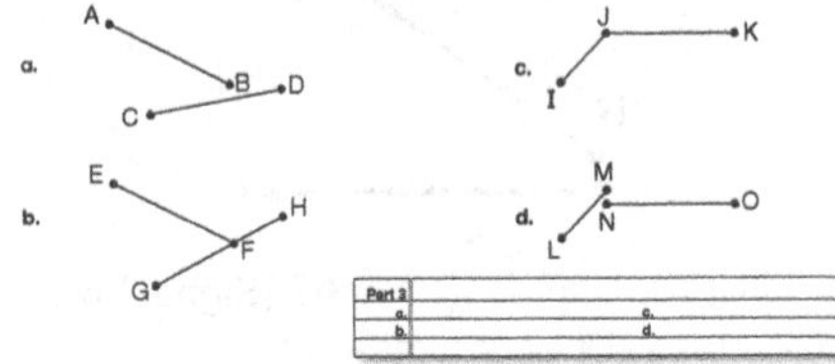

Each problem shows two line segments with a point at each end. You're going to tell me if the line segments in each problem intersect.

- Problem A. Do the line segments intersect? (Signal.) *No.*

(Display:) [120:5J]

Part 3			
a.	not intersect	c.	
b.		d.	

Later you'll write the words **not intersect** to show that the line segments in problem A do not intersect.

- What words will you write for problem A? (Signal.) *Not intersect.*

b. Problem B. Do the line segments intersect? (Signal.) *Yes.*

- Tell me the letter of the point where the line segments intersect. Get ready. (Signal.) *F.*
- What do the line segments do at point F? (Signal.) *Intersect.*

(Add to show:) [120:5K]

Part 3			
a.	not intersect	c.	
b.	intersect, F	d.	

Later you'll write the word intersect and the letter F to show that the line segments intersect at point F in problem B.

- What will you write for problem B? (Signal.) *Intersect, F.*

c. Problem C. Do the line segments intersect? (Signal.) *Yes.*

- Tell me the letter of the point where the line segments intersect. Get ready. (Signal.) *J.*
- What do the line segments do at point J? (Signal.) *Intersect.*
- Tell me what you will write for problem C. Get ready. (Signal.) *Intersect, J.*

d. Problem D. Do the line segments intersect? (Signal.) *No.*

- Tell me what you'll write for problem D. Get ready. (Signal.) *Not intersect.*

(Repeat steps c and d that were not firm.)

e. Write part 3 and the letters A through D below. (Observe students and give feedback.)

- For problems with line segments that intersect, write **intersect** and the letter for the point. For problems with line segments that do not intersect, write the words **not intersect.** (Observe students and give feedback.)

f. Check your work. You'll read what you wrote for each problem.

- Problem A. Get ready. (Signal.) *Not intersect.*
- Problem B. Get ready. (Signal.) *Intersect, F.*
- Problem C. Get ready. (Signal.) *Intersect, J.*
- Problem D. Get ready. (Signal.) *Not intersect.*

(Add to show:) [120:5L]

Part 3			
a.	not intersect	c.	intersect, J
b.	intersect, F	d.	not intersect

Here's what you should have for part 3. Remember what line segments are and what it means when they intersect.

Lesson 120, Exercise 5

Column Subtraction (Lessons 1–80)

Overview of problem types:

- Work with column-subtraction problems begins on Lesson 1 and continues intermittently through Lesson 80.
- In Lessons 1–30, students review borrowing and writing a new place value for the top number.
- The first type of column subtraction requires borrowing from the tens column:

$$\begin{array}{r} 81 \\ -23 \\ \hline \end{array}$$

The new place value for 81 is 70 plus 11. The new problem for the ones column is 11 minus 3.

The new problem for the tens column is 7 minus 2.

Here's the rewritten problem and the answer:

$$\begin{array}{r} {\scriptstyle 7} \\ \not{8}\,{}^{1}1 \\ -23 \\ \hline 58 \end{array}$$

- The next type requires borrowing from the tens **or** hundreds.
- The last reviewed type requires borrowing from both tens and hundreds.
- The problems include dollar-and-cents problems that have 3 to 5 digits.

Starting on Lesson 68, students learn to work new problem types. The first type requires borrowing from zero.

$$\begin{array}{r} 603 \\ -245 \\ \hline \end{array}$$

Students rewrite the top number to show 603 as 500 + 90 + 13.

$$\begin{array}{r} {\scriptstyle 5\,9} \\ \not{6}\,\not{0}\,{}^{1}3 \\ -245 \\ \hline \end{array}$$

The rewritten top value is the same as the original value; however, the new place value permits students to work the problem for the ones (13 – 5), the problem for the tens (9 – 4), and the problem for the hundreds (5 – 2).

Here's the rewritten problem and the answer:

$$\begin{array}{r} {\scriptstyle 5\,9} \\ \not{6}\,\not{0}\,{}^{1}3 \\ -245 \\ \hline 358 \end{array}$$

On Lesson 1, students review how to rewrite the top number.

Here's the first part of the exercise:

a. (Display:) [1:4A]

34 78
53 32

- (Point to **34.**) Read these numbers. Get ready. (Touch each number.) *34, 78, 53, 32.*

b. You're going to do a lot of work with place value.

- (Point to **34.**) My turn: What's the place value for 34? (Touch 3.) 30 plus (touch 4) 4.
- Your turn: Say the place value for 34. (Signal.) *30 + 4.*
- (Point to **78.**) Say the place value for 78. (Signal.) *70 + 8.*
- (Point to **53.**) Say the place value for 53. (Signal.) *50 + 3.*
- (Point to **32.**) Say the place value for 32. (Signal.) *30 + 2.*

(Repeat until firm.)

c. (Point to **34.**) 34 has a tens digit and a ones digit.
What's the tens digit of 34? (Signal.) *3.*

- What's the ones digit of 34? (Signal.) *4.*

d. (Point to **78.**) What's the tens digit of 78? (Signal.) *7.*

- What's the ones digit of 78? (Signal.) *8.*

e. (Point to **53.**) What's the tens digit of 53? (Signal.) *5.*

- What's the ones digit of 53? (Signal.) *3.*

f. (Point to **32.**) What's the tens digit of 32? (Signal.) *3.*

- What's the ones digit of 32? (Signal.) 2.

g. For some subtraction problems, you have to write a **new** place value for the top number. You do that by taking one ten from the tens digit and adding that ten to the ones digit.

- (Point to **34.**) What's the tens digit? (Signal.) *3.*
  I take one from 3. That leaves 2.
  (Add to show:) [1:4B]

  ```
   2
   ꞩ4    78
     53     32
  ```

- I put the ten I took away in front of the 4. Where do I put the one ten I took away? (Signal.) *In front of the 4.*
  (Add to show:) [1:4C]

  ```
   2
   ꞩ¹4    78
     53     32
  ```

  So I now have 14 ones.
- (Point to the **3** in 34.) How many tens did I start with? (Signal.) *3.*
- How many tens do I have now? (Signal.) *2.*
  My turn to say the new place value for 34. (Touch 2.) 20 plus (touch 14) 14.
- Say the new place value for 34. (Signal.) *20 + 14.*
- Say the place value we started with. (Signal.) *30 + 4.*
  (Repeat until firm.)

h. (Point to **78.**) Say the place value for 78. (Signal.) *70 + 8.*
- I want to write the new place value for 78. So I take one from the tens digit. What's one less than 7? (Signal.) *6.*
  (Add to show:) [1:4D]

  ```
   2     6
   ꞩ¹4    ꝥ8
     53     32
  ```

- Where do I write the one ten I took away? (Signal.) *In front of the 8.*
  (Add to show:) [1:4E]

  ```
   2     6
   ꞩ¹4    ꝥ¹8
     53     32
  ```

- What's 10 plus 8? (Signal.) *18.*
- I'll say the new place value for 78: (Touch 6.) 60 plus (touch 18) 18.
- Say the new place value for 78. (Touch.) *60 + 18.*
- Say the place value we started with. (Touch.) *70 + 8.*
  (Repeat step h until firm.)

from Lesson 1, Exercise 4

On Lesson 6, students work problems that require rewriting the top number to show the new place value.

Students work 2-digit and 3-digit problems through Lesson 7. On Lesson 8, students work a mix of problems that require rewriting either the first two digits of the top number or the last two digits.

Here's the first part of the exercise from Lesson 9:

(Teacher reference:)

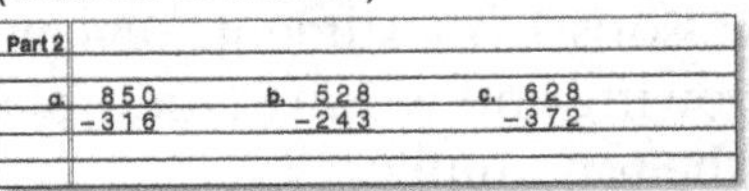

Part 2

a. 850	b. 528	c. 628
−316	−243	−372

For some of these problems, you're going to write the new place value for the first two digits of the top number. For other problems, you'll write the new place value for the last two digits of the top number.
- Copy part 2 on your lined paper. Put your pencil down when you're finished.
  (Observe students and give feedback.)

b. Touch problem A on your lined paper. ✔
- Read the problem. (Signal.) *850 − 316.*
- Say the problem for the ones. (Signal.) *Zero − 6.*
- Can you work that problem? (Signal.) *No.*
- Say the digits you rewrite. (Signal.) *5 and 0.*
- Say the new place value for 50. Get ready. (Signal.) *40 + 10.*
  (Repeat until firm.)

c. Work problem A.
  (Observe students and give feedback.)
- Problem A. Read the problem you started with and the answer. (Signal.) *850 − 316 = 534.*
  (Display:) [9:8A]

  ```
  Part 2
          4
   a.   8 5¹0
      − 3 1 6
        5 3 4
  ```

  Here's what you should have.

d. Read problem B. (Signal.) *528 − 243.*
- Say the problem for the ones. (Signal.) *8 − 3.*
- Can you work that problem? (Signal.) *Yes.*
- Say the problem for the tens. (Signal.) *2 − 4.*
- Can you work that problem? (Signal.) *No.*
- Say the digits you rewrite. (Signal.) *5 and 2.*
- Say the new place value for 52. (Signal.) *40 + 12.*
  (Repeat until firm.)

e. Work problem B.
  (Observe students and give feedback.)
- Problem B. Read the problem you started with and the answer. (Signal.) *528 − 243 = 285.*
  (Display:) [9:8B]

  ```
          4
   b.   5¹2 8
      − 2 4 3
        2 8 5
  ```

  Here's what you should have.

from Lesson 9, Exercise 8

Teaching Note: As part of working each problem, students say the digits they will rewrite. They are to say the digits in the order of hundreds-tens, or tens-ones. Sometimes students will say the digits in reverse order because they work the problem in reverse order (working the ones column first, then the tens . . .).

If students make this mistake, tell them the answer and repeat the step. (You may want to tell them the basis for the correct answer. "The number is 850, so the digits you'll rewrite are 5 and zero.")

The review of double borrowing starts on Lesson 25.

Students determine that they can't work the problem in the ones column, so they rewrite the last two digits of the top number. Students then determine that they can't work the problem in the tens column, so they rewrite the hundreds and new tens digit of the top number.

Here's part of the exercise from Lesson 25:

a. Find part 2 in your workbook. ✔
(Teacher reference:) **R Part N**

a. 270 − 182 b. 613 − 54 c. 582 − 394 d. 461 − 94

You're going to work these problems.
- Read problem A. Get ready. (Signal.) *270 − 182.*
- Read the problem for the ones column. Get ready. (Signal.) *Zero − 2.*
- Can you work that problem? (Signal.) *No.*
- Write the new place value and work the problem in the ones column. Put your pencil down when you've done that much.
(Observe students and give feedback.)

b. Problem A. Read the new problem and the answer for the ones column. Get ready. (Signal.) *10 − 2 = 8.*
- Read the new problem in the tens column. Get ready. (Signal.) *6 − 8.*
- Can you work that problem? (Signal.) *No.*
(Display:) [25:2D]

a. 2 ~~7~~[6] ~~0~~[1]0 − 182 = 8

Here's what you should have.
- You'll write the new place value for 2 digits of the top number. What digits? (Signal.) *2 and 6.*
- Say the new place value for 26. Get ready. (Signal.) *10 + 16.*
- Write the new place value for 26 and work the rest of the problem. Put your pencil down when you've worked problem A.
(Observe students and give feedback.)

c. Check your work for problem A.
- Read the new problem and the answer in the tens column. Get ready. (Signal.) *16 − 8 = 8.*
- Read the new problem and the answer in the hundreds column. Get ready. (Signal.) *1 − 1 = zero.*
(Add to show:) [25:2E]

a. ~~2~~[1] ~~7~~[16] ~~0~~[1]0 − 182 = 88

Here's what you should have.
- Everybody, what's 270 − 182? (Signal.) *88.*

from Lesson 25, Exercise 2

On Lesson 28, students work a mixed set of problems, some of which require double borrowing and others that require only single borrowing.

Here's the set of problems from Textbook Lesson 28:

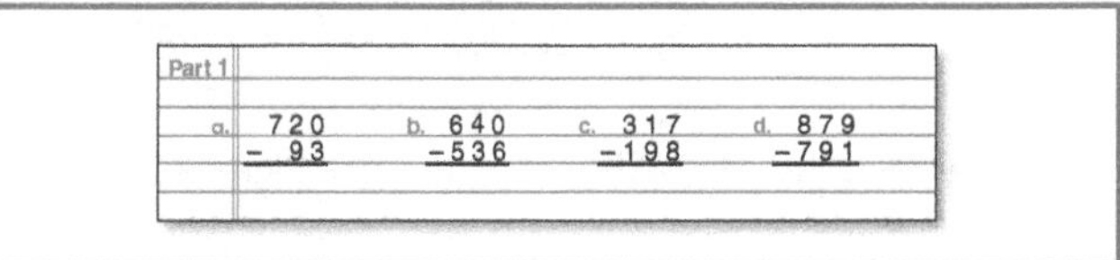
Part 1

a. 720 − 93 b. 640 − 536 c. 317 − 198 d. 879 − 791

Borrowing from Zero

On Lesson 66, students begin work on a preskill that is needed when they borrow from zero. The top number in these problems has three digits. Students rewrite the entire top number to show the new place value.

For 708, the new place value is 600 + 90 + 18.

For 500, the new place value is 400 + 90 + 10.

For 109, the new place value is 90 + 19.

On Lesson 68, students work problems that have a tens digit of zero.

It's not possible to borrow anything from zero so students have to rewrite the whole top number.

Here's the first part of the exercise from Lesson 68:

a. (Display:) [68:7A]

```
 405
-356
```

- Read the problem. (Signal.) *405 – 356.*
- Read the problem for the ones. (Signal.) *5 – 6.*
- Can you work that problem? (Signal.) *No.*
- What's the tens digit of the top number? (Signal.) *Zero.*
- Can you borrow anything from the zero? (Signal.) *No.* So you have to rewrite the top whole number to show the new place-value addition.

b. Say the new place-value addition for 405. Get ready. (Signal.) *300 + 90 + 15.*
(Add to show:) [68:7B]

```
 3 9
 4 0 15
-3 5 6
```

- Say the new problem for the ones. (Signal.) *15 – 6.*
- What's the answer? (Signal.) *9.*
(Add to show:) [68:7C]

```
 3 9
 4 0 15
-3 5 6
     9
```

- Say the new problem for the tens. (Signal.) *9 – 5.*
- What's the answer? (Signal.) *4.*
(Add to show:) [68:7D]

```
 3 9
 4 0 15
-3 5 6
   4 9
```

- Say the new problem for the hundreds. (Signal.) *3 – 3.*
- What's the answer? (Signal.) *Zero.*
- Read the whole problem and the answer. (Signal.) *405 – 356 = 49.*

TEXTBOOK PRACTICE

a. Find part 3 in your textbook. ✔
(Teacher reference:)

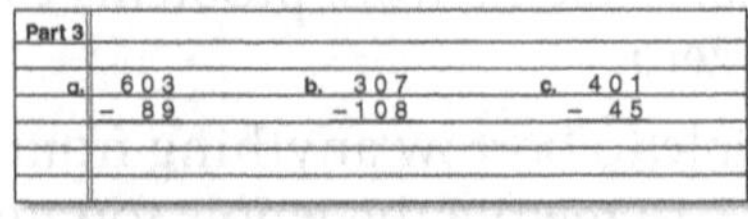

Part 3
a. 603 − 89   b. 307 − 108   c. 401 − 45

- Read problem A. (Signal.) *603 – 89.*
- Read the problem for the ones. (Signal.) *3 – 9.*
- Can you work that problem? (Signal.) *No.* So you have to write the new place value for the whole top number.
- What's the top number? (Signal.) *603.*
- Say the new place-value addition for 603. (Signal.) *500 + 90 + 13.*

b. Copy part 3 on your lined paper. Then work problem A. Put your pencil down when you've finished problem A.
(Observe students and give feedback.)

c. Check your work.
- Problem A. Read the whole problem and the answer. (Signal.) *603 – 89 = 514.*
(Display:) [68:7E]

```
Part 3
      5 9
a.    6 0 13
   -    8 9
      5 1 4
```

Here's what you should have for problem A.

from Lesson 68, Exercise 7

On Lesson 70, students begin work with mixed sets. Some problems require borrowing from zero; some don't.

Here are the problems from Lesson 70:

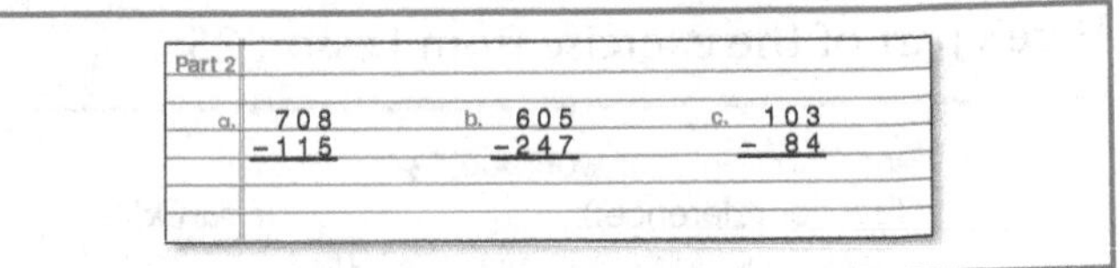

Part 2
a. 708 − 115   b. 605 − 247   c. 103 − 84

DOLLAR-AND-CENTS PROBLEMS

On Lesson 72, students work dollar-and-cents problems, some of which require borrowing from zero.

On Lesson 74, students work problems that require rewriting more than three digits of the top number.

If you bring students to a high level of mastery on the early skills in this track, they will not only learn the rest of the material adequately, they will have an understanding of why they have to rewrite parts of the top number or the whole top number under specific conditions.

a. Find part 4 in your textbook. ✔
(Teacher reference:)

Part 4		
a. $20.00 − 8.49	b. $100.09 − 40.75	c. $200.05 − 73.39

For these subtraction problems, you'll have to write the new place value for at least four digits of the top number.

- Copy part 4. Put your pencil down when you're finished.
(Observe students and give feedback.)
(If students have been 100% for the last two lessons, skip to step c.)

b. Read problem A on your lined paper. Get ready. (Signal.) *$20 – $8.49.*

- Read the problem in the one-cent column. (Signal.) *Zero – 9.*
- Can you work that problem? (Signal.) *No.*
The whole top number needs to be rewritten.
(Display:) [74:7A]

```
Part 4
      1 9 9 1
   a. $ 2̶0̶.̶0̶0
      –   8.49
```

Cross out the digits in the other columns and write the new place value for the top number.
(Observe students and give feedback.)

- Read the new problem in the one-cent column. Get ready. (Signal.) *10 – 9.*
- Can you work that problem? (Signal.) *Yes.*
- Read the new problem in the ten-cent column. (Signal.) *9 – 4.*
- Can you work that problem? (Signal.) *Yes.*
- Read the new problem in the one-dollar column. (Signal.) *9 – 8.*
- Can you work that problem? (Signal.) *Yes.*

c. Work problem A. Put your pencil down when you've finished problem A.
(Observe students and give feedback.)

- Problem A is $20 minus $8.49. Everybody, what's the answer? (Signal.) *$11.51.*
(Add to show:) [74:7B]

```
Part 4
      1 9 9 1
   a. $ 2̶0̶.̶0̶0
      –   8.49
      $ 1 1.5 1
```

Here's what you should have for problem A.
(If students have been 100% for the last two lessons, skip to step e.)

d. Read problem B. Get ready. (Signal.) *$100.09 – $40.75.*

- Read the problem in the one-cent column. (Signal.) *9 – 5.*
- Can you work that problem? (Signal.) *Yes.*
- What's the answer? (Signal.) *4.*

(Display:) [74:7C]

```
   b. $100.09
      – 40.75
            4
```

- Read the problem in the ten-cent column. (Signal.) *Zero – 7.*
- Can you work that problem? (Signal.) *No.*
- Tell me the digits that you'll cross out. Get ready. (Signal.) *1, 0, 0.*
(Add to show:) [74:7D]

```
   b. $ 1̶0̶0̶.09
      – 40.75
            4
```

- What do I write above 1? (Call on a student. Idea:) *Nothing.*
- What do I write above the crossed-out zeros? (Signal.) *9.*
(Add to show:) [74:7E]

```
         9 9
   b. $ 1̶0̶0̶.09
      – 40.75
            4
```

- What do I write in front of the next zero? (Signal.) *1.*
(Add to show:) [74:7F]

```
         9 9 1
   b. $ 1̶0̶0̶.0 9
      – 40.75
            4
```

e. Work problem B. Put your pencil down when you've finished problem B.
(Observe students and give feedback.)

- Problem B is $100.09 minus $40.75. Everybody, what's the answer? (Signal.) *$59.34.*
(Add to show:) [74:7G]

```
         9 9 1
   b. $ 1̶0̶0̶.0 9
      – 40.75
      $  59.34
```

Here's what you should have for problem B.
(If students have been 100% for the last two lessons, skip to step h.)

f. Read problem C. Get ready. (Signal.) *$200.05 – $73.39.*

- Read the problem in the one-cent column. (Signal.) *5 – 9.*
- Can you work that problem? (Signal.) *No.*

from Lesson 74, Exercise 7

Column Multiplication (Lessons 2–130)

In *CMC Level E,* students do a considerable amount of work with column-multiplication problems. The review of multiplying by ten starts on Lesson 2. Students do a limited amount of column multiplication during the first 20 lessons. These lessons emphasize multiplication facts.

Starting on Lesson 21, students review multiplying 2- and 3-digit numbers by one-digit numbers:

$$\begin{array}{r}36\\ \times\ 9\\ \hline\end{array}\qquad\begin{array}{r}324\\ \times\ \ 5\\ \hline\end{array}$$

Starting on Lesson 43, students work on column problems that multiply 2-digit numbers by tens numbers:

$$\begin{array}{r}72\\ \times50\\ \hline\end{array}\qquad\begin{array}{r}48\\ \times20\\ \hline\end{array}$$

To work these problems, students must apply a convention of writing a zero in the ones column.

On Lesson 48, the range is expanded to 3-digit numbers times tens numbers:

$$\begin{array}{r}765\\ \times\ 40\\ \hline\end{array}$$

On Lesson 49, students work problems that multiply 2-digit numbers by 2-digit numbers:

$$\begin{array}{r}36\\ \times91\\ \hline\end{array}\qquad\begin{array}{r}45\\ \times27\\ \hline\end{array}$$

On Lesson 51, students work problems that multiply by hundreds numbers:

$$\begin{array}{r}724\\ \times500\\ \hline\end{array}\qquad\begin{array}{r}14\\ \times300\\ \hline\end{array}$$

As students work on new types, they work problem sets that include the various types they have already reviewed. For instance, on Lesson 55, students work problem sets that multiply multi-digit numbers by multi-digit numbers:

$$\begin{array}{r}13\\ \times502\\ \hline\end{array}\qquad\begin{array}{r}623\\ \times326\\ \hline\end{array}$$

On Lesson 61, students learn a computational shortcut for repeated digits in the number that is multiplied.

For 328 × 676, students are shown that the products for the 6s are the same, except that the product for 600 ends in two zeros.

Problems that involve column multiplication appear in ratios, ratio word problems, area, and area word problems.

Reviewing Multiplying 2-Digit by 1-Digit Numbers

Here's the first review of problems that multiply 2-digit numbers by 1-digit numbers (Lesson 21):

a. (Display:) [21:7A]

$$\begin{array}{r}43\\ \times\ 5\\ \hline\end{array}\qquad\begin{array}{r}25\\ \times\ 9\\ \hline\end{array}\qquad\begin{array}{r}18\\ \times\ 2\\ \hline\end{array}$$

These are multiplication problems.

- (Point to **43.**) This problem is 43 times 5. Read this problem. (Signal.) *43 × 5.*
- (Point to **25.**) Read this problem. (Signal.) *25 × 9.*
- (Point to **18.**) Read this problem. (Signal.) *18 × 2.*

b. (Point to **43.**) Listen: When you work these problems, you first multiply for the ones. Then you multiply for the tens.
My turn to read the problem for the ones: 3 × 5.

- Read the problem for the ones. (Signal.) *3 × 5.*
- Start with 4 and read the problem for the tens. (Signal.) *4 × 5.*

c. (Point to **25.**) Read the problem for the ones. (Signal.) *5 × 9.*

- Read the problem for the tens. (Signal.) *2 × 9.*

d. (Point to **18.**) Read the problem for the ones. (Signal.) *8 × 2.*

- Read the problem for the tens. (Signal.) *1 × 2.*

(Repeat steps b through d until firm.)

e. (Point to **43.**) This time, you'll say the problems and the answers.

- Read the problem for the ones. (Signal.) *3 × 5.*
- What's the answer? (Signal.) *15.*
- Is 15 a two-digit number? (Signal.) *Yes.*
- Do I write the tens digit of 15 in the tens column or in the ones column? (Signal.) *In the tens column.*
(Add to show:) [21:7B]

$$\begin{array}{r}{}^{1}\\43\\ \times\ 5\\ \hline\end{array}\qquad\begin{array}{r}25\\ \times\ 9\\ \hline\end{array}\qquad\begin{array}{r}18\\ \times\ 2\\ \hline\end{array}$$

- Where do I write the ones digit of 15? (Signal.) *In the ones column.*
(Add to show:) [21:7C]

$$\begin{array}{r}{}^{1}\\43\\ \times\ 5\\ \hline 5\end{array}\qquad\begin{array}{r}25\\ \times\ 9\\ \hline\end{array}\qquad\begin{array}{r}18\\ \times\ 2\\ \hline\end{array}$$

f. (Point to **4.**) Listen: First we multiply 4 times 5. (Point to **1.**) Then we add 1.
- Say the multiplication problem for the tens. (Signal.) *4 × 5.*
- What's 4 × 5? (Signal.) *20.*
- Now we add 1. What's 20 + 1? (Signal.) *21.*
I write the digits for 21 in the answer.
(Add to show:) [21:7D]

```
  1
 43      25      18
× 5     × 9     × 2
215
```

- (Point to **43.**) Read the problem we started with and the answer. (Signal.) *43 × 5 = 215.*

g. (Point to **25.**) Read this problem. (Signal.) *25 × 9.*
- Say the multiplication problem for the ones. (Signal.) *5 × 9.*
- What's the answer? (Signal.) *45.*
- Do I write the tens digit of 45 in the tens column or in the ones column? (Signal.) *In the tens column.*
- Where do I write the ones digit of 45? (Signal.) *In the ones column.*
(Add to show:) [21:7E]

```
  1       4
 43      25      18
× 5     × 9     × 2
215       5
```

h. Say the multiplication problem for the tens. (Signal.) *2 × 9.*
- What's 2 × 9? (Signal.) *18.*
- (Point to **4.**) What do we add to 18? (Signal.) *4.*
- What's 18 + 4? (Signal.) *22.*
(Add to show:) [21:7F]

```
  1       4
 43      25      18
× 5     × 9     × 2
215     225
```

- (Point to **25.**) Read the problem we started with and the answer. (Signal.) *25 × 9 = 225.*

i. (Point to **18.**) Read this problem. (Signal.) *18 × 2.*
- Say the multiplication problem for the ones. (Signal.) *8 × 2.*
- What's the answer? (Signal.) *16.*
- Do I write the tens digit of 16 in the tens column or in the ones column? (Signal.) *In the tens column.*
- Where do I write the ones digit of 16? (Signal.) *In the ones column.*
(Add to show:) [21:7G]

```
  1       4       1
 43      25      18
× 5     × 9     × 2
215     225       6
```

j. Say the multiplication problem for the tens. (Signal.) *1 × 2.*
- What's 1 × 2? (Signal.) *2.*
- (Point to **1.**) What do we add to 2? (Signal.) *1.*
- What's 1 + 2? (Signal.) *3.*
(Add to show:) [21:7H]

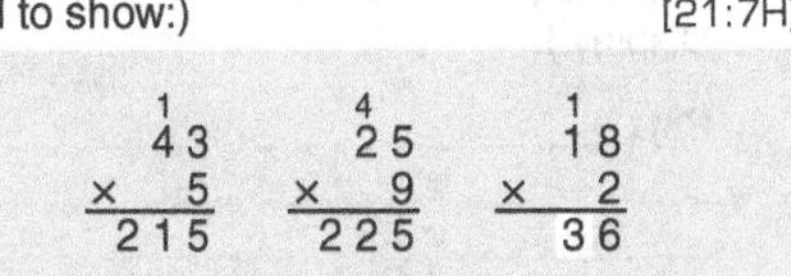

- (Point to **18.**) Read the problem we started with and the answer. (Signal.) *18 × 2 = 36.*

Lesson 21, Exercise 7

Multiplying by Tens

On Lesson 42, students work a new kind of problem: multiplying by a tens number.

Students learn to write zero in the ones column of the answer and then multiply by the tens number. This procedure establishes what students will do later when they work problems like 126 × 43. They work two problems: 126 × 3 and 126 × 40. When they work the second problem, they are multiplying by a tens number. So they first write zero in the ones column and then do the multiplication.

Here's the exercise from Lesson 46:

a. Find part 3 in your textbook. ✔
(Teacher reference:)

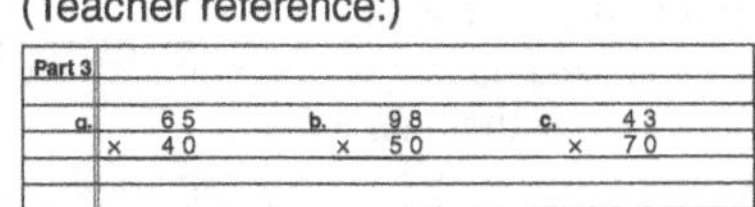
Part 3
a. 65 × 40 b. 98 × 50 c. 43 × 70

These problems multiply by a tens number.
- Copy part 3 on your lined paper.
(Observe students and give feedback.)

b. Touch and read problem A on your lined paper. (Signal.) *65 × 40.*
- What's the first thing you write? (Signal.) *Zero.*
- Where do you write zero? (Signal.) *In the ones column.*
- Write it. ✔

c. Start with 65 and say the problem for the first digit of 40. (Signal.) *65 × 4.*
- Say the problem for the ones digit of 65. (Signal.) *5 × 4.*
- Say the problem for the tens digit of 65. (Signal.) *6 × 4.*
(Repeat until firm.)

d. Again. Start with 65 and say the problem for the first digit of 40. **(Signal.)** *65 × 4.*

- Say the problem for the ones digit of 65. **(Signal.)** *5 × 4.*
- What's the answer? **(Signal.)** *20.*
- Write the digits of 20. ✔

(Display:) [46:6A]

Part 3

```
        2
a.     65
    ×  40
       00
```

Here's what you should have so far.

e. Say the problem for the tens digit of 65. **(Signal.)** *6 × 4.*

- What's the answer? **(Signal.)** *24.*
- What's 24 + 2? **(Signal.)** *26.*

f. Work the rest of problem A.
(Observe students and give feedback.)

g. Check your work.

- Read problem A and the answer. **(Signal.)** *65 × 40 = 2 thousand 600.*

(Add to show:) [46:6B]

Part 3

```
        2
a.     65
    ×  40
     2600
```

Here's what you should have.

h. Touch and read problem B. **(Signal.)** *98 × 50.*

- What's the first thing you write? **(Signal.)** *Zero.*
- Where do you write zero? **(Signal.)** *In the ones column.*
- Write it. ✔

(If students have been 100% for the last two lessons, skip to step k.)

i. Start with 98 and say the problem for the first digit of 50. **(Signal.)** *98 × 5.*

- Say the problem for the ones digit of 98. **(Signal.)** *8 × 5.*
- What's the answer? **(Signal.)** *40.*
- Write the digits of 40.

(Observe students and give feedback.)

(Display:) [46:6C]

```
        4
b.     98
    ×  50
       00
```

Here's what you should have so far.

j. Say the problem for the tens digit of 98. **(Signal.)** *9 × 5.*

- What's the answer? **(Signal.)** *45.*
- What's 45 + 4? **(Signal.)** *49.*

k. Work the rest of problem B.
(Observe students and give feedback.)

- Read problem B and the answer. **(Signal.)** *98 × 50 = 4 thousand 900.*

(Add to show:) [46:6D]

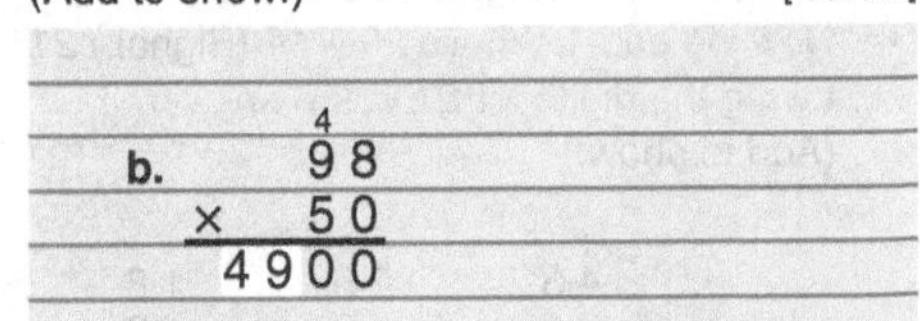

Here's what you should have.

l. Touch and read problem C. **(Signal.)** *43 × 70.*

- What's the first thing you write? **(Signal.)** *Zero.*
- Where do you write zero? **(Signal.)** *In the ones column.*
- Write it. ✔

(If students have been perfect on working 2-digit multiplication problems for at least two lessons, skip to step o.)

m. Start with 43 and say the problem for the first digit of 70. **(Signal.)** *43 × 7.*

- Say the problem for the ones digit of 43. **(Signal.)** *3 × 7.*
- What's the answer? **(Signal.)** *21.*
- Write the digits of 21. ✔

n. Say the problem for the tens digit of 43. **(Signal.)** *4 × 7.*

- What's the answer? **(Signal.)** *28.*
- What's 28 + 2? **(Signal.)** *30.*

o. Work the rest of problem C.
(Observe students and give feedback.)

- Read problem C and the answer. **(Signal.)** *43 × 70 = 3 thousand 10.*

(Display:) [46:6E]

```
        2
c.     43
    ×  70
     3010
```

Here's what you should have.

Lesson 46, Exercise 6

Teaching Note: Praise students who align digits in the tens column and ones column. If students don't keep the digits aligned, serious problems may result, particularly when they multiply by 2-digit numbers and add the products. They will tend to add numbers that aren't in the same column.

On Lesson 48, students work problems that multiply 3-digit numbers. Here's part of the exercise from Lesson 48:

a. Find part 5 in your textbook. ✔
(Teacher reference:)

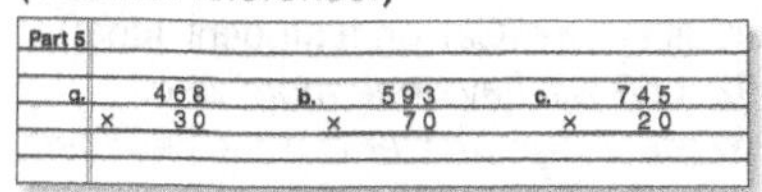

Part 5

a. 468 × 30	b. 593 × 70	c. 745 × 20

These problems multiply 3-digit numbers by tens numbers.

- Read problem A. (Signal.) *468 × 30.*
- What do you write in the answer for the ones column? (Signal.) *Zero.*
- Start with 468 and say the problem for the first digit of 30. (Signal.) *468 × 3.*
- Say the problem for the ones digit of 468. (Signal.) *8 × 3.*
- Say the problem for the tens digit of 468. (Signal.) *6 × 3.*
- Say the problem for the hundreds digit of 468. (Signal.) *4 × 3.*

(Repeat until firm.)
(If students have performed perfectly on working multidigit multiplication problems, skip to step d.)

b. Read problem B. (Signal.) *593 × 70.*
- What do you write in the answer for the ones column? (Signal.) *Zero.*
- Start with 593 and say the problem for the first digit of 70. (Signal.) *593 × 7.*
- Say the problem for the ones digit of 593. (Signal.) *3 × 7.*
- Say the problem for the tens digit of 593. (Signal.) *9 × 7.*
- Say the problem for the hundreds digit of 593. (Signal.) *5 × 7.*

(Repeat until firm.)

c. Read problem C. (Signal.) *745 × 20.*
- What do you write in the answer for the ones column? (Signal.) *Zero.*
- Start with 745 and say the problem for the first digit of 20. (Signal.) *745 × 2.*
- Say the problem for the ones digit of 745. (Signal.) *5 × 2.*
- Say the problem for the tens digit of 745. (Signal.) *4 × 2.*
- Say the problem for the hundreds digit of 745. (Signal.) *7 × 2.*

(Repeat until firm.)

from Lesson 48, Exercise 9

Teaching Note: If students are not fluent in responding to the series of tasks you present for each problem, they may become confused later when they work problems that require multiplying by more than one digit.

Multiplying 2-Digit by 2-Digit Numbers

On Lesson 49, students begin work on problems that multiply by 2-digit numbers.

Here's the first part of the Workbook activity for Lesson 49. Notice that the answer for the ones digit is already written. Students work the problem for the tens digit.

a.	b.	c.
51 × 42 = 102	38 × 24 = 152	63 × 52 = 126

Starting on Lesson 52, students work complete problems that multiply 2-digit numbers by 2-digit numbers.

Multiplying 2-Digit by 3-Digit Numbers

On Lesson 55, students multiply 2-digit numbers by 3-digit numbers. Here's the work for one of the problems:

```
a.       1
         62
   ×    546
        372
       2480
   + 31,000
     33,852
```

Teaching Note: The procedure for the hundreds digit is identical to that for the tens digit except that students write two zeros before writing the first hundreds answer.

When students work problems that multiply by 2-digit or 3-digit numbers, they cross out any numbers that were carried earlier. Students are more likely to make mistakes in problems that multiply by 3-digit numbers, because there may be 3 carried values for one of the digits. If two are not crossed out, students are likely to make mistakes.

Starting on Lesson 58, students work problem sets that multiply by 2-digit and 3-digit numbers.

Here's the set from Lesson 58:

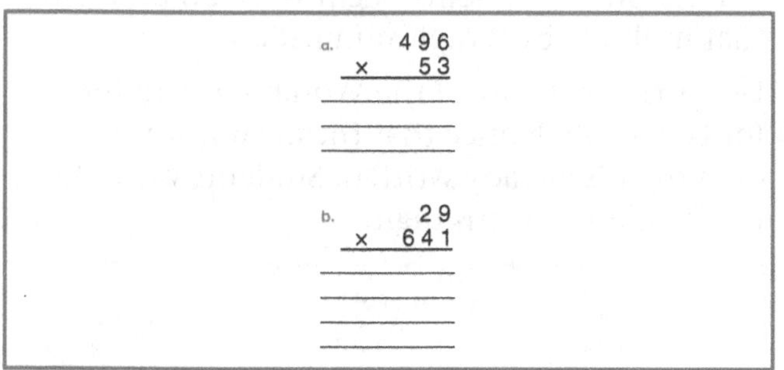

On Lesson 61, students learn a shortcut for problems that multiply by digits that are the same. For these problems, the digits in the answer are the same, except for the number of zeros, so students multiply only once for these problems.

Here's the first part of the exercise on Lesson 61:

a. Find part 5. ✔
(Teacher reference:)

a. 687 × 551 = 34,350 + ...   b. 492 × 366 = ... +147,600

These problems multiply 3-digit numbers. If the digits you multiply by are the same, the digits in the answer are the same. The only difference is the number of zeros you write. Part of the answers are already shown.
(If students have been 100% for the last two lessons, skip to step d.)

b. Read problem A. (Signal.) *687 × 551.*
- Say the 3-digit problem for the ones. (Signal.) *687 × 1.*
- The answer isn't shown, but you know it. What's 687 × 1? (Signal.) *687.*
- Say the 3-digit problem for the tens. Get ready. (Signal.) *687 × 50.*
- What's the answer? (Signal.) *34,350.*
- Say the 3-digit problem for the hundreds. (Signal.) *687 × 500.*
- Who can tell me the answer to 687 × 500? (Call on a student.) *343,500.*
The answer to the 3-digit problem for the hundreds is the same as the answer to the tens with a zero on the end.
(Repeat step b until firm.)

c. Read problem B. (Signal.) *492 × 366.*
- Say the 3-digit problem for the ones. (Signal.) *492 × 6.*
- Say the 3-digit problem for the tens. (Signal.) *492 × 60.*
- Who can tell me how 492 × 6 and 492 × 60 are different? (Call on a student. Idea:) *492 × 60 will have one more zero.*
- Say the 3-digit problem for the hundreds. Get ready. (Signal.) *492 × 300.*
- What's the answer? (Signal.) *147,600.*
(Repeat step c until firm.)

from Lesson 61, Exercise 6

Students work column-multiplication problems as part of their Independent Work.

Division (Lessons 7–75)

This track begins on Lesson 7 and continues intermittently through Lesson 75, after which it is used in ratios, word problems, area problems, and problems that compare fractions.

On Lesson 7, students write division problems from dictation.

For example: "Write the problem, 45 divided by nine."

The only sign for division used early in the program is $\overline{)\ \ }$. (This sign shows the order the numbers appear in the problem.)

The system is not completely consistent, however. 45 divided by 9 shows 9 first. When the problem is read, however, 45 is first.

One of the goals of having students write problems from dictation is to make sure they are very familiar with the conventions of writing division problems when they work them.

Students work with division facts through Lesson 19. On Lesson 20, students work division problems with answers greater than ten. The initial problem sets can be worked a digit at a time. For example: 408 divided by 2. Students work the problem 4 divided by 2, zero divided by 2, and 8 divided by 2.

$$2\overline{)408}\quad 204$$

On Lesson 22, students work problems that require pairing the first two digits of the number under the division sign.

Here's the first part of the exercise from Lesson 23:

(Teacher reference:) R Part M

a. 2)166   c. 3)276   e. 3)690

b. 2)806   d. 9)180   f. 2)148

These are division problems. For some of them you can work them a digit at a time. For the other problems, you'll work the problem for the first two digits.

- Read problem A. Get ready. (Signal.) *166 ÷ 2.*
- Say the problem for the first digit. (Signal.) *1 ÷ 2.*
- Can you work 1 ÷ 2? (Signal.) *No.*
- Say the problem for the first two digits. (Signal.) *16 ÷ 2.*
- Can you work 16 ÷ 2? (Signal.) *Yes.*

b. Read problem B. Get ready. (Signal.) *806 ÷ 2.*
- Say the problem for the first digit. (Signal.) *8 ÷ 2.*
- Can you work 8 ÷ 2? (Signal.) *Yes.*

c. Read problem C. Get ready. (Signal.) *276 ÷ 3.*
- Say the problem for the first digit. (Signal.) *2 ÷ 3.*
- Can you work 2 ÷ 3? (Signal.) *No.*
- Say the problem for the first two digits. (Signal.) *27 ÷ 3.*
- Can you work 27 ÷ 3? (Signal.) *Yes.*

d. Read problem D. Get ready. (Signal.) *180 ÷ 9.*
- Say the problem for the first digit. (Signal.) *1 ÷ 9.*
- Can you work 1 ÷ 9? (Signal.) *No.*
- Say the problem for the first two digits. (Signal.) *18 ÷ 9.*
- Can you work 18 ÷ 9? (Signal.) *Yes.*

e. Read problem E. Get ready. (Signal.) *690 ÷ 3.*
- Say the problem for the first digit. (Signal.) *6 ÷ 3.*
- Can you work 6 ÷ 3? (Signal.) *Yes.*

f. Read problem F. Get ready. (Signal.) *148 ÷ 2.*
- Say the problem for the first digit. (Signal.) *1 ÷ 2.*
- Can you work 1 ÷ 2? (Signal.) *No.*
- Say the problem for the first two digits. (Signal.) *14 ÷ 2.*
- Can you work 14 ÷ 2? (Signal.) *Yes.*

(Repeat problems that were not firm.)

g. Let's do those problems again. This time, for each problem, you'll tell me the 1-digit or 2-digit problem you'll work first. Then you'll tell me about the answer.
- Problem A is 166 ÷ 2. Say the 1-digit or 2-digit problem you'll work first. Get ready. (Signal.) *16 ÷ 2.*
- What's the answer? (Signal.) *8.*
- Do you write the answer above 1 or above 6? (Signal.) *Above 6.*

h. Problem B is 806 ÷ 2. Say the 1-digit or 2-digit problem you'll work first. Get ready. (Signal.) *8 ÷ 2.*
- What's the answer? (Signal.) *4.*
- Where do you write the answer? (Signal.) *Above 8.*

from Lesson 23, Exercise 3

Teaching Note: Make sure that students are firm on responding to both questions for problems that present two questions. The answer to the first question, (Can you work 1 divided by 2?) determines whether students work a problem that divides by a 2-digit number.

The next question confirms that students are able to work the 2-digit problem (Can you work 16 divided by 2?).

In steps G and H, students don't indicate whether they can work 1-digit problems. They simply indicate the problem they'll work first. If they make mistakes, ask if they can work the problem for the first digit. Then direct them to say the problem they'll work.

Remainders

Starting on Lesson 32, students identify the largest multiple they can divide by. For 17÷5 they identify that they can divide 15. The wording the problems use to refer to the largest multiple is "What's the largest part of 17 you can divide by 5?"

Here's the first part of the exercise from Lesson 33:

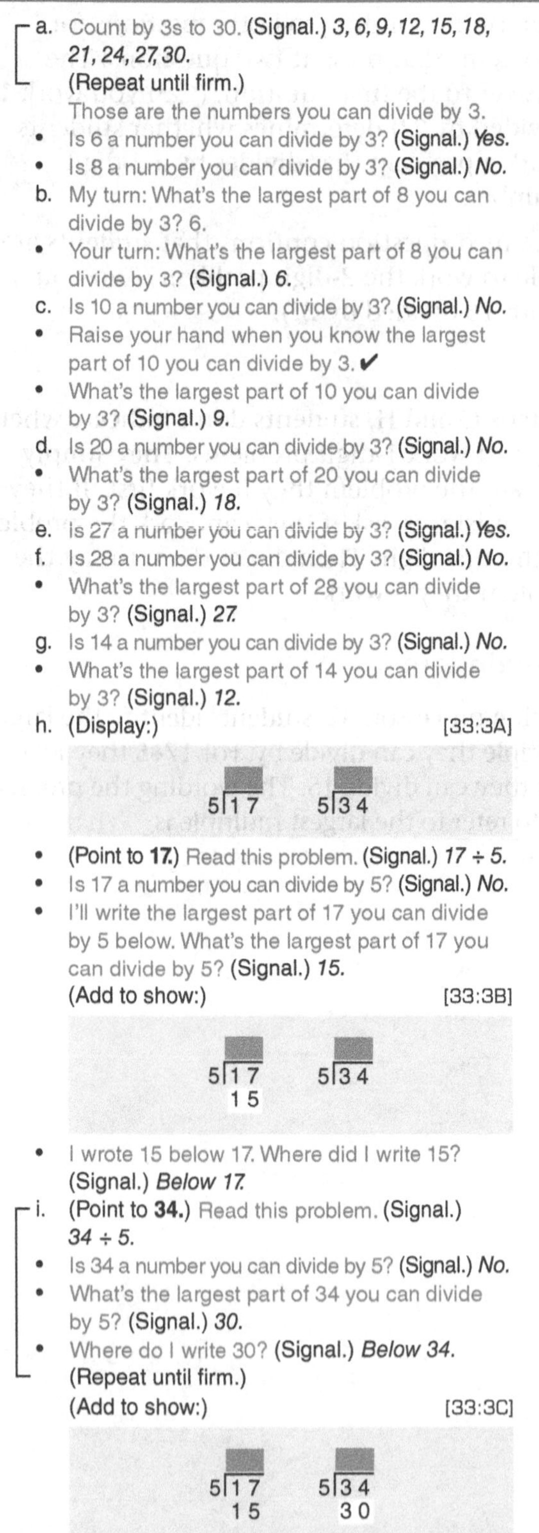

a. Count by 3s to 30. (Signal.) *3, 6, 9, 12, 15, 18, 21, 24, 27, 30.*
(Repeat until firm.)
Those are the numbers you can divide by 3.
- Is 6 a number you can divide by 3? (Signal.) *Yes.*
- Is 8 a number you can divide by 3? (Signal.) *No.*

b. My turn: What's the largest part of 8 you can divide by 3? 6.
- Your turn: What's the largest part of 8 you can divide by 3? (Signal.) *6.*

c. Is 10 a number you can divide by 3? (Signal.) *No.*
- Raise your hand when you know the largest part of 10 you can divide by 3. ✔
- What's the largest part of 10 you can divide by 3? (Signal.) *9.*

d. Is 20 a number you can divide by 3? (Signal.) *No.*
- What's the largest part of 20 you can divide by 3? (Signal.) *18.*

e. Is 27 a number you can divide by 3? (Signal.) *Yes.*

f. Is 28 a number you can divide by 3? (Signal.) *No.*
- What's the largest part of 28 you can divide by 3? (Signal.) *27.*

g. Is 14 a number you can divide by 3? (Signal.) *No.*
- What's the largest part of 14 you can divide by 3? (Signal.) *12.*

h. (Display:) [33:3A]

5⟌17 5⟌34

- (Point to 17.) Read this problem. (Signal.) *17 ÷ 5.*
- Is 17 a number you can divide by 5? (Signal.) *No.*
- I'll write the largest part of 17 you can divide by 5 below. What's the largest part of 17 you can divide by 5? (Signal.) *15.*

(Add to show:) [33:3B]

5⟌17 5⟌34
15

- I wrote 15 below 17. Where did I write 15? (Signal.) *Below 17.*

i. (Point to 34.) Read this problem. (Signal.) *34 ÷ 5.*
- Is 34 a number you can divide by 5? (Signal.) *No.*
- What's the largest part of 34 you can divide by 5? (Signal.) *30.*
- Where do I write 30? (Signal.) *Below 34.*

(Repeat until firm.)
(Add to show:) [33:3C]

5⟌17 5⟌34
15 30

from Lesson 33, Exercise 3

Teaching Note: In steps A–G, students apply the test "Is ___ a number you can divide by 3?" The answers provide information about the multiples of 3 that come into play in this track.

On Lesson 35, students either write the largest multiple of the number that is divided or write the largest part of that number that is a multiple. If the number is a multiple, students write the multiple in the answer.

$$3\overline{)15}\quad\text{answer: }5$$

If the largest number that is divided is not a multiple, students write the largest part multiple below.

$$\begin{array}{r} 3\overline{)17} \\ 15 \end{array}$$

Here's the set of problems from Lesson 36:

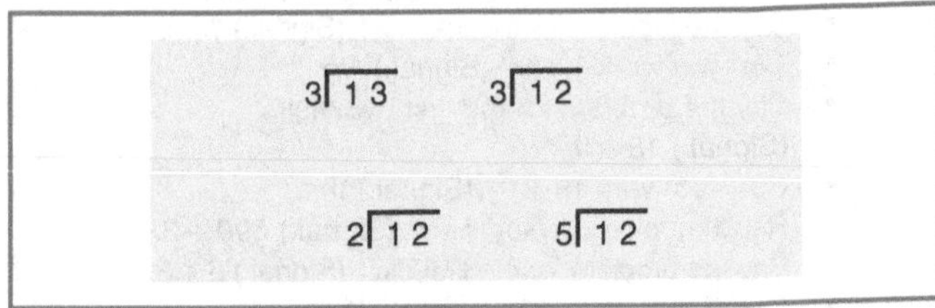

3⟌13 3⟌12
2⟌12 5⟌12

For 13 divided by 3, students write the largest part of 13 that they can divide by 3.

$$\begin{array}{r} 3\overline{)13} \\ 12 \end{array}$$

For 12 divided by 2, students write the answer, 6.

$$2\overline{)12}\quad\text{answer: }6$$

For 12 divided by 3, they write the answer of 4.

For 12 divided by 5, they write 10 below.

Starting on Lesson 42, students work complete problems that have a remainder.

Here's the exercise:

(Teacher reference:)

a. 9⟌37 b. 2⟌11 c. 4⟌15 d. 3⟌22

These are division problems that have leftovers.

- Read problem A. **(Signal.)** *37 ÷ 9.*
- Can you divide 37 by 9? **(Signal.)** *No.*
- Write the largest part below and write the leftovers. Stop when you've done that much.
  (Observe students and give feedback.)
- What's the largest part of 37 you can divide by 9? **(Signal.)** *36.*
- How many leftovers are there? **(Signal.)** *1.*
  (Display:) [42:4A]

```
         1
a. 9⟌37
     36
```

Here's the largest part and the leftovers.

b. Now you have to work a division problem and write the answer above.
- Say the division problem you'll work. **(Signal.)** *36 ÷ 9.*
- Write the answer. ✔
  (Add to show:) [42:4B]

```
      4  1
a. 9⟌37
     36
```

Here's what you should have for problem A. 37 divided by 9 equals 4 and 1 leftover.

c. Read problem B. **(Signal.)** *11 ÷ 2.*
- Can you divide 11 by 2? **(Signal.)** *No.*
- Write the largest part below and write the remainder. The remainder is the number for the leftovers. Stop when you've done that much.
  (Observe students and give feedback.)
- What's the largest part of 11 you can divide by 2? **(Signal.)** *10.*
- How many leftovers are there? **(Signal.)** *1.*
  (Display:) [42:4C]

```
         1
b. 2⟌11
     10
```

d. Now you have to work a division problem and write the answer above.
- Say the division problem you'll work. **(Signal.)** *10 ÷ 2.*
- Write the answer. ✔
  (Add to show:) [42:4D]

```
      5  1
b. 2⟌11
     10
```

Here's what you should have for problem B.

e. Work problem C. First write the largest part and the remainder. Then write the answer to the division problem you work.
(Observe students and give feedback.)

f. Check your work.
- Problem C is 15 ÷ 4. Say the division problem you worked. **(Signal.)** *12 ÷ 4.*
- What's the answer? **(Signal.)** *3.*
- How many leftovers are there? **(Signal.)** *3.*
  (Display:) [42:4E]

```
      3  3
c. 4⟌15
     12
```

Here's what you should have for problem C.

g. Work problem D. First write the largest part and the remainder. Then write the answer to the division problem you work.
(Observe students and give feedback.)

h. Check your work.
- Problem D is 22 ÷ 3. Say the division problem you worked. **(Signal.)** *21 ÷ 3.*
- What's the answer? **(Signal.)** *7.*
- How many leftovers are there? **(Signal.)** *1.*
  (Display:) [42:4F]

```
      7  1
d. 3⟌22
     21
```

Here's what you should have for problem D.

from Lesson 42, Exercise 4

All these problems have "leftover" parts.

The procedure students follow first says the division that confirms there will be leftovers. "37 divided by 9."

Next, students write the largest part of 37 that can be divided by 9 below (36).

Students write the leftover (1).

Finally, students work a division (36 divided by 9) and write the answer.

```
   4  1
9⟌37
  36
```

On later lessons, students work mixed sets of problems. Some have remainders, some don't.

Here's the set of problems from Textbook Lesson 45:

Part 3

a. 4⟌19	c. 9⟌30	e. 5⟌42
b. 4⟌20	d. 9⟌36	f. 9⟌45

Interior Remainders

The next division problem type students work has **interior remainders.** To work these problems, students don't write the remainder as part of the answer. They write the remainder in front of the next digit of the problem.

Here's part of the introduction from Lesson 49:

a. (Display:) [49:6A]

$5\overline{)90}$

Here's a new kind of division problem.

- (Point to **90.**) Read the problem. (Signal.) *90 ÷ 5.*
- Say the problem for the first digit of 90. (Signal.) *9 ÷ 5.*
- What's the largest part of 9 that we can divide by 5? (Signal.) *5.*

(Add to show:) [49:6B]

$5\overline{)90}$
5

b. Say the problem and the answer for the largest part. (Signal.) *5 ÷ 5 = 1.*

(Add to show:) [49:6C]

1
$5\overline{)90}$
5

c. The subtraction problem for the leftovers is 9 minus 5.

- Say the subtraction problem for the leftovers. (Signal.) *9 – 5.*
- What's the answer? (Signal.) *4.*

(Add to show:) [49:6D]

1
$5\overline{)9_40}$
5

We divided 9 tens by 5. The answer is 1 with 4 tens leftover. So we write a little 4 in front of the zero to make a two-digit number for the ones. We don't write it below the 5.

d. The two-digit number for the ones is 40.

- What number? (Signal.) *40.*
- Say the new division problem for the ones. (Signal.) *40 ÷ 5.*
- What's the answer? (Signal.) *8.*

(Add to show:) [49:6E]

18
$5\overline{)9_40}$
5

- Read the problem we started with and the answer. (Signal.) *90 ÷ 5 = 18.*

from Lesson 49, Exercise 6

Starting on Lesson 52, students work problem sets that have two types of problems. Some have an interior remainder (162 ÷ 3), and some have a terminal remainder (207 ÷ 4).

Here's part of the exercise from Lesson 52:

(Teacher reference:)

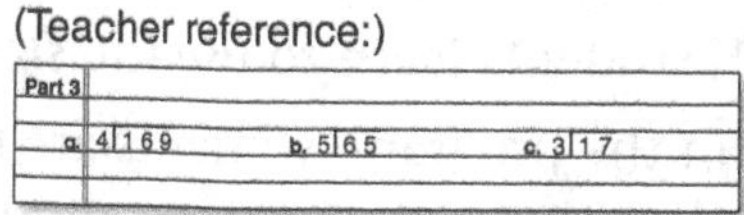

For some of these division problems, you'll get a remainder when you work part of the problem. For one of the problems, you'll write leftovers to make a 2-digit number for the ones. For the rest of the problems, you'll write the leftovers on a line next to the answer.

- Copy part 3.
  (Observe students and give feedback.)

b. Read problem A on your lined paper. **(Signal.)** *169 ÷ 4.*
- Do you work the problem for the first digit or the first two digits? **(Signal.)** *The first two digits.*
- Read the problem for the first two digits. **(Signal.)** *16 ÷ 4.*
- Can you work that problem? **(Signal.)** *Yes.*
- What's the answer? **(Signal.)** *4.*
- Write it. ✔
  (Display:) [52:6I]

Here's what you should have so far for problem A.

c. Read the problem for the ones. Get ready. **(Signal.)** *9 ÷ 4.*
- Can you work that problem? **(Signal.)** *No.*
- What's the largest part of 9 you can divide by 4? **(Signal.)** *8.*
- What's 8 ÷ 4? **(Signal)** *2.*
- Work the problem for the ones. Write the largest part below 9, write the answer above 9, and make a line and write the leftovers.
  (Observe students and give feedback.)

d. Check your work.
- Read problem A and the answer. **(Signal.)** *169 ÷ 4 = 42 and 1 leftover.*
  (Add to show:) [52:6J]

Part 3
42 1
a. 4|169
8

Here's what you should have for problem A.

e. Read problem B. **(Signal.)** *65 ÷ 5.*
- Do you work the problem for the first digit or the first two digits? **(Signal.)** *The first digit.*
- Read the problem for the first digit. **(Signal.)** *6 ÷ 5.*
- Can you work that problem? **(Signal.)** *No.*
- What's the largest part of 6 you can divide by 5? **(Signal.)** *5.*
- What's 5 ÷ 5? **(Signal.)** *1.*
- Work the problem for the first digit. Write the largest part below 6, write the answer above 6, and write the leftovers to make a 2-digit number for the ones.
  (Observe students and give feedback.)
  (Display:) [52:6K]

Here's what you should have so far for problem B.

f. Read the new problem for the ones. Get ready. **(Signal.)** *15 ÷ 5.*
- Can you work that problem? **(Signal.)** *Yes.*
- What's the answer? **(Signal.)** *3.*
- Write it. ✔

g. Check your work.
- Read problem B and the answer. **(Signal.)** *65 ÷ 5 = 13.*
  (Add to show:) [52:6L]

13
b. 5|6,5
5

Here's what you should have for problem B.

from Lesson 52, Exercise 6

Fractional Remainders

On Lesson 52, students learn to express division problems as fractions. (See Fractions page 89.)

On Lesson 54, students write remainders as fractions. The fraction is the leftover above the number the problem divides by.

For the problem 43 divided by 6 the remainder is 1/6.

Here's the introduction from Lesson 54:

a. Here's a rule about leftovers: If you divide by 9, the leftovers are ninths. If you divide by 10, the leftovers are tenths.
- If you divide by 6, what are the leftovers? **(Signal.)** *Sixths.*
- If you divide by 4, what are the leftovers? **(Signal.)** *Fourths.*
- If you divide by 7, what are the leftovers? **(Signal.)** *Sevenths.*
- If you divide by 3, what are the leftovers? **(Signal.)** *Thirds.*
  (Repeat step a until firm.)

from Lesson 54, Exercise 5

Fractional Remainders

Starting on Lesson 56, students work problems that have different combinations of fractional and interior remainders.

On Lesson 57, students work problems that have answers with an interior zero.

For example: $3\overline{)617}$

Students write the answer for the first problem: 6 divided by 3.

The next problem they work is 1 divided by 3. The answer is zero with a remainder of 1. Students underline and work the 2-digit problem: 17 divided by 3.

$$\begin{array}{r} 20 \\ 3\overline{)6\underline{17}} \end{array}$$

Below they show the largest part of 17 that can be divided by 3, (15) and write the last digit of the answer, 5. The remainder is 2/3.

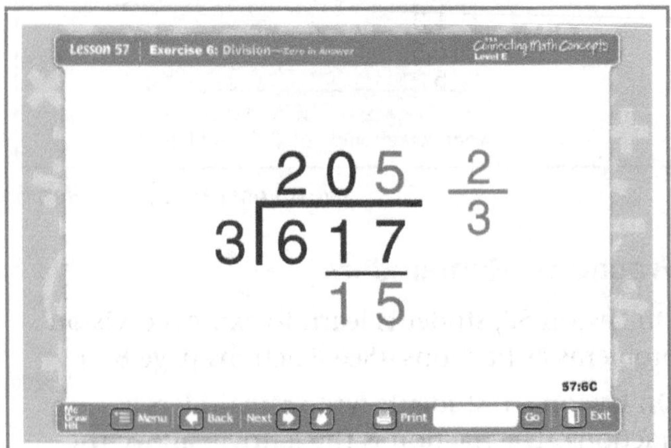

Structured work with different types of division problems continues through Lesson 68.

On Lesson 71, students learn to use the sign ÷.

At this point in the program, students are firm enough with the other sign that they shouldn't get reversals when using the ÷ sign.

Here's the Workbook activity from Lesson 72:

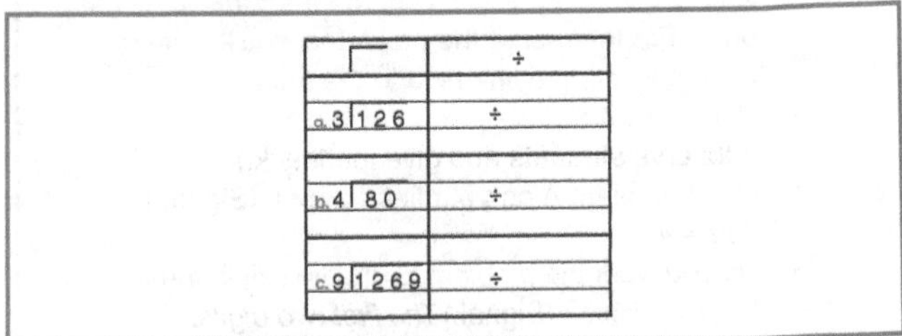

Students use division skills to work a variety of word problems including comparing fractions with percents and working area problems that give the area and ask about the length of a side.

Rounding and Estimation (Lessons 79–110)

This track starts on Lesson 79 and continues through Lesson 110.

Students learn to round numbers to the nearest ten, hundred, and thousand. Students also round dollar values to the nearest dollar; round decimals to the nearest whole number; and round multi-digit numbers to the nearest hundred and thousand. Students use rounded numbers to calculate a variety of addition and subtraction problems.

The first exercise establishes basic rounding conventions. Students learn that they can round 2-digit numbers to the closest ten. If the 2-digit number ends in 5 or more, it rounds up.

Here's the exercise from Lesson 79:

a. You're going to round numbers to the nearest tens. When you round numbers to the nearest tens, you change them so they end in a zero. You round 34 to 30. You round 38 to 40.
My turn to say thirties numbers that round to 30: 31, 32, 33, 34.
- Say thirties numbers that round to 30. (Signal.) *31, 32, 33, 34.*
Thirties numbers that round to 40 are 36, 37, 38, 39.
- Say thirties numbers that round to 40. (Signal.) *36, 37, 38, 39.*

b. Again, say thirties numbers that round to 30. Get ready. (Signal.) *31, 32, 33, 34.*
- Say thirties numbers that round to 40. Get ready. (Signal.) *36, 37, 38, 39.*
(Repeat step b until firm.)

c. Listen: Numbers between 70 and 80 can round to 70 or to 80. What can numbers between 70 and 80 round to? (Signal.) *70 or 80.*
- Say the seventies numbers that round to 70. Get ready. (Signal.) *71, 72, 73, 74.*
- Say the seventies numbers that round to 80. Get ready. (Signal.) *76, 77, 78, 79.*

d. Listen: Numbers between 10 and 20 can round to 10 or to 20. What can teen numbers round to? (Signal.) *10 or 20.*
- Say the teen numbers that round to 10. Get ready. (Signal.) *11, 12, 13, 14.*
- Say the teen numbers that round to 20. Get ready. (Signal.) *16, 17, 18, 19.*
(Repeat steps c and d that were not firm.)

e. (Display:) [79:5A]

a. 48	d. 11
b. 76	e. 87
c. 23	

Each number is between two tens numbers. You'll tell me the tens numbers.
- Read number A. (Signal.) *48.*
- Tell me the tens numbers 48 is between. Get ready. (Signal.) *40 and 50.*
- Does 48 round to 40 or 50? (Signal.) *50.*

f. Read number B. (Signal.) *76.*
- Tell me the tens numbers 76 is between. Get ready. (Signal.) *70 and 80.*
- Does 76 round to 70 or 80? (Signal.) *80.*

g. Read number C. (Signal.) *23.*
- Tell me the tens numbers 23 is between. Get ready. (Signal.) *20 and 30.*
- Does 23 round to 20 or 30? (Signal.) *20.*

h. Read number D. (Signal.) *11.*
- Tell me the tens numbers 11 is between. Get ready. (Signal.) *10 and 20.*
- Does 11 round to 10 or 20? (Signal.) *10.*

i. Read number E. (Signal.) *87.*
- Tell me the tens numbers 87 is between. Get ready. (Signal.) *80 and 90.*
- Does 87 round to 80 or 90? (Signal.) *90.*

j. (Add to show:) [79:5B]

Part A			
a.	d.	a. 48	d. 11
b.	e.	b. 76	e. 87
c.		c. 23	

- Write part A on your lined paper with the letters A through E below. After each letter, write the number rounded to the nearest tens.
(Observe students and give feedback.)

k. Check your work.
- Problem A: What's 48 rounded to the nearest tens? (Signal.) *50.*
- Problem B: What's 76 rounded to the nearest tens? (Signal.) *80.*
- Problem C: What's 23 rounded to the nearest tens? (Signal.) *20.*
- Problem D: What's 11 rounded to the nearest tens? (Signal.) *10.*
- Problem E: What's 87 rounded to the nearest tens? (Signal.) *90.*

(Display:) [79:5C]

Part A		
a.	50	d. 10
b.	80	e. 90
c.	20	

Here's what you should have for part A.

Lesson 79, Exercise 5

Teaching Note: The key to rounding is to know the numbers that round down and those that round up. In step B, students review the numbers that are closer to 30 and those that are closer to 40. Make sure that students are very firm in this discrimination and apply it correctly to the early examples. If students are firm on the idea that the first four values in a decade round down and the last four values round up, basic rounding conventions present far fewer problems.

In step J, students round different numbers to the nearest tens number. If students make mistakes, show them how to test the numbers. For example, 76.

If students respond weakly tell them, "Say 70s numbers that round to 70."

Ask, "Did you say 76?"

"So does 76 round to 70 or 80?"

Practice this correction. You'll probably have opportunities to use it.

On Lesson 81, students round values in problems and write the estimated answer, which is "close to" the actual answer.

For example:

$$\begin{array}{r} 87 \\ 51 \\ +73 \\ \hline \end{array}$$

rounds to

$$\begin{array}{r} 90 \\ 50 \\ +70 \\ \hline \end{array}$$

Students add the values to determine the rounded answer, 210, which is close to the actual answer, which is 211.

On Lesson 83, students round numbers that have an arrow pointing to one of the digits. Students determine the place of the arrowed digit. They learn that if the digit that follows the arrowed digit is 5 or more, the arrowed digit rounds up.

m. (Point to **1857.**) What digit is the arrow pointing to? (Signal.) *Hundreds.*
- What are the choices? (Signal.) *8 or 9.*

n. (Point to **7450.**) What digit is the arrow pointing to? (Signal.) *Hundreds.*
- What are the choices? (Signal.) *4 or 5.*

o. (Point to **2015.**) What digit is the arrow pointing to? (Signal.) *Tens.*
- What are the choices? (Signal.) *1 or 2.*

(Repeat steps k through o until firm.)

We'll learn more about rounding to any digit in the next lessons.

from Lesson 83, Exercise 9

In this lesson, students practice identifying the arrowed digit. Looking at the arrowed digit will indicate the two rounding choices. The arrowed digit 8 will either round up to 9 or remain at 8. The arrowed digit 4 will either round up to 5 or remain at 4.

Rounding to Tens, Hundreds, and Thousands

Starting on Lesson 87, students round numbers to the tens, hundreds, and thousands.

Here's the exercise:

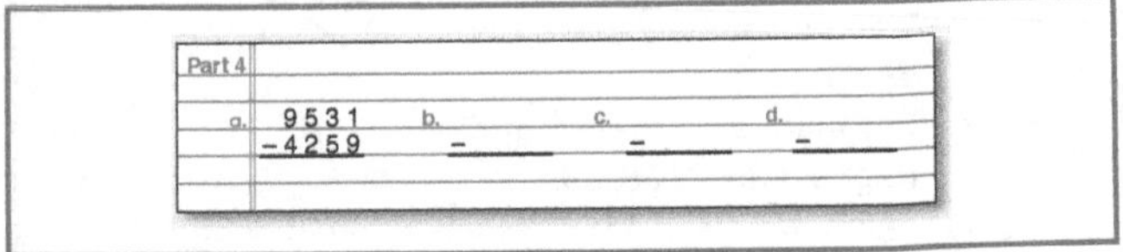

Teaching Note: In steps B and C students verbally identify what 9531 is when rounded to the tens, the hundreds, and the thousands. Make sure that students are firm on this step. Students later show the problems with rounded numbers as part of their Independent Work.

On Lesson 91, students round 2-digit numbers to either 0 or 100. They also round 3-digit numbers to either 0 or 100.

Students learn that a number like 80 is either 100 or zero when rounded to the hundred.

Initially, problems show zeros for the hundreds and thousands digit.

84 is written as 0084.

Students follow the conventions they have learned to determine whether they round up. To round to thousands, they note that the digit after the thousands digit is zero. That's not 5 or more, so the number rounds to zero.

To round to hundreds, students note that the digit after the hundreds is more than 5 so they round the number to 100.

Here's part of the exercise from Lesson 92:

a. You learned that some numbers round to zero. (Display:) [92:6A]

36

- Read this number. (Signal.) *36.*
- If you round 36 to the tens, what number do you get? (Signal.) *40.*
- If you round 36 to the hundreds, do you get 100 or zero? (Signal.) *Zero.*
- Listen: If you round 36 to the thousands, do you get 1000 or zero? (Signal.) *Zero.*

b. (Display:) [92:6B]

53

- Read this number. (Signal.) *53.*
- If you round 53 to the tens, what number do you get? (Signal.) *50.*
- If you round 53 to the hundreds, do you get 100 or zero? (Signal.) *100.*

c. (Add to show:) [92:6C]

↓
0053

If you show the zeros, it's easy to figure out if you round the hundreds or the thousands digits up. The thousands digit is zero and the hundreds digit is zero.

- Is the arrow pointing to the thousands or hundreds? (Signal.) *Hundreds.*
- What's the digit after the hundreds? (Signal.) *5.*
- So do you round up? (Signal.) *Yes.*
- What's 53 rounded to the hundreds? (Signal.) *100.*

d. (Change to show:) [92:6D]

↓
0053

- Is the arrow pointing to the thousands or hundreds? (Signal.) *Thousands.*
- What's the digit after the thousands? (Signal.) *Zero.*
- So do you round up? (Signal.) *No.* Yes, when you round 53 to thousands, you get zero.
- What's 53 rounded to the thousands? (Signal.) *Zero.*

from Lesson 92, Exercise 6

On later lessons, students don't write zeros, but they assume that there are zeros in front of the numbers that they are rounding.

Here's part of the exercise from Lesson 95:

(Teacher reference:)

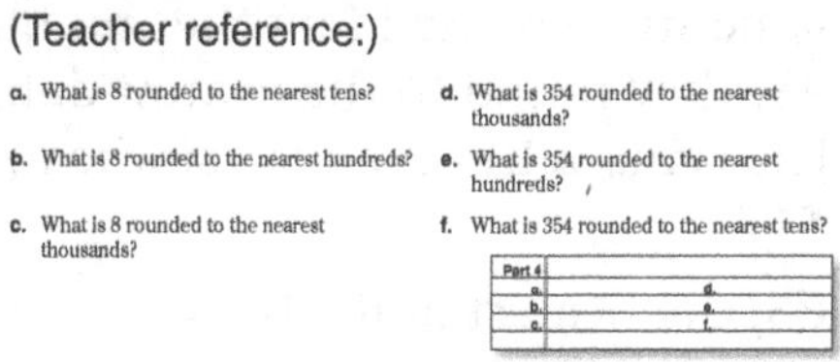
a. What is 8 rounded to the nearest tens?
b. What is 8 rounded to the nearest hundreds?
c. What is 8 rounded to the nearest thousands?
d. What is 354 rounded to the nearest thousands?
e. What is 354 rounded to the nearest hundreds?
f. What is 354 rounded to the nearest tens?

These are questions that ask you to round numbers to digits. You'll round some of the numbers to digits that aren't shown, but you know what those digits are.

- Write part 4 on your lined paper with the letters A through F below. (Observe students and give feedback.)

b. Read question A. (Call on a student.) *What is 8 rounded to the nearest tens?*

- Raise your hand when you know the answer. ✔
- Everybody, what's 8 rounded to the nearest tens? (Signal.) *10.*

c. Read question B. (Call on a student.) *What is 8 rounded to the nearest hundreds?*

- Raise your hand when you know the answer. ✔
- Everybody, what's 8 rounded to the nearest hundreds? (Signal.) *Zero.* (If students were 100% on questions A and B, skip to step h.)

d. Read question C. (Call on a student.) *What is 8 rounded to the nearest thousands?*

- Raise your hand when you know the answer. ✔
- Everybody, what's 8 rounded to the nearest thousands? (Signal.) *Zero.*

e. Read question D. (Call on a student.) *What is 354 rounded to the nearest thousands?*

- Raise your hand when you know the answer. ✔
- Everybody, what's 354 rounded to the nearest thousands? (Signal.) *Zero.*

f. Read question E. (Call on a student.) *What is 354 rounded to the nearest hundreds?*

- Raise your hand when you know the answer. ✔
- Everybody, what's 354 rounded to the nearest hundreds? (Signal.) *400.*

g. Read question F. (Call on a student.) *What is 354 rounded to the nearest tens?*

- Raise your hand when you know the answer. ✔
- Everybody, what's 354 rounded to the nearest tens? (Signal.) *350.* (Repeat steps b through g that were not firm.)

from Lesson 95, Exercise 7

ROUNDING DOLLARS-AND-CENTS/ DECIMALS

Starting on Lesson 100, students round dollars-and-cents amounts to the nearest dollar. Students use the same strategy used for other numbers. For \$17.48, they round to the nearest dollar—\$17 or \$18. The cents is less than 50, so the rounded value is \$17.

For \$42.83, the nearest dollar is \$43.

For \$0.29, the nearest dollar is \$0.

Starting on Lesson 103, students apply the same strategy used to round dollar-and-cents amounts to decimal numbers. They round each to the nearest whole number.

.141 rounds to zero.

17.65 rounds to 18.

3.09 rounds to 3.

Throughout the rounding track, students work estimation problems that parallel what they have recently learned about rounding.

Estimation Problems

For instance, after students learn to round 2-digit numbers to thousands, they work estimation problems that require this type of rounding.

For example:

$$\begin{array}{r} 9806 \\ -\quad 73 \\ \hline \end{array}$$

The directions tell students to round both numbers to thousands. The estimation problem they work is:

$$\begin{array}{r} 10{,}000 \\ -\qquad 0 \\ \hline \end{array}$$

After students learn to round dollars-and-cents amounts to the nearest dollar, they work estimation problems like:

$$\begin{array}{r} \$\quad .29 \\ 18.56 \\ +\ \ 1.03 \\ \hline \end{array}$$

Students work the estimation problem:

$$\begin{array}{r} \$\qquad 0 \\ 19.00 \\ +\ \ 1.00 \\ \hline \end{array}$$

After students learn to round decimal values, they work estimation problems that round decimal numbers (not dollars and cents) to the nearest whole number.

For example:

$$\begin{array}{r} 1.24 \\ .80 \\ +10.09 \\ \hline \end{array}$$

The estimation problem they work is:

$$\begin{array}{r} 1.00 \\ 1.00 \\ +10.00 \\ \hline \end{array}$$

Coefficients and Inverse Operations (Lessons 85–130)

This track begins on Lesson 85 and continues intermittently through the end of the program.

The track first teaches students that terms like 2/3 × M are also expressed as 2/3M and 3 × 4J = 12J. Students use these equivalences when they work problems like 4J = 36. Students convert 4J into 4 × J.

COEFFICIENTS

Students later use multiplication equivalences to complete function tables that have missing X or Y values. (See Coordinate System page 155.) For example, 3K = P. If the value for K is 5, the function converts into 3 × K = P. Students substitute 5 for K and work the problem 3 × 5. P = 15.

Here's the first part of the exercise from Lesson 85. It introduces the multiplication equivalence:

a. Listen: Another way to say **6 times X** is **6X**.
- What's another way to say 6 times X? (Signal.) *6X.*
- What's another way to say 52 times X? (Signal.) *52X.*
- What's another way to say 1/2 times X? (Signal.) *1/2X.*
- What's another way to say 7/4 times X? (Signal.) *7/4X.*
- What's another way to say 1 times X? (Signal.) *1X.*

b. Yes, 1 times X is 1X or just X.
- Tell me one of the other ways to say 1 times X. (Call on a student.) *1X* or *X.*
- Everybody, tell me the other way to say 1 times X. Get ready. (Signal.) *(Students respond with remaining answer.)*

c. Everybody, what's another way to say 5/2 times X? (Signal.) *5/2X.*
- What's another way to say 3/8 times M? (Signal.) *3/8M.*
- What's another way to say 5 times Q? (Signal.) *5Q.*
- What's another way to say Y times X? (Signal.) *YX.*

(Repeat step c until firm.)

d. (Display:) [85:7A]

35 × X $\qquad \frac{5}{9} \times R$

.9 × W

You're going to read each problem and tell me what it equals. I'll complete the equations.
- (Point to **35.**) Read the problem. (Signal.) *35 times X.*
- What does 35 times X equal? (Signal.) *35X.*

(Add to show:) [85:7B]

35 × X = 35X $\qquad \frac{5}{9} \times R$

.9 × W

e. (Point to $\frac{5}{9}$.) Read the problem. (Signal.) *5/9 times R.*
- What does 5/9 times R equal? (Signal.) *5/9R.*

(Add to show:) [85:7C]

35 × X = 35X $\qquad \frac{5}{9} \times R = \frac{5}{9}R$

.9 × W

f. (Point to **.9.**) Read the problem. (Signal.) *9 tenths times W.*
- What does 9 tenths times W equal? (Signal.) *9 tenths W.*

(Add to show:) [85:7D]

35 × X = 35X $\qquad \frac{5}{9} \times R = \frac{5}{9}R$

.9 × W = .9W

from Lesson 85, Exercise 7

On Lesson 89, students complete equations that express terms like 5R as 5 × R and R × 5.

Here are the sample sets from Workbook Lesson 89:

a.	b.	c.
________ = 1.2R	________ = 8Y	________ $= \frac{3}{2}G$
________ = 1.2R	________ = 8Y	________ $= \frac{3}{2}G$

On Lesson 91, students work problems of the types: 4 × 5L and 2P × 7. Students say the multiplication for the numbers (4 × 5) (2 × 7), then say the complete product (20L) (14P).

INVERSE OPERATIONS

CMC Level E introduces inverse operations on Lesson 104.

Students first learn to analyze addition-subtraction problems that have a missing number. Students identify whether the missing number is the big number or a small number. If the problem subtracts, it starts with the big number.

If the problem adds, it ends with the big number.

With this information, students figure out whether they add or subtract to find the missing number.

Here's the exercise from Lesson 105:

a. (Display:) [105:4A]

a. ___ − 11 = 41 **d.** ___ − 66 = 7

b. 36 − ___ = 12 **e.** 310 + 75 = ___

c. 13 + ___ = 50 **f.** 96 − ___ = 51

Here are some addition and subtraction problems with missing numbers. I'll read each problem. You'll tell me what the big number is and if the missing number is a small number or the big number.

- Remember the rule. A problem that subtracts starts with the big number. What does a problem that subtracts start with? (Signal.) *The big number.*
(Repeat until firm.)

b. Problem A: What number − 11 = 41.
- Does problem A subtract? (Signal.) *Yes.*
- So does it start with the big number? (Signal.) *Yes.*
- So is the big number missing? (Signal.) *Yes.*

c. Problem B: 36 − what number = 12.
- Does problem B subtract? (Signal.) *Yes.*
- So does it start with the big number? (Signal.) *Yes.*
- What's the big number? (Signal.) *36.*
- Is the missing number the big number or a small number? (Signal.) *A small number.*

d. Problem C: 13 + what number = 50.
- Does problem C subtract? (Signal.) *No.*
- So does it start with the big number? (Signal.) *No.*
- Does it end with the big number? (Signal.) *Yes.*
- What's the big number? (Signal.) *50.*
- Look at the missing number. What kind of number is the missing number? (Signal.) *A small number.*

e. Problem D: What number − 66 = 7.
- Does problem D subtract? (Signal.) *Yes.*
- So does it start with the big number? (Signal.) *Yes.*
- What's the big number? (Signal.) *The missing number.*
(If students are firm, skip to step h.)

f. Problem E: 310 + 75 = what number.
- Does problem E subtract? (Signal.) *No.*
- So does it start with the big number? (Signal.) *No.*
- So does it end with the big number? (Signal.) *Yes.*
- What's the big number? (Signal.) *The missing number.*

g. Problem F: 96 − what number = 51.
- Does problem F subtract? (Signal.) *Yes.*
- So does it start with the big number? (Signal.) *Yes.*
- What's the big number? (Signal.) *96.*
- What kind of number is the missing number? (Signal.) *A small number.*
(Repeat steps b through g that were not firm.)

h. We'll do those problems again. This time you'll read the problems and tell me if the missing number is the big number or a small number.
- Read problem A. (Signal.) *What number − 11 = 41.*
- Tell me what kind of number the missing number is. Get ready. (Signal.) *The big number.*

i (Repeat the following tasks for problems B through F:)

	Read problem ___.	Tell me what kind of number the missing number is. Get ready.
B	*36 − what number = 12*	*A small number*
C	*13 + what number = 50*	*A small number*
D	*What number − 66 = 7*	*The big number*
E	*310 + 75 = what number*	*The big number*
F	*96 − what number = 51*	*A small number*

(Repeat problems that were not firm.)

Lesson 105, Exercise 4

On Lesson 110, missing numbers in problems are shown with letters. Students read each problem and identify whether the letter is the big number or a small number. They work problems the same way they did earlier. They show what each letter equals with an equation (i.e., M = 45).

Here's the problem set from Textbook Lesson 110:

a. M − 103 = 83	c. 103 − X = 83	e. Q + 56 = 196
b. 64 + U = 289	d. J − 64 = 289	f. R − 25 = 121

Also on Lesson 110, students apply what they know about missing numbers in problems that add or subtract to problems that multiply or divide.

If a problem divides, it starts with the big number. If a problem multiplies, it ends with the big number.

On Lesson 112, students work multiplication and division problems that have letters. R ÷ 2 = 86. The problem divides, so R is the big number. Students work the multiplication problem: 86 × 2.

R = 172.

On Lesson 115, students work the first mixed set of problems that include problems that add or subtract, multiply or divide.

Here's the exercise from Lesson 115:

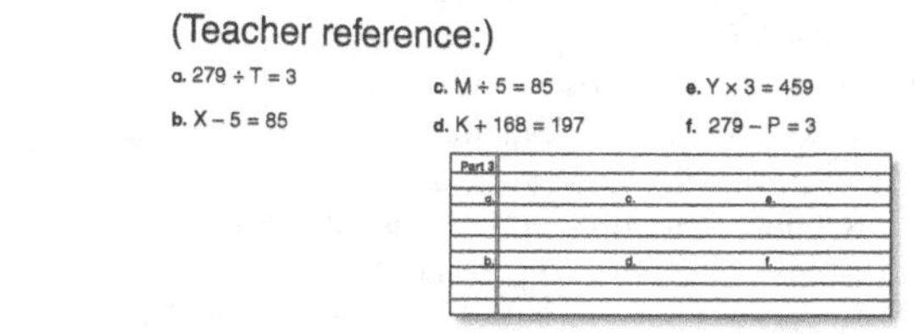
(Teacher reference:)

a. 279 ÷ T = 3   c. M ÷ 5 = 85   e. Y × 3 = 459

b. X – 5 = 85   d. K + 168 = 197   f. 279 – P = 3

These are problems that have a letter for the missing number. The numbers for some of the problems are from addition number families. The numbers for the rest of the problems are from multiplication families. For each problem, you'll tell me if the big number or a small number is missing and say the problem for finding the letter.

b. Touch and read problem A. **(Signal.)** *279 ÷ T = 3.*
- Is T the big number or a small number? **(Signal.)** *A small number.*
- Say the problem for finding T. Get ready. **(Signal.)** *279 ÷ 3.*

c. **(Repeat the following tasks for problems B through F:)**

Touch and read problem __.		Is __ the big number or a small number?		Say the problem for finding __.
B	*X – 5 = 85*	X	*The big number.*	X	*85 + 5*
C	*M ÷ 5 = 85*	M	*The big number.*	M	*85 × 5*

(If students are firm, skip to step d.)

D	*K + 168 = 197*	K	*A small number.*	K	*197 – 168*
E	*Y × 3 = 459*	Y	*A small number.*	Y	*459 ÷ 3*
F	*279 – P = 3*	P	*A small number.*	P	*279 – 3*

(Repeat problems that were not firm.)

d. Write part 3 and the letters A through F on your lined paper. Then work problem A. Put your pencil down when you've written an equation for T.
(Observe students and give feedback.)

e. Check your work.
- Problem A: 279 ÷ T = 3. Read the problem you worked and the answer. **(Signal.)** *279 ÷ 3 = 93.*
- Read the equation for T. **(Signal.)** *T = 93.*
(Display:) [115:5A]

Part 3	
	93 T = 93
a.	3⟌279

Here's what you should have for problem A.

f. Work the rest of the problems. Put your pencil down when you've written an equation for each letter in part 3.
(Observe students and give feedback.)

g. Check your work.
- Problem B: X – 5 = 85. Read the problem you worked and the answer. **(Signal.)** *85 + 5= 90.*
- Read the equation for X. **(Signal.)** *X = 90.*

h. **(Repeat the following tasks for problems C through F:)**

Problem __. Read the problem you worked and the answer.		Read the equation for __.	
C: M ÷ 5 = 85	*85 × 5 = 425*	M	*M = 425*
D: K + 168 = 197	*197 – 168 = 29*	K	*K = 29*
E: Y × 3 = 459	*459 ÷ 3 = 153*	Y	*Y = 153*
F: 279 – P = 3	*279 – 3 = 276*	P	*P = 276*

Lesson 115, Exercise 5

Teaching Note: Students should be fluent in answering the questions and in working the problems correctly.

If students are not fluent, go back and work these exercises in the following order: Lesson 108, Exercise 6; Lesson 110, Exercise 5; Lesson 109, Exercise 7; Lesson 111, Exercise 6; Lesson 110, Exercise 3; Lesson 112, Exercise 3 so they are presented in order of addition, subtraction, multiplication, and division.

Mixed-Number Computations (Lessons 76–110)

This track starts on Lesson 76 and continues intermittently through Lesson 110. The problem types introduced are:

- Adding and subtracting mixed numbers with like denominators.
- Working addition and subtraction mixed-number problems that require carrying or borrowing.
- Multiplying mixed numbers by whole numbers.

ADDING-SUBTRACTING

On Lesson 76, students apply what they know about adding and subtracting whole numbers or fractions to adding and subtracting mixed numbers.

Students follow these steps: First they work the problem for the fractions and write the answer in the fraction column. Then they work the whole-number problem and combine the answer with the fraction.

The first problem type presents fractions with the same denominator. The fraction students end up with is less than 1.

Here's part of the introduction from Lesson 76:

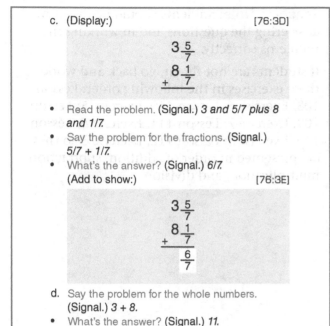

c. (Display:) [76:3D]

$3\frac{5}{7}$
$+ 8\frac{1}{7}$

- Read the problem. (Signal.) *3 and 5/7 plus 8 and 1/7.*
- Say the problem for the fractions. (Signal.) *5/7 + 1/7.*
- What's the answer? (Signal.) *6/7.*

(Add to show:) [76:3E]

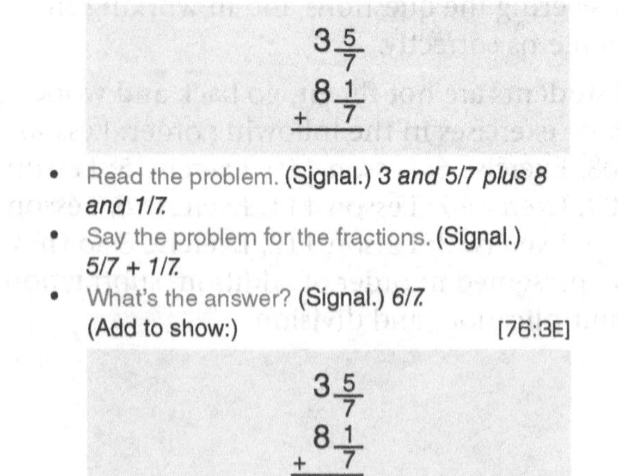

d. Say the problem for the whole numbers. (Signal.) *3 + 8.*
- What's the answer? (Signal.) *11.*

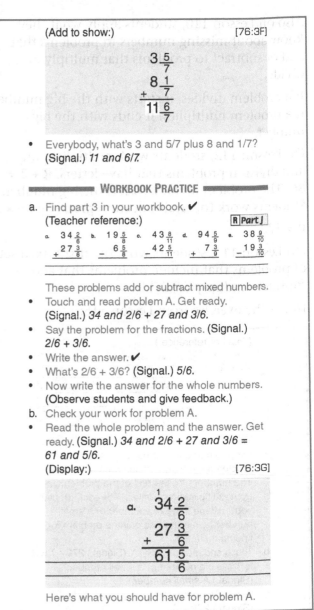

(Add to show:) [76:3F]

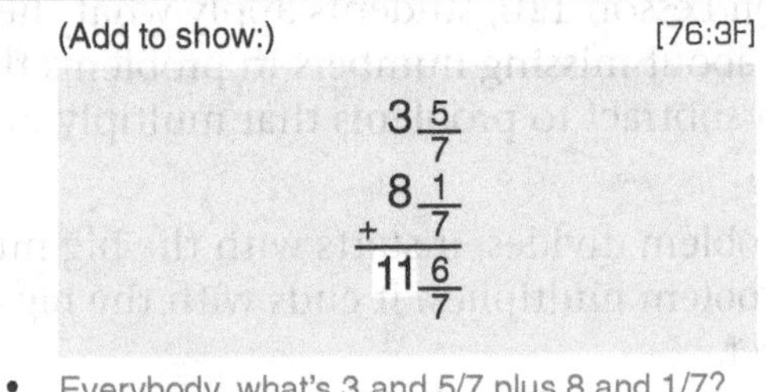

- Everybody, what's 3 and 5/7 plus 8 and 1/7? (Signal.) *11 and 6/7.*

WORKBOOK PRACTICE

a. Find part 3 in your workbook. ✔
(Teacher reference:) R Part J

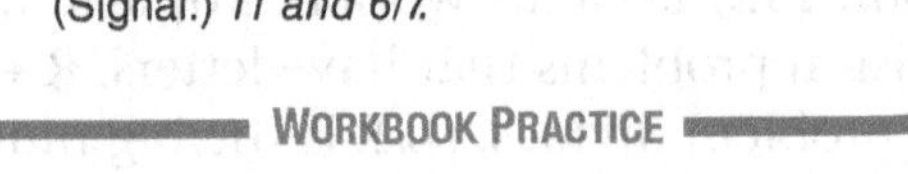

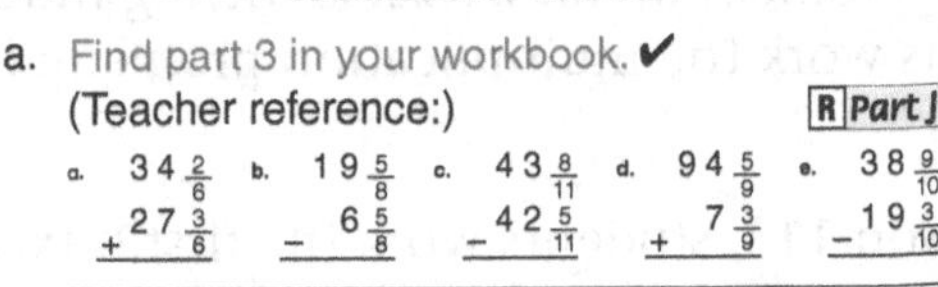

These problems add or subtract mixed numbers.

- Touch and read problem A. Get ready. (Signal.) *34 and 2/6 + 27 and 3/6.*
- Say the problem for the fractions. (Signal.) *2/6 + 3/6.*
- Write the answer. ✔
- What's 2/6 + 3/6? (Signal.) *5/6.*
- Now write the answer for the whole numbers. (Observe students and give feedback.)

b. Check your work for problem A.
- Read the whole problem and the answer. Get ready. (Signal.) *34 and 2/6 + 27 and 3/6 = 61 and 5/6.*

(Display:) [76:3G]

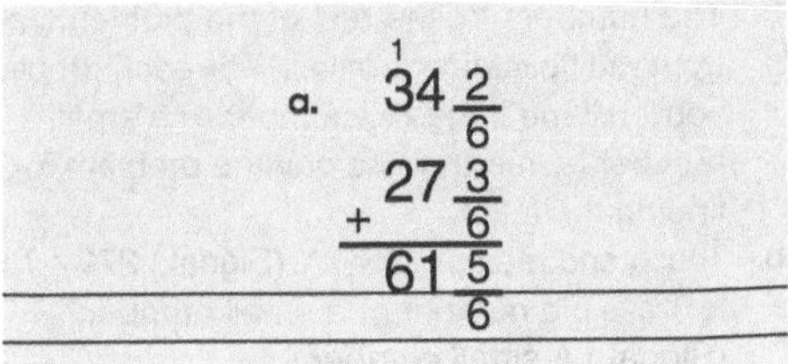

Here's what you should have for problem A.

from Lesson 76, Exercise 3

Teaching Notes: Make sure that students read the problems correctly. One of the more serious mistakes students tend to make is ignoring the plus or minus sign. Working the problems is fairly simple. Students first work the problem for the fractions and write the answer in the fraction column. Then they work the problem for the whole numbers and write the answer in the whole number column.

Carrying

On Lesson 78, students work problems that require carrying from the fraction column to the whole-number column. Students follow the rule that they must have less than 1 in the fraction column.

If the total in the fraction column is more than 1, students treat it as a mixed number. For example, 9/6. Students write it as a mixed number, 1 3/6. They carry the 1 to the whole-number column. They write 3/6 in the fraction column.

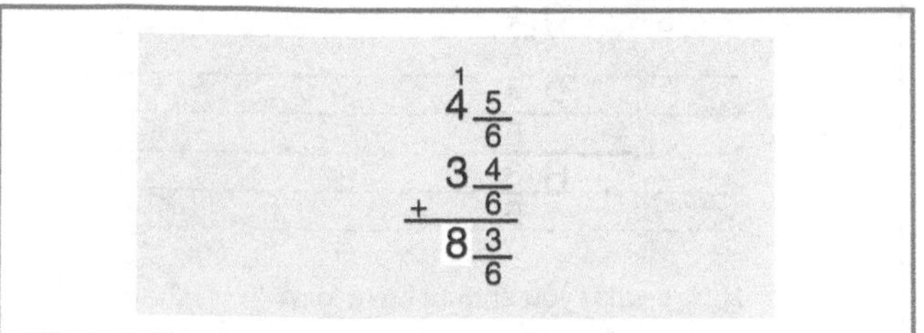

If the total in the fraction column equals 1 (8/8 for instance), students convert it into 1 and carry it to the whole-number column. You will probably have to remind them of this convention.

$$\begin{array}{r} {}^{1} \\ 7\frac{1}{8} \\ +\ 1\frac{7}{8} \\ \hline 9\phantom{\frac{1}{8}} \end{array}$$

Mixed Number–Borrowing

On Lesson 80, students learn a preskill that they use when they borrow to find the answers to some problems that subtract. Students convert the mixed numbers into two fractions that are added, the fraction for the whole number plus the fraction that is more than 1.

For example:

$$5\frac{1}{7} = 4 + \text{—}$$

The missing fraction = 1 1/7. That's 8/7.

$$5\frac{1}{7} = 4 + \frac{8}{7}$$

Here's the set of problems students work in Lesson 80:

a. $3\frac{4}{9} = 2 + \text{—}$

b. $7\frac{3}{8} = 6 + \text{—}$

c. $9\frac{1}{2} = 8 + \text{—}$

d. $3\frac{2}{5} = 2 + \text{—}$

On Lesson 82, students work mixed-number problems that require borrowing.

For example:

$$\begin{array}{r} 9\frac{2}{5} \\ -\ 2\frac{4}{5} \\ \hline \end{array}$$

Students first determine that they can't work the problem for the fractions:

$$\begin{array}{r} \frac{2}{5} \\ -\ \frac{4}{5} \\ \hline \end{array}$$

So they borrow 1 from the whole number 9. They treat that 1 as a 5/5 and add it to 2/5.

$$\begin{array}{r} {}^{8}\ {}^{7} \\ \cancel{9}\frac{\cancel{2}}{5} \\ -\ 2\frac{4}{5} \\ \hline \end{array}$$

That leaves 8 in the whole-number column and 1 and 2/5 in the fraction column.

Students compute the new fraction: 1 + 2/5 is 7/5.

Part 3

a. $\begin{array}{r} {}^{8}\ {}^{7} \\ \cancel{9}\frac{\cancel{2}}{5} \\ -\ 2\frac{4}{5} \\ \hline \end{array}$

Now students are able to work the problem for the fractions (7/5 – 4/5) and the problem for the whole numbers: 8 – 2.

$$\begin{array}{r} \overset{8}{\cancel{9}}\,\frac{\overset{7}{\cancel{2}}}{5} \\ -\;2\,\frac{4}{5} \\ \hline 6\,\frac{3}{5} \end{array}$$

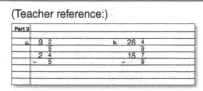

The answer is 6 3/5.

Here's the first part of the Textbook practice from Lesson 82:

(Teacher reference:)

Part 3

a. $9\frac{2}{5} - 2\frac{4}{5}$ b. $26\frac{4}{9} - 15\frac{7}{9}$

- Copy part 3 on your lined paper. (Observe students and give feedback.)

b. Read problem A on your lined paper. Get ready. (Signal.) *9 and 2/5 – 2 and 4/5.*

- Read the problem for the fractions. (Signal.) *2/5 – 4/5.*
- Can you work that problem? (Signal.) *No.*

c. So you rewrite the top number. You subtract 1 from the whole number and add it to the fraction.

- What's 9 minus 1? (Signal.) *8.*
- Cross out 9 and write 8 above it. ✔
- Tell me what 1 plus 2/5 is. Get ready. (Signal.) *7/5.*
- Cross out 2 and write the new numerator. ✔ (Display:) [82:7F]

Part 3

a. $\overset{8}{\cancel{9}}\,\frac{\overset{7}{\cancel{2}}}{5} - 2\frac{4}{5}$

Here's what you should have for A.

d. Say the new problem for the fractions. (Signal.) *7/5 – 4/5.*

- What's the answer? (Signal.) *3/5.*
- Say the new problem for the ones. (Signal.) *8 – 2.*
- What's the answer? (Signal.) *6.*
- Work the rest of problem A. (Observe students and give feedback.)

e. Check your work for problem A.

- We worked the problem 9 and 2/5 minus 2 and 4/5. What's the whole answer? (Signal.) *6 and 3/5.* (Add to show:) [82:7G]

Part 3

a. $\overset{8}{\cancel{9}}\,\frac{\overset{7}{\cancel{2}}}{5} - 2\frac{4}{5} = 6\frac{3}{5}$

Here's what you should have for A.

from Lesson 82, Exercise 7

On Lesson 86, students first work mixed sets of problems. Some require carrying; others require borrowing.

Here's the set of problems from Textbook Lesson 86:

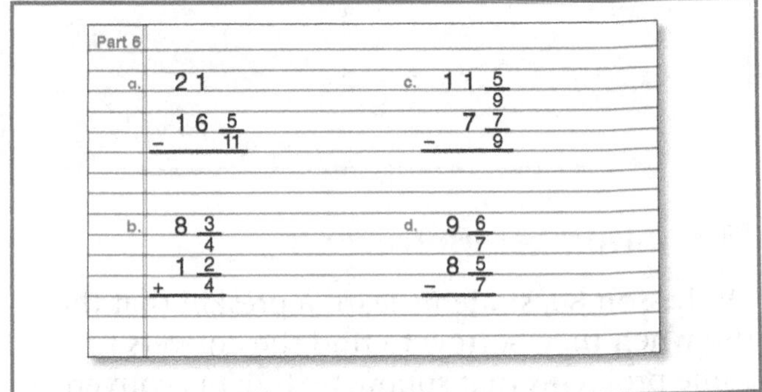
Part 6

a. $21 - 16\frac{5}{11}$

c. $11\frac{5}{9} - 7\frac{7}{9}$

b. $8\frac{3}{4} + 1\frac{2}{4}$

d. $9\frac{6}{7} - 8\frac{5}{7}$

Multiplying Mixed Numbers by Whole Numbers

Multiplication of mixed numbers by whole numbers begins on Lesson 98 and continues through 110. Students continue to use the analysis to work area problems that present mixed numbers.

a. You're going to find mixed-number answers to multiplication problems. Remember, a mixed number is a whole number and a fraction less than 1.
- Do mixed numbers have a fraction less than 1? (Signal.) *Yes.*
- Do mixed numbers have a fraction equal to 1? (Signal.) *No.*
- Do mixed numbers have a fraction more than 1? (Signal.) *No.*

(Repeat until firm.)

b. (Display:) [98:3A]

$$\begin{array}{r} 6\frac{4}{5} \\ \times\ 9 \\ \hline \end{array}$$

This problem shows a mixed number times a whole number.
- Read the problem. Get ready. (Signal.) *6 and 4/5 × 9.*

You work this problem in much the same way you work problems with 2 digits. The problem for the fraction is 4/5 × 9. The problem for the whole numbers is 6 × 9.
- Say the problem for the fraction. Get ready. (Signal.) *4/5 × 9.*
- Say the problem for the whole numbers. Get ready. (Signal.) *6 × 9.*

(Repeat until firm.)

c. Again, say the problem for the fraction. (Signal.) *4/5 × 9.*
- Raise your hand when you can tell me the fraction 4/5 × 9 equals. ✔
- What does 4/5 × 9 equal? (Signal.) *36/5.*

(Add to show:) [98:3B]

$$\begin{array}{r} 6\frac{4}{5} \\ \times\ 9 \\ \hline \frac{36}{5} \end{array}$$

- Say the problem **and** the answer for the whole numbers. Get ready. (Signal.) *6 × 9 = 54.*

(Add to show:) [98:3C]

$$\begin{array}{r} 6\frac{4}{5} \\ \times\ 9 \\ \hline 54\frac{36}{5} \end{array}$$

- Read the problem and the answer. (Signal.) *6 and 4/5 × 9 = 54 and 36/5.*
- Why isn't 54 and 36/5 a mixed number? (Call on a student. IDEA:) *36/5 is more than 1. The numerator of the fraction is more than the denominator.*
- 54 and 36/5 is equal to the answer, but is it a mixed number? (Signal.) *No.*

d. Raise your hand when you know the mixed number 36/5 equals. ✔
- Everybody, what mixed number does 36/5 equal? (Signal.) *7 and 1/5.*

(Add to show:) [98:3D]

$$\begin{array}{r} 6\frac{4}{5} \\ \times\ 9 \\ \hline 54\frac{36}{5} = 54 + 7\frac{1}{5} \end{array}$$

54 and 36/5 = 54 + 7 and 1/5.
- Tell me what 54 + 7 and 1/5 equals. Get ready. (Signal.) *61 and 1/5.*

(Add to show:) [98:3E]

$$\begin{array}{r} 6\frac{4}{5} \\ \times\ 9 \\ \hline 54\frac{36}{5} = 54 + 7\frac{1}{5} = 61\frac{1}{5} \end{array}$$

- So what mixed number does 6 and 4/5 × 9 equal? (Signal.) *61 and 1/5.*

from Lesson 98, Exercise 3

Starting on Lesson 98, students work problems that multiply mixed numbers by whole numbers. The steps students follow to work these problems that involve mixed numbers parallel those for working other multiplication problems that have a mixed number and a whole number. The only difference is that after the computation is completed, the answer often has a fraction that is not less than 1. Students rewrite the answer as a mixed number.

In Lesson 106, at the end of the sequence, students work problems that have 2-digit mixed numbers.

Here's the set of problems students work:

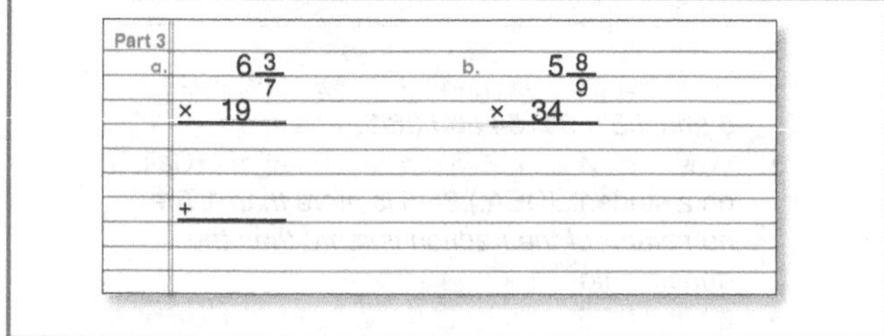

For problem A, students work the problem for 6 3/7 × 9.

Then they work the problem 6 3/7 × 1.

$$
\begin{array}{r}
6\frac{3}{7} \\
\times \quad 19 \\
\hline
54\frac{27}{7} \\
+ \quad 60\frac{30}{7} \\
\hline
114\frac{57}{7}
\end{array}
\qquad
\begin{array}{r}
8\;\frac{1}{7} \\
7\overline{)57} \\
56
\end{array}
$$

$$114\frac{57}{7} = 114 + 8\frac{1}{7} = 122\frac{1}{7}$$

Coordinate System and Functions (Lessons 81–130)

This track coordinates two closely related sets of operations—those that have to do with the coordinate system used to graph relationships between X and Y values, and those used to derive functions such as X + 5 and 3M. These functions are applied to the coordinate system but are not limited to those applications. The first work with the coordinate system occurs on Lesson 81. Students create X and Y descriptions for points on the coordinate system.

Coordinate Systems

They learn that the X direction is to the right and the Y direction is up. The distance is measured by the number of units on the coordinate grid. The early grids do not have numbers on the axes. The reason is that we want to establish the procedure of starting at zero and counting the squares in the X direction and the Y direction.

Note: The standard notation for a point on the coordinate system is (X, Y). If point A on a coordinate system has an X value of 2 and a Y value of 9, the coordinates are (2, 9).

Here's the first part of the exercise from Lesson 83:

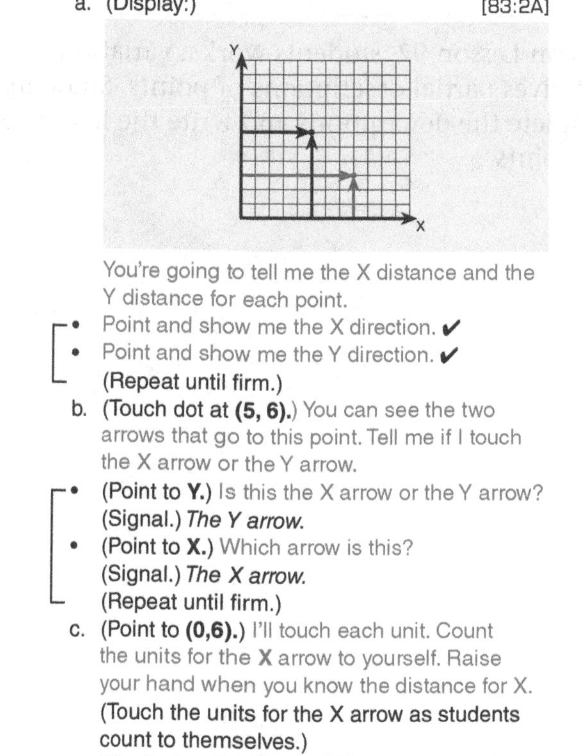

a. (Display:) [83:2A]

You're going to tell me the X distance and the Y distance for each point.

- Point and show me the X direction. ✔
- Point and show me the Y direction. ✔

(Repeat until firm.)

b. **(Touch dot at (5, 6).)** You can see the two arrows that go to this point. Tell me if I touch the X arrow or the Y arrow.

- **(Point to Y.)** Is this the X arrow or the Y arrow? (Signal.) *The Y arrow.*
- **(Point to X.)** Which arrow is this? (Signal.) *The X arrow.*

(Repeat until firm.)

c. **(Point to (0,6).)** I'll touch each unit. Count the units for the **X** arrow to yourself. Raise your hand when you know the distance for X. (Touch the units for the X arrow as students count to themselves.)

- What's the distance for X? (Signal.) *5 units.*
- So what equation do I write above the X arrow? (Signal.) *X = 5.*

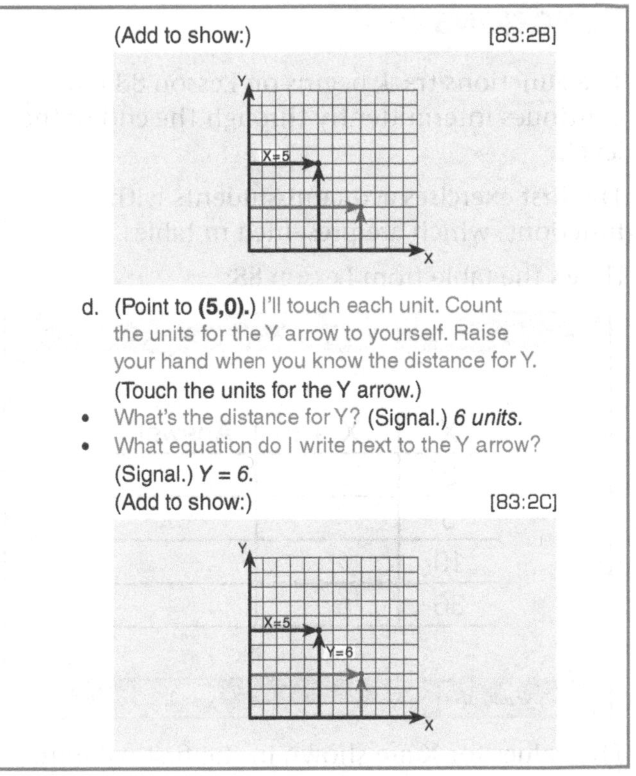

(Add to show:) [83:2B]

d. **(Point to (5,0).)** I'll touch each unit. Count the units for the Y arrow to yourself. Raise your hand when you know the distance for Y. (Touch the units for the Y arrow.)

- What's the distance for Y? (Signal.) *6 units.*
- What equation do I write next to the Y arrow? (Signal.) *Y = 6.*

(Add to show:) [83:2C]

from Lesson 83, Exercise 2

On Lesson 86, students use a notation system that is closer to the traditional one. They continue to write equations for X and Y, but each pair of equations appears next to the letter for each point.

Here's the set of points students complete for Workbook Lesson 86:

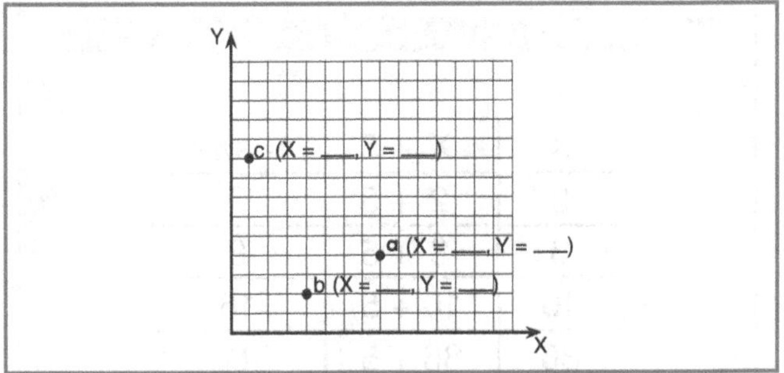

Students start at the origin, count in the X direction to the vertical line the point is on, record the X distance, then count the Y direction to the point and record the Y distance.

For example: Point A is (X = 8, Y = 4).

FUNCTIONS

The Functions track begins on Lesson 88 and continues intermittently through the end of the level.

The first exercises acquaint students with functions, which are presented in tables.

Here's the table from Lesson 88:

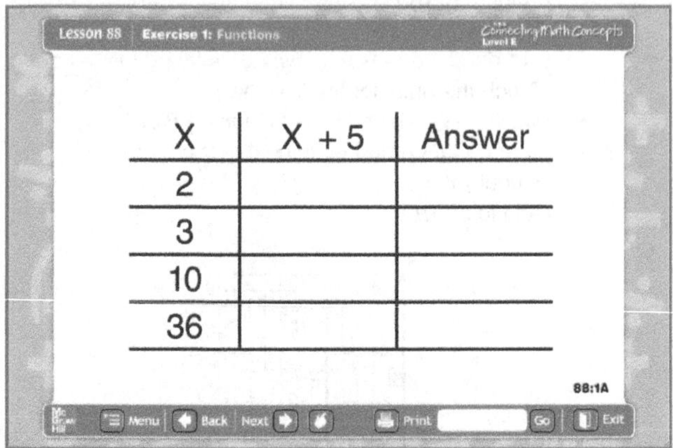

X	X + 5	Answer
2		
3		
10		
36		

The values for X are shown in the first column.

The functions are shown in the middle column.

The answers are shown in the last column.

For each row, students say the number X equals (for the first row X = 2); the problem with the number for X (for the first row, 2 + 5) and the answer (for the first row, 7).

Here's the completed table:

Lesson 88 Exercise 1: Functions — Connecting Math Concepts Level E

X	X + 5	Answer
2	2 + 5	7
3	3 + 5	8
10	10 + 5	15
36	36 + 5	41

88:1I

Menu Back Next Print Go Exit

Starting on Lesson 88, students record the coordinates on lined paper.

Here's the Textbook material for Lesson 88:

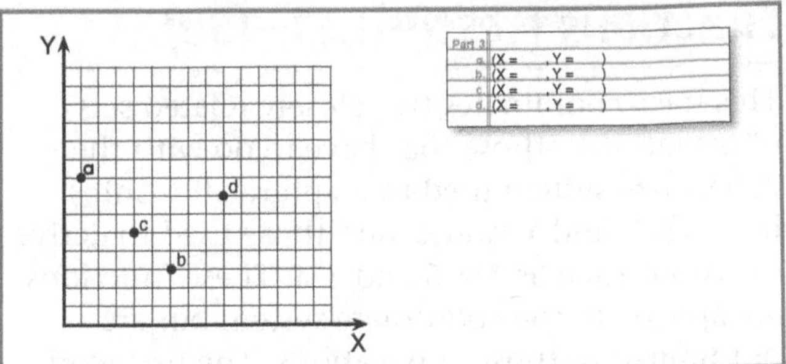

Teaching Note: Students follow the same steps they performed earlier, but the mechanics are more demanding because they write the equations for X and Y on lined paper instead of next to the point. Remind students not to write anything in their Textbook.

On Lesson 92, the function table is changed so that the third column does not refer to the answer—but rather to another letter.

Here's the table from Lesson 92:

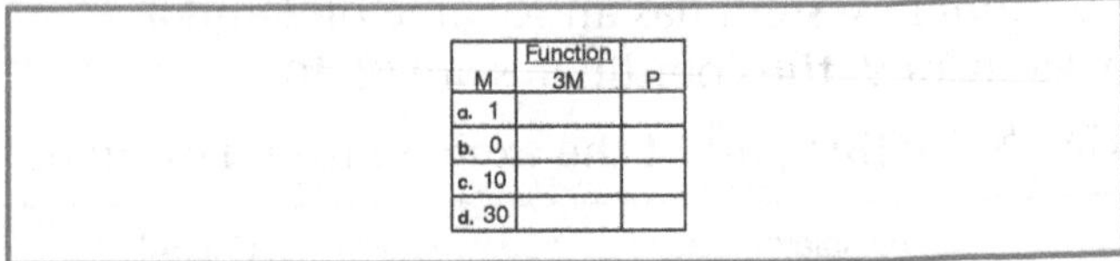

	M	Function 3M	P
a.	1		
b.	0		
c.	10		
d.	30		

Also on Lesson 92, students work a variation that gives partial descriptions of points. Students complete the descriptions and write the letters for the points.

Here's the exercise from Lesson 92:

(Teacher reference:)

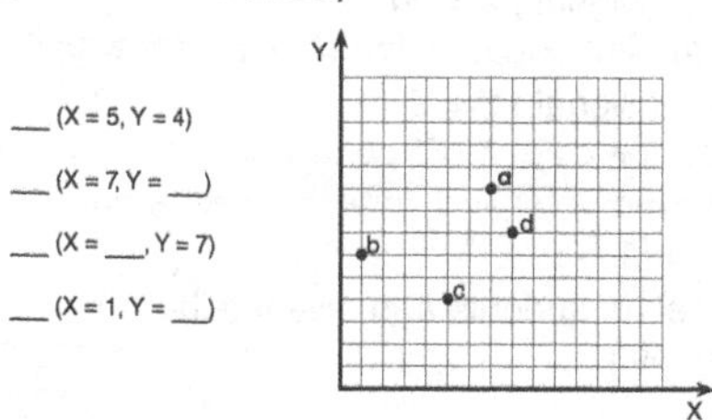

The grid shows points A through D. The rows next to the grid are supposed to show the letters and the equations for the points. The letters and some of the numbers are missing. You're going to write the letters and complete the equations for each row.

- Touch where you'll write the letter for the first row. ✔
- What does X equal for that point? **(Signal.)** *5.*
- What does Y equal? **(Signal.)** *4.*

b. Touch the point on the grid with the X distance of 5 and the Y distance of 4. Raise your hand when you know the letter for that point. ✔

- What's the letter for the point with an X value of 5 and a Y value of 4? **(Signal.)** *C.*
- Write the letter for point C. ✔

c. Touch the row for the next point. ✔

- What does X equal? **(Signal.)** *7.*
- Count 7 on the X arrow. Then touch the point with the X distance of 7 on the grid. Raise your hand when you know the letter and the Y distance for that point. ✔
- What's the letter for the point with an X value of 7? **(Signal.)** *A.*
- Tell me the Y distance of point A. Get ready. **(Signal.)** *9.*

(Display:) [92:2A]

c. (X = 5, Y = 4)
a. (X = 7, Y = 9)
__ (X = __, Y = 7)
__ (X = 1, Y = __)

- Write the letter and complete the Y equation for point A. ✔

d. Touch the row for the next point. ✔

- What does Y equal? **(Signal.)** *7.*
- Count 7 on the Y arrow. Then touch the point with the Y distance of 7. Raise your hand when you know the letter and the X distance for that point. ✔
- What's the letter for the point with a Y value of 7? **(Signal.)** *D.*
- Tell me the X distance of point D. Get ready. **(Signal.)** *8.*
- Write the letter and complete the X equation for point D. ✔

(Add to show:) [92:2B]

c. (X = 5, Y = 4)
a. (X = 7, Y = 9)
d. (X = 8, Y = 7)
__ (X = 1, Y = __)

Here's what you should have.

e. Touch the row for the last point. ✔

- What does X equal? **(Signal.)** *1.*
- Touch the point with the X distance of 1. Raise your hand when you know the letter and the Y distance for that point. ✔
- What's the letter for the point with an X value of 1? **(Signal.)** *B.*
- Tell me the Y distance of point B. Get ready. **(Signal.)** *6.*
- Write the letter and complete the Y equation for point B. ✔

(Add to show:) [92:2C]

c. (X = 5, Y = 4)
a. (X = 7, Y = 9)
d. (X = 8, Y = 7)
b. (X = 1, Y = 6)

Here's what you should have for part 2.

Lesson 92, Exercise 2

On Lesson 94, students learn that the conventions for the coordinate system are related to function tables with X and Y values. All of the functions students work with in *CMC Level E* generate points on a straight line.

Here's the first part of the exercise from Lesson 95:

(Teacher reference:) **R Part B**

Table 1

X	Function X − 3	Y
a. 10		
b. 7		
c. 4		

Table 2

X	Function 9 − X	Y
a. 9		
b. 4		
c. 0		

- Touch table 1. ✔
  Table 1 shows a function and two letters. The first column shows what X equals for each row. The last column will show what Y equals for each row.
- Read the function for table 1. Get ready. **(Signal.)** *X − 3.*

b. Write the problem and what Y equals for all of the rows. Put your pencil down when you've completed table 1.
(Observe students and give feedback.)

c. Check your work. For each row, you'll say the equation for X, the problem, and the equation for Y.
- Row A. Say the equation for X. **(Signal.)** *X = 10.*
- Read the problem. **(Signal.)** *10 − 3.*
- Say the equation for Y. **(Signal.)** *Y = 7.*

d. **(Repeat the following tasks for rows B through D:)**

Row ___. Say the equation for X.		Read the problem.	Say the equation for Y.
B	*X = 7*	*7 − 3*	*Y = 4*
C	*X = 4*	*4 − 3*	*Y = 1*

(Display:) [95:3A]

Table 1

X	Function X − 3	Y
a. 10	10 − 3	7
b. 7	7 − 3	4
c. 4	4 − 3	1

Here's what you should have.
- Make sure your table 1 is correct because you're going to use it to make the point for each row. ✔

e. Touch the row for point A in table 1. ✔
- Tell me the equation for the X distance. Get ready. **(Signal.)** *X = 10.*
- Tell me the equation for the Y distance. Get ready. **(Signal.)** *Y = 7.*
- Touch the point X = 10 and Y = 7 on the grid. ✔
- Make the point and write the letter A on your grid.
  (Observe students and give feedback.)
  (Display:) [95:3B]

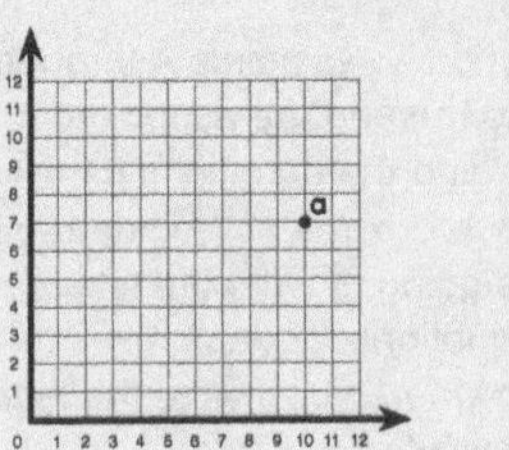

Here's the point for row A.

f. Make the points for B and C. Put your pencil down when you've made the points for all of the rows of table 1.
(Observe students and give feedback.)
(Add to show:) [95:3C]

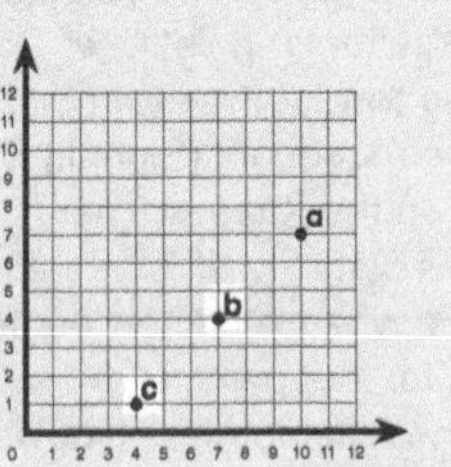

Here are the points for table 1.

from Lesson 95, Exercise 3

Teaching Note: The numbers are shown for the X and Y axes. With this addition, students can find and describe points faster.

Fractional Coefficients

On Lesson 98, students work with functions that show a fraction times X.

Here's the table from Workbook Lesson 98:

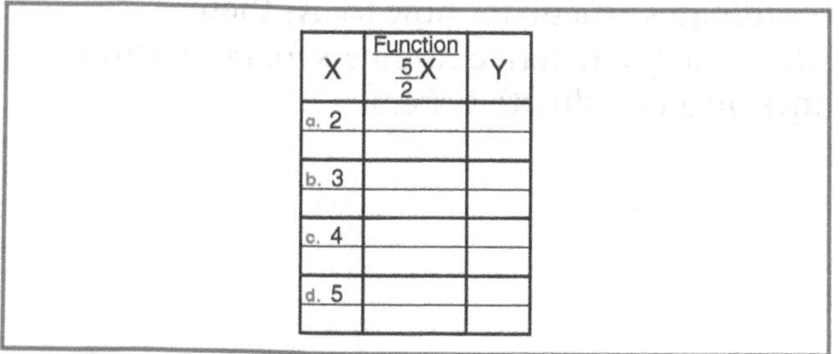

X	Function $\frac{5}{2}X$	Y
a. 2		
b. 3		
c. 4		
d. 5		

Students write the function for each row and indicate the Y value.

For row A, the function is 5/2 × 2. The Y value is 10/2.

In later lessons, students simplify fractions (showing 10/2 = 5) and make the points on a coordinate grid.

On Lesson 100, students plot two points, draw the line that connects the points, and then identify X or Y values by inspecting the line. Students confirm the location of the points by working the problems in the function table.

Here's the table and the coordinate system for Workbook Lesson 102:

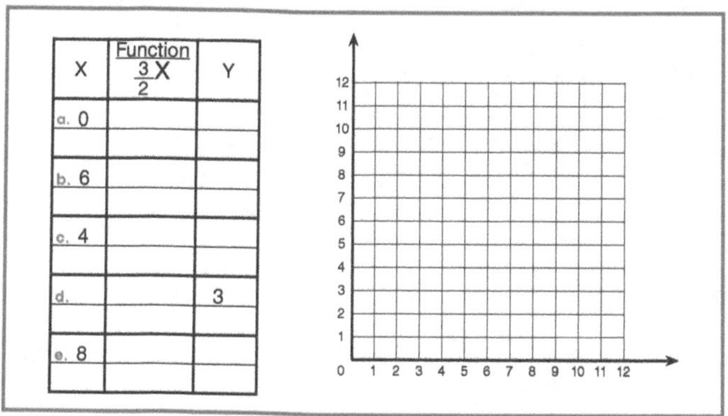

X	Function $\frac{3}{2}X$	Y
a. 0		
b. 6		
c. 4		
d.		3
e. 8		

Here's the completed table and the line on the coordinate grid with the points.

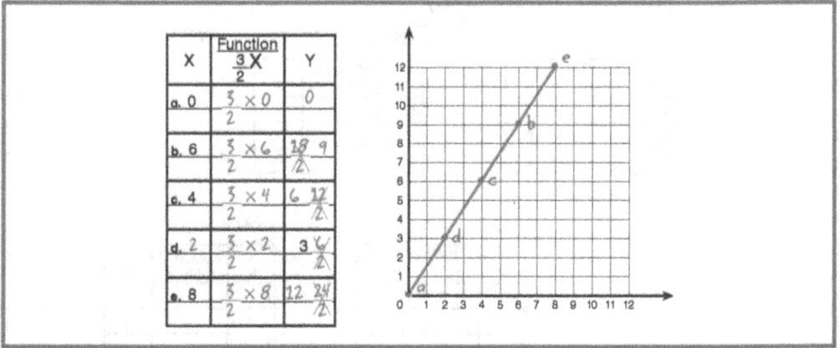

X	Function $\frac{3}{2}X$	Y
a. 0	$\frac{3}{2} \times 0$	0
b. 6	$\frac{3}{2} \times 6$	$\frac{18}{2}$ 9
c. 4	$\frac{3}{2} \times 4$	$\frac{12}{2}$ 6
d. 2	$\frac{3}{2} \times 2$	$\frac{6}{2}$ 3
e. 8	$\frac{3}{2} \times 8$	$\frac{24}{2}$ 12

Starting on Lesson 113, students solve for missing X values without referring to lines on a coordinate system.

Students say the equation for each row. For example, if the function for a table is 20 – X, and Y equals 8 for a row, then the equation for the row is 20 – X = 8. Students solve for X using inverse operations. Students infer that 20 is the big number and X is a small number. So they subtract to find the missing small number. The problem they work is 20 – 8. So the value for X is 12.

Here's the first part of the exercise from Lesson 116:

(Teacher reference:)

Table 1

X	Function 3X	Y
a.		27
b.		6
c.		0

Table 2

X	Function X ÷ 5	Y
d.		5
e.		10
f.		0

Table 3

X	Function 27 – X	Y
g.		20
h.		10
i.		0

Table 4

X	Function 24 ÷ X	Y
j.		1
k.		4
l.		8

You're going to say the problem and the answer to the missing X value for the first row of each table.

- Read the function for table 1. Get ready. (Signal.) *3X.*
- Say the problem for 3X. (Signal.) *3 times X.*
- For row A, what does 3 times X equal? (Signal.) *27.*
- Say the equation for row A. (Signal.) *3 × X = 27.*
- Is X the big number or a small number? (Signal.) *A small number.*
- Say the problem for finding the X value of row A. Get ready. (Signal.) *27 ÷ 3.*
- What's X for row A? (Signal.) *9.*

b. Read the function for table 2. Get ready. (Signal.) *X ÷ 5.*
- Say the equation for row D. Get ready. (Signal.) *X ÷ 5 = 5.*
- Is X the big number or a small number? (Signal.) *The big number.*
- Say the problem for finding the X value of row D. Get ready. (Signal.) *5 × 5.*
- What's X for row D? (Signal.) *25.*

c. Read the function for table 3. Get ready. (Signal.) *27 – X.*
- Say the equation for row G. Get ready. (Signal.) *27 – X = 20.*
- Is X the big number or a small number? (Signal.) *A small number.*
- Say the problem for finding the X value of row G. Get ready. (Signal.) *27 – 20.*
- What's X for row G? (Signal.) *7.*

d. Read the function for table 4. Get ready. (Signal.) *24 ÷ X.*
- Say the equation for row J. Get ready. (Signal.) *24 ÷ X = 1.*
- Is X the big number or a small number? (Signal.) *A small number.*
- Say the problem for finding the X value of row J. Get ready. (Signal.) *24 ÷ 1.*
- What's X for row J? (Signal.) *24.*

(Repeat until firm.)

from Lesson 116, Exercise 1

This track is complicated, but the various operations that students are to perform are developed carefully and make sense. The functions show how to express relationships between X and Y values. The work with the coordinates articulates how to use these relationships to derive numbers for points and lines on a coordinate system.

Ratio Word Problems (Lessons 76–130)

This track begins on Lesson 76 and continues intermittently through the end of the program. Exercises teach students to solve problems that refer to ratios.

For example: The ratio of corn to wheat was 5 to 3. If there were 30 pounds of corn, how many pounds of wheat were there?

The problem gives two names—*corn* and *wheat.* Students write a ratio with those names:

$$\begin{matrix}\text{Corn}\\\text{Wheat}\end{matrix}$$

Students write a fraction with the ratio numbers 5 and 3:

$$\begin{matrix}\text{Corn}\\\text{Wheat}\end{matrix}\ \frac{5}{3}$$

The problem gives a number for a second fraction (30 pounds of corn):

$$\begin{matrix}\text{Corn}\\\text{Wheat}\end{matrix}\ \frac{5}{3} \times \frac{}{} = \frac{30}{}$$

The problem has two numbers in the numerators. Students work the problem 5 times what number equals 30 and complete the fraction that equals 1:

$$\begin{matrix}\text{Corn}\\\text{Wheat}\end{matrix}\ \frac{5}{3} \times \frac{6}{6} = \frac{30}{}$$

Students are now able to work the problem in the denominators:

$$3 \times 6 = \_\_$$

The problem asks, "... how many pounds of wheat were there?"

The answer is 18 pounds:

$$\begin{matrix}\text{Corn}\\\text{Wheat}\end{matrix}\ \frac{5}{3} \times \frac{6}{6} = \frac{30}{18}$$

Each/Every

On Lesson 76, students learn about writing fractions from sentences of the form "There were 4 dogs for every 7 cats." All sentences in the early exercises of this track refer to *each* or *every.*

Here's the first part of the exercise on Lesson 76. It introduces the procedures for identifying names and the first fraction in ratio equations:

a. We're going to write fractions for sentences that name two things.
- Listen: There were 3 girls for every 2 boys. Say that sentence. (Signal.) *There were 3 girls for every 2 boys.*
- The sentence names girls and boys. What does the sentence name? (Signal.) *Girls and boys.*
(Display:) [76:6A]

girls
boys

b. The sentence gives a number for girls and a number for boys. Listen: There were 3 girls for every 2 boys.
- What's the number for girls? (Signal.) *3.*
- What's the number for boys? (Signal.) *2.*
(Add to show:) [76:6B]

$$\begin{matrix}\text{girls}\\\text{boys}\end{matrix}\ \frac{3}{2}$$

c. 3/2 shows that there were 3 girls for every 2 boys. The fraction doesn't tell you that there were 3 girls. It just tells you if there were 3 girls, there would be 2 boys.
If there were **6** girls, there wouldn't be 2 boys. There would be 4 boys.
If there were 9 girls, there would be 6 boys.

d. Your turn: If there were **3** girls, how many boys would there be? (Signal.) *2.*
- If there were **6** girls, how many boys would there be? (Signal.) *4.*
- If there were **9** girls, how many boys would there be? (Signal.) *6.*
- If there were **12** girls, how many boys would there be? (Signal.) *8.*
- If there were **15** girls, how many boys would there be? (Signal.) *10.*
(Repeat step d until firm.)

e. Listen: If there were **2 boys,** how many girls would there be? (Signal.) *3.*
- If there were **4** boys, how many girls would there be? (Signal.) *6.*
- If there were **6** boys, how many girls would there be? (Signal.) *9.*
(Repeat step e until firm.)

f. New sentence: There were 17 trucks for every 5 cars. Say that sentence. (Signal.) *There were 17 trucks for every 5 cars.*
- The sentence names two things. What are the names? (Signal.) *Trucks and cars.*

(Display:) [76:6C]

trucks
cars

The sentence gives a number for trucks and cars. Listen again: There were 17 trucks for every 5 cars.

- What's the number for trucks? (Signal.) *17.*
- What's the number for cars? (Signal.) *5.*

(Add to show:) [76:6D]

trucks $\frac{17}{5}$
cars

- What's the fraction for trucks to cars? (Signal.) *17/5.*

g. If there were 17 trucks, how many cars would there be? (Signal.) *5.*

- If there were **10 cars**, would there be 17 or more than 17 trucks? (Signal.) *More than 17.*

from Lesson 76, Exercise 6

Teaching Notes: For each sentence, students answer two questions—one that identifies the top number of the fraction; the other that identifies the bottom number.

For example:

There were 8 girls for every 2 boys.

The names appear in the fraction in the order they occur in the sentence. So the fraction with names is:

Girls
Boys

The names are followed by the ratio numbers:

Girls $\frac{8}{2}$
Boys

Repeat examples that are not firm.

Students work complete problems starting on Lesson 81.

Here's part of the exercise from Lesson 82:

k. Touch problem C. ✔

- Read the problem. (Call on a student.) *There were 4 wheels on each truck. There were 8 trucks. How many wheels were there?*
- Write the names and the numbers for the first fraction. Then stop.

(Observe students and give feedback.)

l. Check your work for problem C.

- What are the names? (Signal.) *Wheels and trucks.*
- What is the first fraction? (Signal.) *4 over 1.*

(Display:) [82:6I]

c. wheels $\frac{4}{1}$
trucks

Here's what you should have.

m. Now write the signs, the fraction bars, and the number for the fraction that equals 4 over 1.

(Observe students and give feedback.)

- Did you write a number for wheels or trucks? (Signal.) *Trucks.*
- What number? (Signal.) *8.*

(Add to show:) [82:6J]

c. wheels trucks $\frac{4}{1} \times \frac{\ }{\ } = \frac{\ }{8}$

Here's what you should have for C.

n. Now complete the equation. Box the answer to the question.

(Observe students and give feedback.)

o. Check your work for problem C.

- Everybody, what's the fraction that equals 1? (Signal.) *8/8.*
- If there were 8 trucks, tell me how many wheels there were. Get ready. (Signal.) *32.*

(Add to show:) [82:6K]

c. wheels trucks $\frac{4}{1} \times \frac{8}{8} = \frac{\boxed{32}}{8}$

Here's what you should have for problem C.

p. Touch problem D. ✔

- Read the problem. (Call on a student.) *For every 5 houses, there were 3 garages. There were 30 houses. How many garages were there?*
- Write the names and the numbers for the first fraction. Then stop.

(Observe students and give feedback.)

q. Check your work for problem D.

- What are the names? (Signal.) *Houses and garages.*
- What is the first fraction? (Signal.) *5/3.*

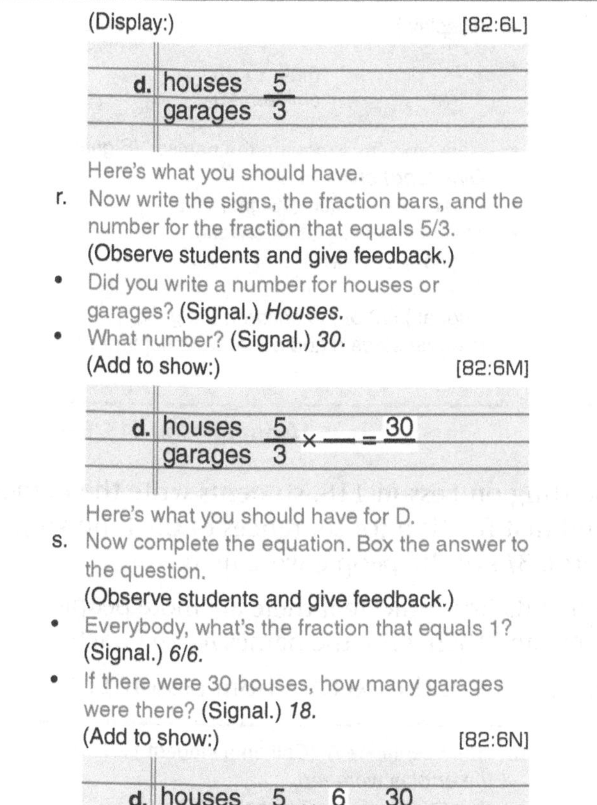

(Display:) [82:6L]

d.	houses	5
	garages	3

Here's what you should have.

r. Now write the signs, the fraction bars, and the number for the fraction that equals 5/3.
(Observe students and give feedback.)

- Did you write a number for houses or garages? (Signal.) *Houses.*
- What number? (Signal.) *30.*

(Add to show:) [82:6M]

d. houses/garages $\frac{5}{3} \times \frac{\ }{\ } = \frac{30}{\ }$

Here's what you should have for D.

s. Now complete the equation. Box the answer to the question.
(Observe students and give feedback.)

- Everybody, what's the fraction that equals 1? (Signal.) *6/6.*
- If there were 30 houses, how many garages were there? (Signal.) *18.*

(Add to show:) [82:6N]

d. houses/garages $\frac{5}{3} \times \frac{6}{6} = \frac{30}{\boxed{18}}$

Here's what you should have for problem D.

from Lesson 82, Exercise 6

Teaching Notes: The only new part of working these problems involves setting up the last part of the ratio equation so it corresponds to the information the problem gives. Students identify the sentence that tells how to make the number family: *There were 4 wheels on every truck.* Students set up the names and the first fraction according to the procedures they have practiced:

$$\frac{\text{Wheels}}{\text{Trucks}} \ \frac{4}{1}$$

They write one number for the last fraction. It comes from the sentence "There were 8 trucks."

$$\frac{\text{Wheels}}{\text{Trucks}} \ \frac{4}{1} \times \frac{\ }{\ } = \frac{\ }{8}$$

Students complete the equation by first figuring out the fraction that equals one (8/8) and then by multiplying to complete the last fraction (32/8). Finally, they answer the question the problem asks with a number and a unit name, 32 wheels.

$$\frac{\text{Wheels}}{\text{Trucks}} \ \frac{4}{1} \times \frac{8}{8} = \frac{32}{8}$$

Students are introduced to sentences that have "new" wording on Lesson 85. This wording refers to ratios.

For example:

The ratio of tires to bolts was 9 to 4.

The first number named goes in the numerator of the first fraction.

$$\frac{\text{Tires}}{\text{Bolts}} \ \frac{9}{4}$$

Here are the sentences from Textbook Lesson 85:

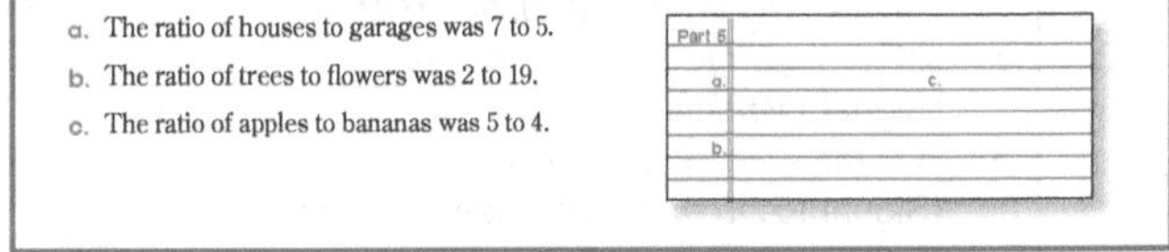

a. The ratio of houses to garages was 7 to 5.
b. The ratio of trees to flowers was 2 to 19.
c. The ratio of apples to bananas was 5 to 4.

On Lesson 86, students work problems that have 3-digit and 4-digit numbers.

Here are the problems from Textbook Lesson 86:

Problems

a. An MP3 player contains 954 songs. The songs are divided into sections. There are 9 songs in every 2 sections. How many sections are the songs divided into?

b. Aluminum oxide contains 2 aluminum atoms for every 3 oxygen atoms. There are 7326 oxygen atoms. How many aluminum atoms are there?

Part 7	
a.	
b.	

On Lesson 88, students work a mixed set of ratio word problems. Some have the word *ratio*. Others refer to *each* or *every*.

FRACTION WORDING

On Lesson 115, students create sentences that refer to the ratio numbers of two groups. For instance, the names presented are *robins* and *birds*. Students answer the question, "Are there more robins or birds?" Students then start with a fraction (4/7) and say the sentence about robins and birds.

"4/7 of the birds are robins."

Here's part of the exercise from Lesson 115:

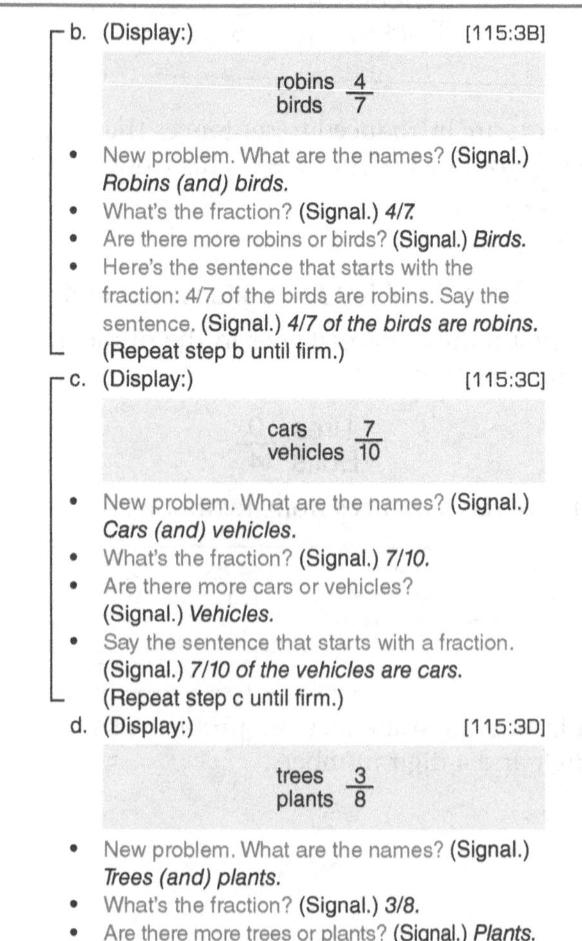

b. (Display:) [115:3B]

robins $\frac{4}{7}$
birds

- New problem. What are the names? (Signal.) *Robins (and) birds.*
- What's the fraction? (Signal.) *4/7.*
- Are there more robins or birds? (Signal.) *Birds.*
- Here's the sentence that starts with the fraction: 4/7 of the birds are robins. Say the sentence. (Signal.) *4/7 of the birds are robins.*

(Repeat step b until firm.)

c. (Display:) [115:3C]

cars $\frac{7}{10}$
vehicles

- New problem. What are the names? (Signal.) *Cars (and) vehicles.*
- What's the fraction? (Signal.) *7/10.*
- Are there more cars or vehicles? (Signal.) *Vehicles.*
- Say the sentence that starts with a fraction. (Signal.) *7/10 of the vehicles are cars.*

(Repeat step c until firm.)

d. (Display:) [115:3D]

trees $\frac{3}{8}$
plants

- New problem. What are the names? (Signal.) *Trees (and) plants.*
- What's the fraction? (Signal.) *3/8.*
- Are there more trees or plants? (Signal.) *Plants.*
- Say the sentence that starts with the fraction. (Signal.) *3/8 of the plants are trees.*

e. (Display:) [115:3E]

girls $\frac{5}{9}$
children

- New problem. What are the names? (Signal.) *Girls (and) children.*
- What's the fraction? (Signal.) *5/9.*
- Are there more girls or children? (Signal.) *Children.*
- Say the sentence that starts with a fraction. (Signal.) *5/9 of the children are girls.*

(Repeat steps d and e that were not firm.)

from Lesson 115, Exercise 3

Starting on Lesson 118, students write the names and first fraction for sentences of the following form: 3/4 of the people were men.

Students first indicate if there are more people or men, and then write the names for the fraction 3/4.

Here's part of the exercise from Lesson 118.

b. Read sentence A. (Call on a student.) *2/7 of the apples were red.*

- What's the fraction? (Signal.) *2/7.*
- What are the names? (Signal.) *Apples (and) red.*
- Are there more apples or red apples? (Signal.) *Apples.*
- So is apples the name for 2 or 7? (Signal.) *7.*

(Display:) [118:3I]

Part 2	
a.	red $\frac{2}{7}$ apples

Here are the names and the first fraction for sentence A.

- Write the names and the first fraction for sentence A. ✔

(If students are 100% on this exercise, skip to step d.)

c. Read sentence B. (Call on a student.) *3/4 of the people were men.*

- What's the fraction? (Signal.) *3/4.*
- What are the names? (Signal.) *People (and) men.*
- Are there more people or men? (Signal.) *People.*
- So is people the name for 3 or 4? (Signal.) *4.*

from Lesson 118, Exercise 3

Students work complete problems on Lesson 122. They write ratio problems and figure out the missing values.

> b. Read problem A to yourself and set up the ratio equation. Do not complete the ratio equation. Put your pencil down when you've set up the ratio equation for problem A. **(Observe students and give feedback.)**
>
> c. Check your setup of ratio equation A.
>
> - What are the names? **(Signal.)** *Dirty (and) windows.*
> - What's the first fraction? **(Signal.)** *2/9.*
> - You can work the problem for the numerators **or** denominators. Read the problem you can work. Get ready. **(Signal.)** *2 × what number = 36.*
>
> **(Display:)** [122:5A]
>
> **Part 3**
>
> a. dirty / windows $\frac{2}{9} \times \frac{}{} = \frac{36}{}$
>
> Here's the setup for problem A.
>
> **(If students are 100%, skip to step h.)**

from Lesson 122, Exercise 5

Here's the set up for the ratio equation:

$$\begin{matrix}\text{Dirty} \\ \text{Windows}\end{matrix} \quad \frac{2}{9} \times \frac{}{} = \frac{36}{}$$

To complete the problem, students figure out that the fraction of 1 is 18/18. Then they multiply 9 x 18 in the denominators to complete the equation and answer the question about how many windows there were.

$$\begin{matrix}\text{Dirty} \\ \text{Windows}\end{matrix} \quad \frac{2}{9} \times \frac{18}{18} = \frac{36}{162}$$

Starting on Lesson 124, students work a mix of ratio problems. Some refer to *each* and *every;* some have the word *ratio;* some name a fraction.

Here's the set of problems from Lesson 124:

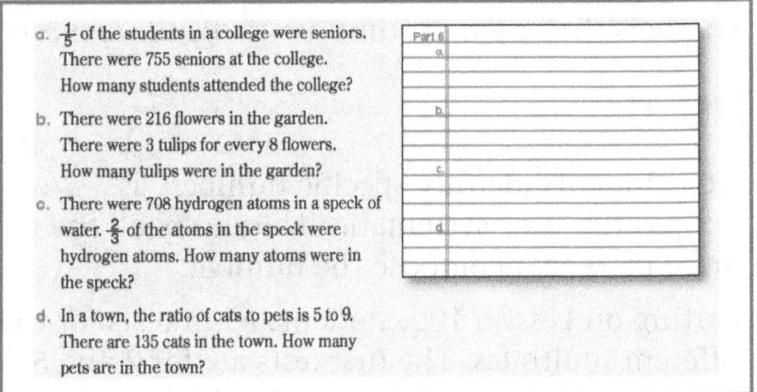

a. $\frac{1}{5}$ of the students in a college were seniors. There were 755 seniors at the college. How many students attended the college?

b. There were 216 flowers in the garden. There were 3 tulips for every 8 flowers. How many tulips were in the garden?

c. There were 708 hydrogen atoms in a speck of water. $\frac{2}{3}$ of the atoms in the speck were hydrogen atoms. How many atoms were in the speck?

d. In a town, the ratio of cats to pets is 5 to 9. There are 135 cats in the town. How many pets are in the town?

Students continue to work mixed-problem sets through the end of *CMC Level E.*

Prime Factors (Lessons 100–130)

This track begins on Lesson 100 and continues through the end of the program. Students first learn about multiples of 2, 3, 5, and 7. Next, they learn that *prime numbers have only two multiplication facts—1 times the number and the number times 1.*

Students use knowledge of 2, 3, 5, and 7 to identify prime numbers to 100.

If a number (79 for instance) is not a multiple of 2, 3, 5, or 7, it is a prime.

Numbers that are not prime are *composite numbers.*

MULTIPLES

After students identify specific numbers as composites, they systematically identify all the factor pairs that compose the number.

Starting on Lesson 100, students learn tests for the different multiples. The first tests are for 2 and 5. Even numbers are multiples of 2 (which means that dividing these numbers by 2 yields whole-number answers). Numbers that end in 5 or zero are multiples of 5.

Students indicate whether specific numbers are divisible by 2, by 5, by 2 and 5, or by neither 2 nor 5.

Here's part of the Workbook exercise from Lesson 101:

a. Find part 2 in your workbook. ✔
(Teacher reference:) **R Part D**

a. 80 d. 74 f. 91
b. 35 e. 110 g. 145
c. 93

2	
5	

To get whole numbers, you can divide some of these numbers by 2. You can divide some of these numbers by 5. You can divide some of these numbers by 2 **and** by 5.

- Read number A. (Signal.) *80.*
- Can you divide it by 2 and get a whole number? (Signal.) *Yes.*
- So write 80 in the row for 2. ✔

b. Does 80 end in zero or 5? (Signal.) *Yes.*
- So can you divide 80 by 5 and get a whole number? (Signal.) *Yes.*
  Yes, it ends in zero, so you can divide it by 5.
- Write 80 in the row for 5. ✔
  (Display:) [101:2F]

2	80
5	80

c. Read number B. (Signal.) *35.*
- Can you divide 35 by 2? (Signal.) *No.*
- Can you divide 35 by 5? (Signal.) *Yes.*
- Write it in the row for 5. ✔

d. Read number C. (Signal.) *93.*
- Can you divide 93 by 2? (Signal.) *No.*
- Can you divide 93 by 5? (Signal.) *No.*
  So you don't write 93 in either row.

from Lesson 101, Exercise 2

Students complete the table to show which numbers are multiples of 2, 5, or both.

a. 80 d. 74 f. 91
b. 35 e. 110 g. 145
c. 93

2	
5	

On Lesson 105, students learn the test for whether numbers are multiples of 3: You add the digits in the answer. If the answer is a multiple of 3, the number is divisible by 3.

Example:

417. The sum of the digits is 12.

12 is a multiple of 3.

So 417 is a multiple of 3.

Multiples Divisible by 7

Starting on Lesson 112, students identify whether numbers are divisible by 7. There is no shorthand method for identifying multiples of 7. Students apply their knowledge of multiplication facts involving 7 × 1 = 7 through 7 × 10 = 70. Students also learn to count by 7 from 77 through 98. Then they test whether numbers through 100 are divisible by 7.

Here's the first part of the exercise:

a. (Display:) [112:7A]

7	14	21	28
77	84	91	98

These are some of the numbers for counting by 7.

- Read the numbers in the top row. Get ready. (Signal.) *7, 14, 21, 28.*

The numbers in the bottom row are the numbers for counting by 7 from 77 to 98.

- Read the numbers in the bottom row. Get ready. (Signal.) *77, 84, 91, 98.*

b. The ones digits are the same as the ones digits in the top row.

- What's the ones digit of 77? (Signal.) *7.*
- What's the ones digit of 84? (Signal.) *4.*
- What's the ones digit of 91? (Signal.) *1.*
- What's the ones digit of 98? (Signal.) *8.*

Remember the pattern: 7, 4, 1, 8.

c. Start with 7 and say the numbers for counting by 7 to 28. Get ready. (Signal.) *7, 14, 21, 28.*

- Start with 77 and say the numbers for counting by 7 from 77 to 98. Get ready. (Signal.) *77, 84, 91, 98.*

(Repeat step c until firm.)

from Lesson 112, Exercise 7

On Lesson 113, students test whether different numbers are multiples of 7, 2, 3, or 5.

Here are the numbers they test and the table used for recording results from Workbook Lesson 113:

a. 48 c. 64 e. 35 g. 77 i. 91
b. 21 d. 84 f. 93 h. 81

7	
2	
3	
5	

Students write all the multiples for the number. The multiples they write for 7 are 21, 84, 35, 77 and 91.

Prime Numbers

On Lesson 118, students learn about prime numbers. The rule they learn is that *there are only two whole-number multiplication equations that equal a prime number.* If more than two whole-number equations equal a whole number, the number is not a prime.

For example:

11 is a prime number. The two facts are
11 × 1 = 11 and 1 × 11 = 11.

On Lesson 120, students learn the primes through 10 (2, 3, 5, and 7).

On Lesson 121, students learn the test for prime numbers through 100. The test students use is that if the number is not divisible by 2, 3, 5, or 7, it is a prime.

Here's part of the exercise from Lesson 121:

b. You'll read each number and tell me if the number is divisible by 2, 3, 5, or 7. Then you'll tell me if the number is prime.
- Read number A. Get ready. (Signal.) *83.*
- Tell me if 83 is divisible by 2. Get ready. (Signal.) *No.*
- By 3. Get ready. (Signal.) *No.*
- By 5. Get ready. (Signal.) *No.*
- By 7. Get ready. (Signal.) *No.*
- 83 is not divisible by 2, 3, 5, or 7, so is 83 prime? (Signal.) *Yes.*
- Start with 1 and say the whole-number multiplication equation that equals 83. (Signal.) *1 × 83 = 83.*
- Say the other equation. (Signal.) *83 × 1 = 83.*
(If students are 100%, skip to step k.)

c. Read number B. (Signal.) *84.*
- Tell me if 84 is divisible by 2. Get ready. (Signal.) *Yes.*
- 84 is divisible by 2, so is 84 prime? (Signal.) *No.* I'll say the whole-number multiplication equation that equals 84 and does not have 1 in it. 2 × 42 = 84.

d. Read number C. (Signal.) *29.*
- Tell me if 29 is divisible by 2. Get ready. (Signal.) *No.*
- By 3. Get ready. (Signal.) *No.*
- By 5. Get ready. (Signal.) *No.*
- By 7. Get ready. (Signal.) *No.*
- 29 is not divisible by 2, 3, 5, or 7, so is 29 prime? (Signal.) *Yes.*

e. Read number D. Get ready. (Signal.) *91.*
- Tell me if 91 is divisible by 2. Get ready. (Signal.) *No.*
- By 3. Get ready. (Signal.) *No.*
- By 5. Get ready. (Signal.) *No.*
- By 7. Get ready. (Signal.) *Yes.*
- 91 is divisible by 7, so is 91 prime? (Signal.) *No.*

from Lesson 121, Exercise 4

Note that the two-equation test discloses that 1 *is not* a prime number because there is only one equation (not 2) that equals 1: 1 × 1 = 1.

Composite Numbers

On Lesson 124, students learn another name for numbers that are not prime—*composite numbers.*

Starting on Lesson 122, students learn the procedure for identifying all the factor pairs for composite numbers.

Students first identify the prime factors.

For 84, there are four prime factors:

$$2 \times 2 \times 3 \times 7 = 84$$

Next students write all possible combinations of factors.

$$2 \times 2 \times 3 \times 7 = 84$$

a. $2 \times (\qquad) = 84$ d. $2 \times 3 \times (\qquad) = 84$

b. $3 \times (\qquad) = 84$ e. $7 \times (\qquad) = 84$

c. $2 \times 2 \times (\qquad) = 84$ f. $2 \times 2 \times 3 \times (\ ) = 84$

Students do the multiplication for each group, which shows the factor pairs.

$$2 \times 2 \times 3 \times 7 = 84$$

a. $2 \times \overset{42}{(2 \times 3 \times 7)} = 84$ d. $\overset{6}{2 \times 3} \times \overset{14}{(2 \times 7)} = 84$

b. $3 \times \overset{28}{(2 \times 2 \times 7)} = 84$ e. $7 \times \overset{12}{(2 \times 2 \times 3)} = 84$

c. $\overset{4}{2 \times 2} \times \overset{42}{(3 \times 7)} = 84$ f. $\overset{12}{2 \times 2 \times 3} \times (7) = 84$

Later students figure out all the factor pairs for specified numbers.

Students first divide by prime factors 2, 3, 5, or 7 to identify prime factors.

Students then write all possible combinations of factors. Students express the combinations as whole numbers, which show all possible combinations of factors more than 1.

For example:

21, the prime factors are 3 and 7.

The prime-factor multiplication is 3 × 7.

CMC Level E and Common Core State Standards for Mathematics

According to the Common Core State Standards for Mathematics:

> In Grade 4, instructional time should focus on three critical areas: (1) developing understanding and fluency with multi-digit multiplication, and developing understanding of dividing to find quotients involving multi-digit dividends; (2) developing an understanding of fraction equivalence, addition and subtraction of fractions with like denominators, and multiplication of fractions by whole numbers; (3) understanding that geometric figures can be analyzed and classified based on their properties, such as having parallel sides, perpendicular sides, particular angle measures, and symmetry.

CMC Level E meets all the Grade 4 standards. A comprehensive listing of the standards and where they are met in the program appears in Presentation Book 1, and Presentation Book 2. Note that most of the Common Core State Standards are covered in the major tracks discussed on pages 46–168. Parts of the standards that have not already been addressed are discussed in this section. Note that text of the Common Core State Standards below has been abridged to highlight the parts of the standards that are discussed here. For a correlation to all of the standards with full text, see backs of Presentation Books 1 and 2.

Operations and Algebraic Thinking (OA)

Common Core State Standards

4.OA 1: Interpret a multiplication equation as a comparison, e.g., interpret 35 = 5 × 7 as a statement that 35 is 5 times as many as 7 and 7 times as many as 5. Represent verbal statements of multiplicative comparisons as multiplication equations.

Students interpret multiplication equations as a comparison and represent verbal statements of multiplicative comparisons of multiplication equations when they work multiplication number-family word problems. See **Multiplication Number Family,** Word Problems pages 68–72 for a detailed description.

Common Core State Standards

4.OA 2 Multiply or divide to solve word problems involving multiplicative comparison, e.g., by using drawings and equations with a symbol for the unknown number to represent the problem, distinguishing multiplicative comparison from additive comparison.

Students multiply or divide to solve word problems involving multiplicative comparisons when they work multiplication number-family word problems. See **Multiplication Number Family,** Word Problems pages 68–72 for a detailed description.

Common Core State Standards

4.OA 3	Solve multistep word problems posed with whole numbers and having whole-number answers using the four operations, including problems in which remainders must be interpreted. Represent these problems using equations with a letter standing for the unknown quantity. Assess the reasonableness of answers using mental computation and estimation strategies including rounding.

Students solve multistep word problems posed with whole numbers and having whole-number answers using the four operations, and interpret remainders. On Lesson 112, students make and work multiplication number-family word problems like:

There were 96 ounces of juice.
Each cup holds 5 ounces.

The cups were as full as possible.
How many ounces were in the cup that was not full?

Students make the number family for the multiplication comparison:

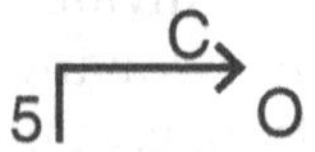

5 ⌐ C → O

5 times the number of cups equals the number of ounces.

Then students substitute for the number of ounces:

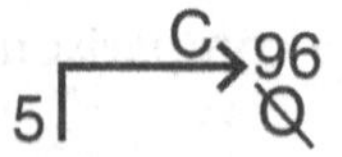

5 ⌐ C → 96 (O crossed out)

C stands for the number of full cups. Students work the division problem 96 ÷ 5 to find the answer:

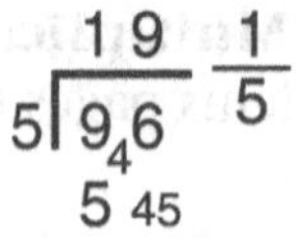

$$5\overline{)9_46}\quad 19\,\tfrac{1}{5}$$

5 45

There are 19 cups filled with juice. 1/5 of a cup is left over. So there is one ounce in the cup that is not full. Students can work the multiplication problem 19 × 5 to figure out how many ounces of juice are in the full cups. Students can add the 1 ounce that is left over to determine that there are in fact 96 total ounces of juice.

On Lesson 130, students are presented with ratio word problems or number-family word problems that show a student's work. See Ratio Word Problems on pages 161–165 for a more detailed description of ratio word problems.

Here's the Textbook material for problem B of Lesson 130 Part 7:

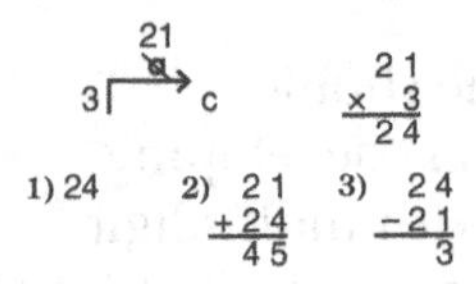

3 ⌐ 21 (a crossed out) → c

21 × 3 = 24

1) 24 2) 21 + 24 = 45 3) 24 − 21 = 3

b. There were 21 adults at a park. At the park, there were 3 times as many children as adults.

1) How many children were at the park?

2) How many people were there at the park altogether?

3) How many fewer adults were there than children?

For this multiplication number-family word problem, students first determine if the family has the letters that stand for unknowns in the right place. For the work that's shown, the letters are in the right place. The students identify if the correct letter was replaced with what it equals. For the work that's shown, the correct letter, a, was replaced with the number for adults, 21.

Students identify the problem the student should work to find the number for the unknown: C = 21 × 3. Students confirm that the correct problem is shown but then identify that the answer is NOT reasonable. Later, students work the problems correctly and write the correct answers to the questions.

Common Core State Standards

4.OA 4	Find all factor pairs for a whole number in the range 1–100. Recognize that a whole number is a multiple of each of its factors. Determine whether a given whole number in the range 1–100 is a multiple of a given one-digit number. Determine whether a given whole number in the range 1–100 is prime or composite.

Students learn to identify all of the prime factors for numbers 100 or less and identify numbers less than 100 as prime or composite. Students learn to identify all of the factor pairs for numbers 100 or less, so they also identify 1-digit multiples of numbers 100 or less. See **Prime Factors** on pages 166–168 for a more detailed description of prime and composite numbers and how students use them.

On Lesson 128, students divide 96 by its prime factors to determine the prime factor equation for 96. Students can determine from inspection that 96 is not divisible by 7 or 5, but is divisible by 3 because the sum of the digits is 15, which is divisible by 3. Students divide 96 by 3 and end up with 32. Students identify 2 as the only prime factor of 32, so they divide 32 by 2. The answer is 16. Students divide 16 by 2, and then divide the quotient 8 by 2 and the quotient 4 by 2.

Later in the lesson students rewrite all possible variations of the prime factor multiplication equations arranged in two groups. Here are the equations with the values for the groups with more than one factor written above it.

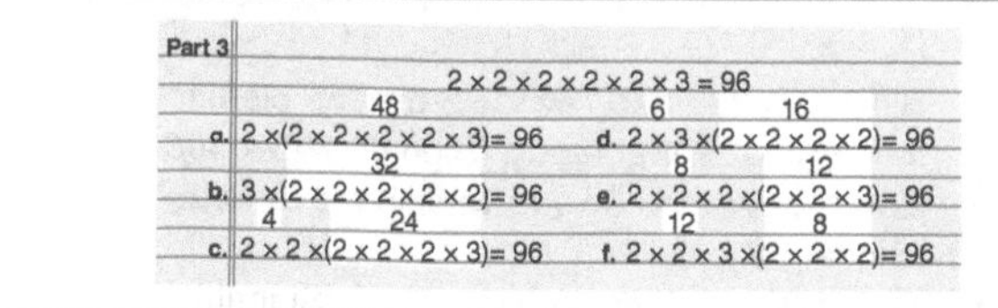

Part 3

2 × 2 × 2 × 2 × 2 × 3 = 96

	48		6 16
a.	2 ×(2 × 2 × 2 × 2 × 3)= 96	d.	2 × 3 ×(2 × 2 × 2 × 2)= 96
	32		8 12
b.	3 ×(2 × 2 × 2 × 2 × 2)= 96	e.	2 × 2 × 2 ×(2 × 2 × 3)= 96
	4 24		12 8
c.	2 × 2 ×(2 × 2 × 2 × 3)= 96	f.	2 × 2 × 3 ×(2 × 2 × 2)= 96

For equation a, students say the equation for both factor pairs: 2 × 48 = 96 and 48 × 2 = 96. Students say the equations for the factor pairs in equations b, c, d, and e. After identifying all of the factor pairs from the prime-number equations, students say the equations for the final factor pairs: 1 × 96 = 96 and 96 × 1 = 96.

Common Core State Standards

4.OA 5 Generate a number or shape pattern that follows a given rule. Identify apparent features of the pattern that were not explicit in the rule itself. *For example, given the rule "Add 3" and the starting number 1, generate terms in the resulting sequence and observe that the terms appear to alternate between odd and even numbers. Explain informally why the numbers will continue to alternate in this way.*

Students work two- or three-part problems in which they create patterns that follow a given rule. After students correctly fill in the spaces for a pattern, they are asked to identify apparent features of the pattern that were not explicit in the rule itself.

Here are all three parts of an example from Level E Practice Software Block 5, Activity 2:

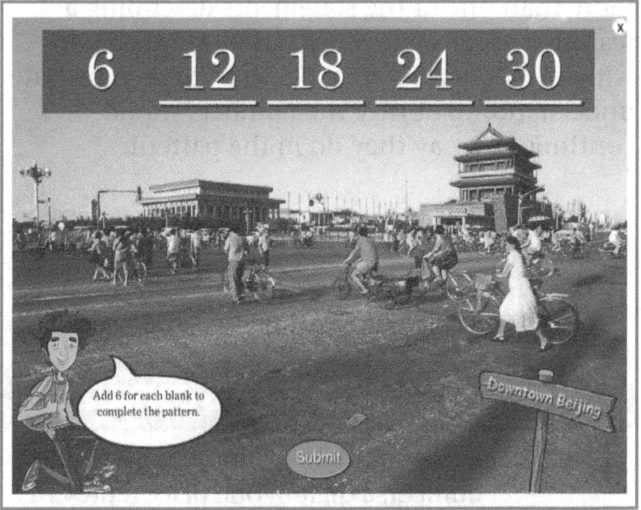

Note that students complete patterns using all four operations and that, for some of the patterns, more than one of the statements describing a feature is true. Some items have a third step in which students answer a question to demonstrate understanding of why the numbers would continue the way they do in the pattern.

Number and Operations in Base Ten (NBT)

Common Core State Standards

4.NBT 1	Recognize that in a multi-digit whole number, a digit in one place represents ten times what it represents in the place to its right. *For example, recognize that 700 ÷ 70 = 10 by applying concepts of place value and division.*

Beginning on Lesson 2, students review multiplying numbers by 10 or 100.

On Lesson 60, the relationship between numbers that are multiplied and the number of zeros is taught.

Here's part of the exercise that focuses on teaching that relationship:

c. Copy part 4. Then, for problem A, work the 3-digit problem for the ones. Cross out the numbers you carry. Then work the 3-digit problem for the tens. Then stop. Remember to write the commas if you need to.
(Observe students and give feedback.)

d. Check your work.
- What's 295 × 4? **(Signal.)** *1180.*
- What's 295 × 60? **(Signal.)** *17,700.*

(Display:) [60:7A]

Part 4

a.

```
    5 3
    3 2
    2 9 5
 ×  4 6 4
  1 1 8 0
 17,7 0 0
```

Here's what you should have so far.

e. Start with 295 and say the 3-digit problem for the hundreds. Get ready. **(Signal.)** *295 × 400.*
- How many zeros do you write in the answer? **(Signal.)** *2.*
- Then what problem do you work? **(Signal.)** *295 × 4.*
- You've already worked that problem. What's 295 × 4? **(Signal.)** *1180.*

So the answer to the 3-digit problem for the hundreds is the same number with two more zeros on the end. 295 × 400 = 118,000.

f. Write the answer to the 3-digit problem for the hundreds. Then add the numbers and figure out the answer to the whole problem. Put your pencil down when you've worked problem A. (Observe students and give feedback.)

- What's 295 × 464? (Signal.) *136,880.*

(Add to show:) [60:7B]

Part 4

```
       5 3
       3 2
a.     2 9 5
   ×   4 6 4
     1 1 1 8 0
     1 7,7 0 0
   + 1 1 8,0 0 0
     1 3 6,8 8 0
```

Here's what you should have for problem A.

g. Touch and read problem B on your lined paper. (Signal.) *178 × 334.*

- Say the 3-digit problem for the ones. (Signal.) *178 × 4.*
- Say the 3-digit problem for the tens. (Signal.) *178 × 30.*
- Say the 3-digit problem for the hundreds. (Signal.) *178 × 300.*
- You don't know what the answers to 178 × 30 and 178 × 300 are, but who can tell me how they are different? (Call on a student. Ideas:) *178 × 300 will have one more zero; you write one zero for 178 × 30 and you write two zeros for 178 × 300.*
  The answer to the 3-digit problem for the hundreds is the same as the answer to the tens, but it has one more zero at the end.
- The answer to the hundreds has how many more zeros than the answer to the tens? (Signal.) *One.*

from Lesson 60, Exercise 7

Common Core State Standards

4.NBT 2 Read and write multi-digit whole numbers using base-ten numerals, number names, and expanded form. Compare two multi-digit numbers based on meanings of the digits in each place, using >, =, and < symbols to record the results of comparison.

Students review reading and writing up to 6-digit numbers in the first thirty lessons. See **Whole Numbers** on pages 101–103 for a more detailed description of how students read and write multi-digit whole numbers using base-ten numerals, number names, and expanded forms.

After comparing pairs of tenths, decimal numbers and hundredths decimal numbers, students compare more than two whole numbers using the signs <, >, and = to make statements. See **Whole Numbers** on pages 101–103 for a more detailed description on how students compare two multi-digit numbers to make statements for a group of whole numbers in order from largest to smallest or from smallest to largest.

Common Core State Standards

4.NBT 3 Use place value understanding to round multi-digit whole numbers to any place.

See **Rounding and Estimation,** Rounding to 10s, 100s, 1000s, on pages 143–146 for a more detailed description of how students round different whole numbers to any place.

Common Core State Standards

4.NBT 4 Fluently add and subtract multi-digit whole numbers using the standard algorithm.

Mastery of the standard column addition algorithm is a prerequisite skill for placement in the *CMC Level E* program. Students use the standard addition algorithm to solve addition number families/word problems, perimeter problems, and missing angle problems. Students also learn to apply this algorithm to column multiplication problems.

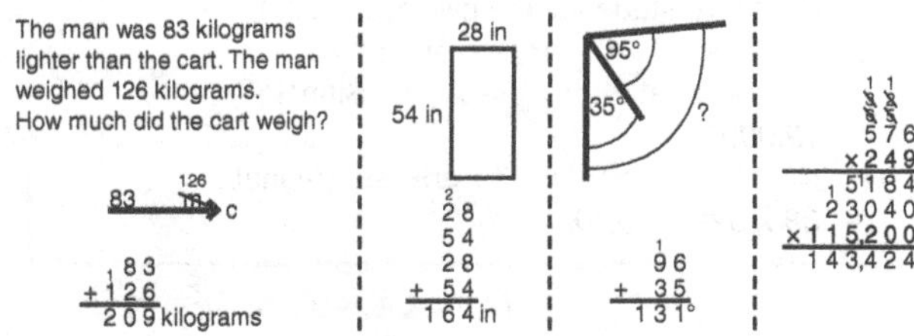

See Column Subtraction on pages 127–131 for a more detailed description of how students learn to subtract multi-digit whole numbers, fluently.

Common Core State Standards

4.NBT 5 Multiply a whole number of up to four digits by a one-digit whole number, and multiply two two-digit numbers, using strategies based on place value and the properties of operations. Illustrate and explain the calculation by using equations, rectangular arrays, and/or area models.

See **Column Multiplication** on pages 132–136 for a more detailed description of how students multiply a whole number of up to four digits by a one-digit whole number and multiply two multi-digit numbers, using strategies based on place value.

Students also learn that area problems relate multiplication and division. Students use this relationship to further illustrate the calculations. See 4.NBT 6 for part of an exercise from Lesson 94 that introduces sketches to illustrate multiplication and division calculation.

Students routinely answer questions about the calculations from multiplication problems they've worked.

Here's part of an exercise that illustrates students' understanding of the component calculations for the column multiplication operation:

d. Touch and read problem B. (Signal.) *38 × 529.*
- Start with 38 and say the 2-digit problem for the ones. (Signal.) *38 × 9.*
- What does 38 × 9 equal? (Signal.) *342.*
- Say the 2-digit problem for the tens. (Signal.) *38 × 20.*
- What does 38 × 20 equal? (Signal.) *760.*
- Say the 2-digit problem for the hundreds. (Signal.) *38 × 500.*
- Work 38 × 500. Remember a comma. Then write the addition sign and the equals bar and figure out the whole answer. Put your pencil down when you've completed problem B. (Observe students and give feedback.)

e. Check your work for problem B.
- Tell me what 38 × 500 equals. (Signal.) *19,000.*
- Read problem B and the answer. (Signal.) *38 × 529 = 20,102.*

from Lesson 56, Exercise 2

Students also write equations for the components calculations.

Here's part of an exercise in which students write equations for the component calculations for a 1-digit by 4-digit multiplication problem.

g. Find part 6 in your textbook. ✔
(Teacher reference:)

Part 6	
	5 × 4168

You're going to work this multiplication problem and write equations for the rows.
- Read the multiplication problem. Get ready. (Signal.) *5 × 4168.*
- Copy part 6 on your lined paper and work the problem. (Observe students and give feedback.)

h. Read the multiplication problem and the answer for part 6. Get ready. (Signal.) *5 × 4168 = 20,840.*
(Display:) [130:6D]

Part 6	
	5
	× 4168
	40
	300
	500
	+ 20,000
	20,840

Here's what you should have for part 6 so far.

i. Touch the row that equals 5 × 8 on your lined paper. ✔
- What's the value for 5 × 8? (Signal.) *40.*
- Say the equation for 40. (Signal.) *40 = 5 × 8.*

(Repeat step i until firm.)

j. Touch and say the value for 5 × 60. Get ready. (Signal.) *300.*
- Say the equation for 300. (Signal.) *300 = 5 × 60.*

k. Touch and say the value for 5 × 100. Get ready. (Signal.) *500.*
- Say the equation for 500. (Signal.) *500 = 5 × 100.*

l. Touch and say the value for 5 × 4000. Get ready. (Signal.) *20,000.*
- Say the equation for 20,000. (Signal.) *20,000 = 5 × 4000.*

(Repeat steps j through l that were not firm.)

m. Complete the equations for 40, 300, 500, and 20,000.
(Observe students and give feedback.)
(Add to show:) [130:6E]

Part 6	
	5
	× 4168
	40 = 5 × 8
	300 = 5 × 60
	500 = 5 × 100
	+ 20,000 = 5 × 4000
	20,840

Here's what you should have for part 6.

from Lesson 130, Exercise 6

Common Core State Standards

4.NBT 6 Find whole-number quotients and remainders with up to four-digit dividends and one digit divisors, using strategies based on place value, the properties of operations, and/or the relationship between multiplication and division. Illustrate and explain the calculation by using equations, rectangular arrays, and/or area models.

See **Division** on pages 137–142 for a more detailed description of how students learn to find whole-number quotients and remainders with up to four-digit dividends, using strategies based on place value and the properties of division.

Students also learn that area problems relate multiplication and division. Students use this relationship to further illustrate the calculations.

Here's part of the exercise from Lesson 94 that introduces sketches to illustrate the division calculation:

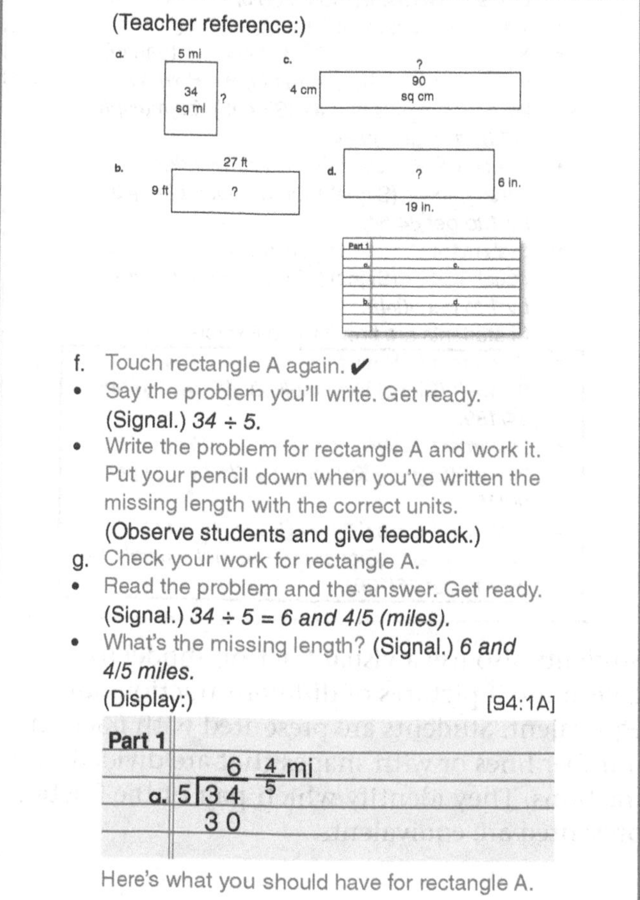

(Teacher reference:)

f. Touch rectangle A again. ✔
- Say the problem you'll write. Get ready. (Signal.) *34 ÷ 5.*
- Write the problem for rectangle A and work it. Put your pencil down when you've written the missing length with the correct units. (Observe students and give feedback.)

g. Check your work for rectangle A.
- Read the problem and the answer. Get ready. (Signal.) *34 ÷ 5 = 6 and 4/5 (miles).*
- What's the missing length? (Signal.) *6 and 4/5 miles.*

(Display:) [94:1A]

Here's what you should have for rectangle A.

h. Here's a picture of rectangle A with the square miles shown.

(Display:) [94:1B]

- What's the length of the side with a question mark. (Signal.) *6 and 4/5 miles.*
- (Point to the bottom row.)The bottom row does not show squares. They are 1 mile in the X direction, but they are not 1 mile in the Y direction. Tell me the distance in the Y direction of the bottom row. Get ready. (Signal.) *4/5 mile.*
Yes, the Y distance for the bottom row is 4/5 of a mile.

from Lesson 94, Exercise 1

See Geometry, Area, and Perimeter on pages 115–126 for a more detailed description of how students learn to use rectangles and the relationship between area and perimeter to illustrate the relationship between multiplication and division.

Students explain division calculations. The work for a division problem with a 4-digit dividend is shown. Students answer questions about the calculations from the worked problem.

Here's the student material from Workbook Lesson 130:

Part 2

$$3\overline{)7{,}_15\,2\,2}\quad 2\,5\,0\,7\tfrac{1}{3}$$
6 21

a. 7 ÷ 3 = ___ with a remainder of ___.

b. 15 ÷ 3 = ___ with a remainder of ___.

c. 2 ÷ 3 = ___ with a remainder of ___.

d. 22 ÷ 3 = ___ with a remainder of ___.

Number and Operations—Fractions (NF)

Common Core State Standards

4.NF 1 Explain why a fraction *a/b* is equivalent to a fraction *(n × a)/(n × b)* by using visual fraction models, with attention to how the number and size of the parts differ even though the two fractions themselves are the same size. Use this principal to recognize and generate equivalent fractions.

Students learn how to identify equivalent fractions from fraction-multiplication equations. Students also learn how to create equivalent fractions from a fraction and a multiple of the numerator or denominator. See **Fractions,** Fraction Multiplication, Equivalent Fractions on pages 92–100 for a detailed description of what students learn about fraction-multiplication equations and how they identify and generate equivalent fractions.

Students explain why a fraction *a*/*b* is equivalent to a fraction *(n × a)/(n × b)* and why a fraction *a*/*b* is not equivalent to a fraction *(n × a)/(m × b)*.

Here's the Workbook practice from an exercise in which students complete statements and explain how they know if the fractions are equal or not equal.

(Teacher reference:)

a. $\frac{5}{7} \times \frac{12}{12} = \frac{60}{84}$ $\frac{5}{7}\ \ \frac{60}{84}$

c. $\frac{2}{3} \times \frac{15}{16} = \frac{30}{48}$ $\frac{2}{3}\ \ \frac{30}{48}$

e. $\frac{6}{5} \times \frac{15}{15} = \frac{90}{75}$ $\frac{6}{5}\ \ \frac{90}{75}$

b. $\frac{3}{2} \times \frac{28}{27} = \frac{84}{54}$ $\frac{3}{2}\ \ \frac{84}{54}$

d. $\frac{4}{9} \times \frac{21}{21} = \frac{84}{189}$ $\frac{4}{9}\ \ \frac{84}{189}$

f. $\frac{1}{8} \times \frac{16}{15} = \frac{16}{120}$ $\frac{1}{8}\ \ \frac{16}{120}$

You're going to complete the statement for the starting fraction and the fraction after the equals. Then you're going to explain how you know the statements are correct.

- Complete the statement for each problem to show if the fractions are equal or not equal. Put your pencil down when you've completed all of the statements for part 2.

(Observe students and give feedback.)

(Teacher reference:)

a. $\frac{5}{7} \times \frac{12}{12} = \frac{60}{84}$ $\frac{5}{7} = \frac{60}{84}$

c. $\frac{2}{3} \times \frac{15}{16} = \frac{30}{48}$ $\frac{2}{3} \neq \frac{30}{48}$

e. $\frac{6}{5} \times \frac{15}{15} = \frac{90}{75}$ $\frac{6}{5} = \frac{90}{75}$

b. $\frac{3}{2} \times \frac{28}{27} = \frac{84}{54}$ $\frac{3}{2} \neq \frac{84}{54}$

d. $\frac{4}{9} \times \frac{21}{21} = \frac{84}{189}$ $\frac{4}{9} = \frac{84}{189}$

f. $\frac{1}{8} \times \frac{16}{15} = \frac{16}{120}$ $\frac{1}{8} \neq \frac{16}{120}$

b. Check your work for part 2. You'll read the statement for each problem.

- Problem A. Get ready. (Signal.) *5/7 = 60/84.*
- (Repeat for:) B, *3/2 ≠ 84/54;* C, *2/3 ≠ 30/48;* D, *4/9 = 84/189;* E, *6/5 = 90/75;* F, *1/8 ≠ 16/120.*

c. Touch the statement for problem A again.

- You know 5/7 equals 60/84 because (pause) you multiply 5/7 by 1 to get 60/84. How do you know 5/7 equals 60/84? (Signal.) *You multiply 5/7 by 1 to get 60/84.*
- Statement B. How do you know 3/2 does not equal 84/54? (Signal.) *You do not multiply 3/2 by 1 to get 84/54.*
- Statement C. How do you know 2/3 does not equal 30/48? (Signal.) *You do not multiply 2/3 by 1 to get 30/48.*

(If students are firm, skip to Exercise 3.)

- Statement D. How do you know 4/9 equals 84/189? (Signal.) *You multiply 4/9 by 1 to get 84/189.*
- Statement E. How do you know 6/5 equals 90/75? (Signal.) *You multiply 6/5 by 1 to get 90/75.*
- Statement F. How do you know 1/8 does not equal 16/120? (Signal.) *You do not multiply 1/8 by 1 to get 16/120.*

Students also use a visual fraction model to determine if pictures of different fractions are equivalent. Students are presented with fraction number lines or with shapes that are divided into fractions. They identify which pair of the fractions presented are equivalent.

Here are three examples from the *Level E* Practice Software, Block 2 Activity 4.

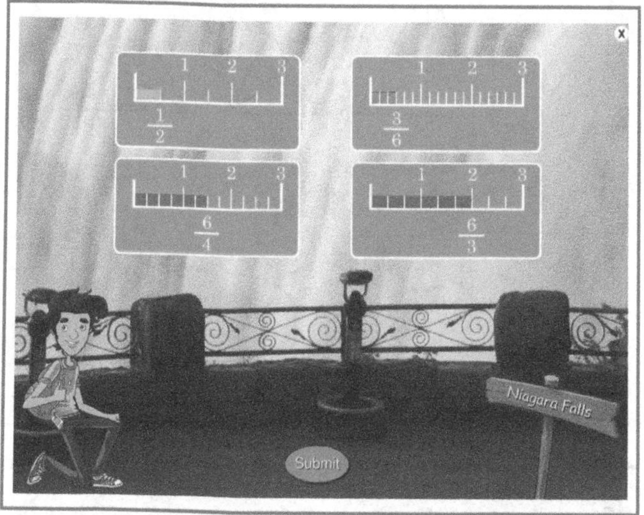

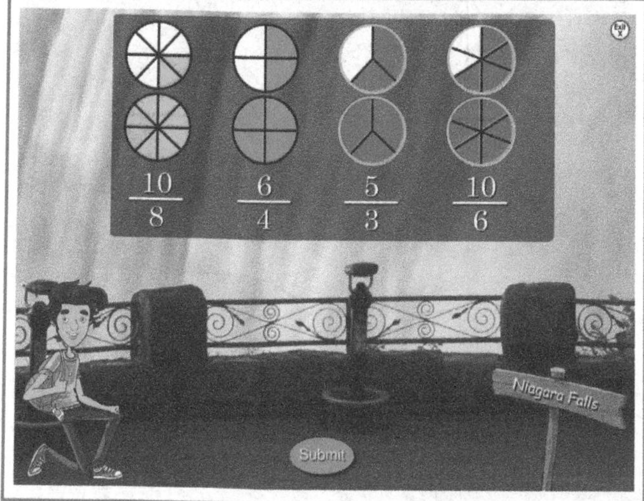

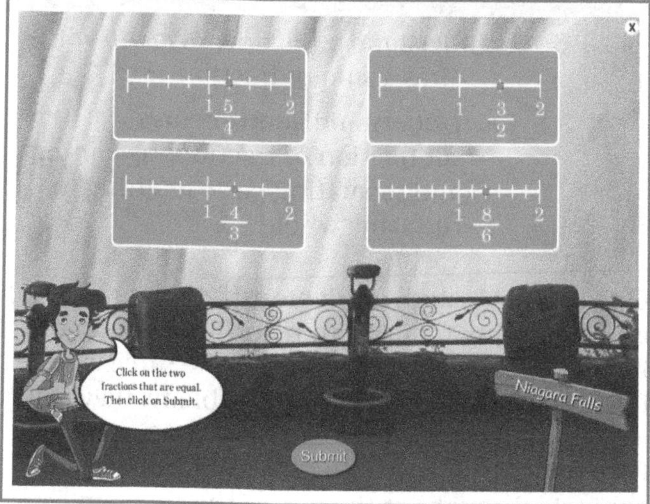

Note that the first example shows a number line with three units, and the fraction is represented by the shaded part of the number line. The second example shows two units represented by the shaded portion of circles. The third example shows a number line with two units, and the fraction is represented by a point on the number line.

Common Core State Standards

4.NF 2	Compare two fractions with different numerators and different denominators, e.g., by creating common denominators or numerators, or by comparing to a benchmark fraction such as 1/2. Recognize that comparisons are valid only when the two fractions refer to the same whole. Record the results of comparisons with symbols >, =, or <, and justify the conclusions, e.g., by using a visual fraction model.

Students learn four strategies for comparing fractions. See **Fractions,** Equivalent Fractions on pages 95–100 for a detailed description of the strategies students learn for writing greater than, less than, or equal statements for pairs of fractions.

Students recognize that comparisons are only valid when fractions refer to the same whole. Students are presented with four pictures of fractions, but only two of the pictures refer to the same whole. Students indicate which two of the fractions refer to the same whole and can be compared.

Here are two examples of pictures of fractions students are presented with in *Level E* Practice Software, Block 2 Activity 6.

Note that the pictures for the first example are number lines and the pictures for the other example are triangles.

After students indicate the fractions that can be compared, they are presented with just those pictures. Students select the sign between the fractions to compare them.

Here are the second parts of the examples shown above with the signs shown that compare the fractions:

Common Core State Standards

4.NF 3 Understand a fraction with a/b with $a > 1$ as a sum of fractions $1/b$.

a. Understand addition and subtraction of fractions as joining and separating parts referring to the same whole.

On Lesson 31, students demonstrate their understanding that addition and subtraction of fractions refers to fractions from the same whole by discriminating identifying problems they can work and problems they can't work. Students then apply this discrimination to multiple contexts throughout the rest of the program. See **Fractions,** Fractions Lessons 31–130 on pages 86–91 for a detailed description of the discriminations and procedures students learn for working addition and subtraction problems with fractions and mixed numbers.

Common Core State Standards

4.NF 3 Understand a fraction with *a/b* with $a > 1$ as a sum of fractions $1/b$.

b. Decompose a fraction into a sum of fractions with the same denominator in more than one way, recording each decomposition by an equation. Justify decompositions, e.g., by using a visual fraction model. *Examples: 3/8 = 1/8 + 1/8 + 1/8; 3/8 = 1/8 + 2/8; 2 1/8 = 1 + 1 + 1/8 = 8/8 + 8/8 + 1/8.*

Students apply the fraction skills they've learned for composing and decomposing fractions.

Here's part of an exercise that directs students to decompose fractions and mixed numbers:

j. For the rest of the problems, you'll tell me the decomposition equation.

- Problem C. Read the decomposition equation for 3/8. (Signal.) *3/8 = 1/8 + 1/8 + 1/8.*

(Display:) [129:1B]

a. $91 - 10 + 15 = 9 \times \underline{9} + 15$

b. $200 + \underline{50} + 6 = 356 - 100$

c. $\frac{3}{8} = \frac{1}{8} + \frac{1}{8} + \frac{1}{8}$

Here's what you should have for problems A, B, and C.

k. Problem D. Read the decomposition equation for 4 and 2/5. (Signal.) *4 and 2/5 = 1 + 1 + 1 + 1 + 1/5 + 1/5.*

- Problem E. Read the decomposition equation for 5/2. (Signal.) *5/2 = 1/2 + 1/2 + 1/2 + 1/2 + 1/2.*
- Problem F. Read the decomposition equation for 3 and 4/7. (Signal.) *3 and 4/7 = 1 + 1 + 1 + 1/7 + 1/7 + 1/7 + 1/7.*

(Display:) [129:1C]

d. $4\frac{2}{5} = 1 + \underline{1} + \underline{1} + \underline{1} + \frac{1}{5} + \frac{1}{5}$

e. $\frac{5}{2} = \frac{1}{2} + \frac{1}{2} + \frac{1}{2} + \frac{1}{2} + \frac{1}{2}$

f. $3\frac{4}{7} = 1 + 1 + 1 + \frac{1}{7} + \frac{1}{7} + \frac{1}{7} + \frac{1}{7}$

Here's what you should have for problems D, E, and F.

from Lesson 129, Exercise 1

Students are also presented with problems that require them to select decompositions of fractions.

Here are two examples of problems students work in the *Level E* Practice Software, Block 4 Activity 1.

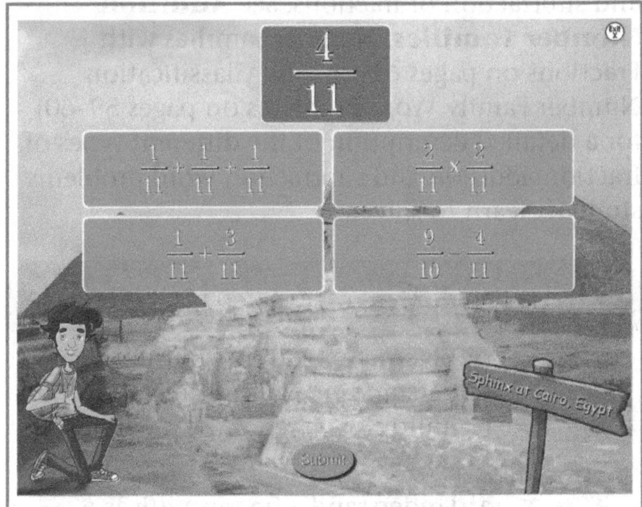

Note that the first example shows two decompositions for 6/10. There is only one correct decomposition shown for the second example.

Common Core State Standards

4.NF 3 Understand a fraction with *a/b* with $a > 1$ as a sum of fractions $1/b$.

c. Add and subtract mixed numbers with like denominators, e.g., by replacing each mixed number with an equivalent fraction, and/or by using properties of operations and the relationship between addition and subtraction.

See **Fractions,** Mixed Numbers on pages 86–91 for a detailed description of the addition and subtraction operations students learn to work with mixed numbers.

Common Core State Standards

4.NF 3 Understand a fraction with *a*/*b* with *a* > 1 as a sum of fractions 1/*b*.

d. Solve word problems involving addition and subtraction of fractions referring to the same whole and having like denominators, e.g., by using visual fraction models and equations to represent the problem.

Students solve word problems involving addition and subtraction of fractions. See **Addition Number Families,** Number Families with Fractions on pages 61–64 (and Classification Number Family Word Problems on pages 59–60) for a detailed description of the different types of fraction addition and subtraction word problems students learn to solve.

Common Core State Standards

4.NF 4 Apply and extend previous understandings of multiplication to multiply a fraction by a whole number.

a. Understand a fraction *a*/*b* as a multiple of 1/*b*. *For example, use a visual fraction model to represent 5/4 as the product 5 × (1/4), recording the conclusion by the equation 5/4 = 5 × (1/4).*

Students demonstrate their understanding by saying the multiplication for fractions that do not have a numerator or denominator of 1 and completing the equation to show the relationship.

Here's an exercise that directs students to perform those tasks:

b. Now you're going to read each fraction and say the whole number times the fraction with a numerator of one that it equals.
- Read fraction A on your lined paper. Get ready. **(Signal.)** *2/6.*
- What's the fraction with a numerator of 1? **(Signal.)** *1/6.*
- What whole number do you multiply 1/6 by to get 2/6? **(Signal.)** *2.*
- Yes, 2/6 equals 2 times 1/6. Say the equation for 2/6. **(Signal.)** *2/6 = 2 × 1/6.*

c. Read fraction B. **(Signal.)** *8/5.*
- Say the equation for 8/5. Get ready. **(Signal.)** *8/5 = 8 × 1/5.*

d. Read fraction C. **(Signal.)** *4/9.*
- Say the equation for 4/9. Get ready. **(Signal.)** *4/9 = 4 × 1/9.*

e. Read fraction A again. **(Signal.)** *2/6.*
- Say the equation for 2/6. **(Signal.)** *2/6 = 2 × 1/6.*

(Repeat steps b through e until firm.)

f. Complete the equations for part 7. Put your pencil down when you're finished.
(Observe students and give feedback.)

g. Check your work.
(Display:) [96:9A]

Part 7

a. $\frac{2}{6} = 2 \times \frac{1}{6}$ c. $\frac{4}{9} = 4 \times \frac{1}{9}$

b. $\frac{8}{5} = 8 \times \frac{1}{5}$

Here are the equations for the fractions.

from Lesson 96, Exercise 9

Common Core State Standards

4.NF 4 Apply and extend previous understandings of multiplication to multiply a fraction by a whole number.

b. Understand a multiple of *a/b* as a multiple of *1/b*, and use this understanding to multiply a fraction by a whole number. *For example, use a visual fraction model to express 3 × (2/5) as 6 × (1/5), recognizing this product as 6/5. (In general n × (a/b) = (n × a)/b.)*

Students demonstrate this understanding by working problems that multiply whole numbers and fractions. See **Fractions,** Fractions Lesson 31–130, Fraction Multiplication on pages 86–91 for a detailed description of what students understand and what they apply with respect to whole numbers times fractions.

Common Core State Standards

4.NF 4 Apply and extend previous understandings of multiplication to multiply a fraction by a whole number.

c. Solve word problems involving multiplication of a fraction by a whole number, e.g., by using visual fraction models and equations to represent the problem. *For example, if each person at a party will eat 3/8 of a pound of roast beef, and there will be 5 people at the party, how many pounds of roast beef will be needed? Between what two whole numbers does your answer lie?*

Students work area problems for rectangles with a fractional length and a whole number length.

Here is one of these sets of problems students work In Textbook Lesson 125:

a. The grass strip between a street and sidewalk is 28 yards long and $\frac{3}{4}$ of a yard wide. What is the area of the grass strip?

b. The cover for a florescent light is $\frac{8}{3}$ inches wide and 37 inches long. What is the area of the light cover?

Part 1

a.

b.

Common Core State Standards

4.NF 5 Express a fraction with denominator 10 as an equivalent fraction with denominator 100, and use this technique to add two fractions with respective denominators 10 and 100. *For example, express 3/10 as 30/100 and add 3/10 + 4/100 = 34/100.*

Students learn to write decimals that are equivalent to fractions and fractions that are equivalent to decimals. Students also learn to complete tables that show rows with an equivalent whole number, tenth and hundredth decimal. Students learn to work decimal addition and subtraction problems involving whole numbers, tenths and hundredths decimals. See **Decimals** on pages 104–110 for a detailed description of the decimal skills students learn.

Later, students are taught how to apply their computation skills and knowledge of equivalent decimals to convert decimal problems into fraction problems.

Here's the introduction to working fraction addition and subtraction problems with denominators that are powers of 10:

a. (Display:) [121:3A]

.8 – .47 =

- Read this problem. Get ready. (Signal.) *8/10 – 47/100.*
  You need to change one of the numbers before you can work this problem.
- Which number do you change? (Signal.) *8/10.*
- Tell me what you change 8 tenths into. Get ready. (Signal.) *80/100.*
  (Repeat until firm.)

b. (Display:) [121:3B]

$\frac{8}{10} - \frac{47}{100} =$

Here's the same problem with fractions. It works the same way. You need to change one of the fractions before you can work this problem.

- Which fraction do you change? (Signal.) *8/10.*
- Tell me what you can change 8/10 into. Get ready. (Signal.) *80/100.*
  (Add to show:) [121:3C]

$\frac{80}{100} - \frac{47}{100} =$

- Read the problem. (Signal.) *80/100 – 47/100.*
- Can you work that problem? (Signal.) *Yes.*

c. (Display:) [121:3D]

$\frac{28}{100} + \frac{6}{10} =$ $\frac{9}{10} - \frac{5}{100} =$

$\frac{34}{100} + 5 + \frac{7}{10} =$

To work each of these problems, you have to change at least one of the values. You'll tell me the values you'd change and what you'd change them into to work the problems.

- (Point to **28.**) Read the problem. (Signal.) *28/100 + 6/10.*
- Would you change 28/100 or 6/10 to work the problem? (Signal.) *6/10.*
- Tell me what you'd change 6/10 into. Get ready. (Signal.) *60/100.*

d. (Point to **9.**) Read the problem. (Signal.) *9/10 – 5/100.*

- Tell me the value you'd change. Get ready. (Signal.) *9/10.*
- Tell me what you'd change 9/10 into. Get ready. (Signal.) *90/100.*

e. (Point to **34.**) Read the problem. (Signal.) *34/100 + 5 + 7/10.*

- You'd change more than one value to work this problem. Tell me the values you'd change. Get ready. (Signal.) *5 (and) 7/10.*
- Tell me what you'd change 5 into. Get ready. (Signal.) *500/100.*
- Tell me what you'd change 7/10 into. Get ready. (Signal.) *70/100.*
  (Repeat problems that were not firm.)

from Lesson 121, Exercise 3

Common Core State Standards

4.NF 6 Use decimal notation for fractions with denominators 10 or 100. *For example, rewrite 0.62 as 62/100; describe a length as 0.62 meters; locate 0.62 on a number line diagram.*

See **Decimals** on pages 104–110 for a detailed description of the equivalence skills students learn between fractions and decimals.

Here's a table students complete to show a fraction and the equivalent decimal for each row:

Part 3

	Fraction	Decimal
a.	$\frac{9}{100}$	
b.		61.7
c.	$\frac{617}{100}$	
d.		.04
e.		78.0

Common Core State Standards

4.NF 7 Compare two decimals to hundredths by reasoning about their size. Recognize that comparisons are valid only when the two decimals refer to the same whole. Record the results of comparisons with the symbols >, =, or <, and justify the conclusions, e.g., by using a visual model.

Students learn to compare whole numbers and decimals with the same number of decimal places. Then students use what they've learned about equivalent decimals to compare pairs of numbers having different decimal places.

Here's the first part of the exercise that teaches students a strategy for comparing numbers with different decimal places:

a. (Display:) [111:1A]

$$\begin{array}{r} 20.6 \\ +20.06 \\ \hline \end{array}$$

You've worked decimal problems like this one that add or subtract whole numbers, tenths, or hundredths.

- Read the problem. Get ready. (Signal.) *20 and 6/10 + 20 and 6/100.*
- To work this problem, do we change both numbers to whole numbers, tenths, or hundredths? (Signal.) *Hundredths.*

b. (Display:) [111:1B]

20.6 20.06

Here are the same decimals. We're going to compare them to show which is more or if they're equal. To compare them, we change the decimals the same way you change them when you work addition and subtraction problems.

- Do we change both numbers to whole numbers, tenths, or hundredths? (Signal.) *Hundredths.*
- What do I add after 20 and 6 tenths to make it hundredths? (Signal.) *Zero.*

(Add to show:) [111:1C]

20.60 20.06

c. Read the decimals now. Get ready. (Signal.) *20 and 60/100 (and) 20 and 6/100.*

- Which is more? (Signal.) *20 and 60/100.*

(Add to show:) [111:1D]

20.60 > 20.06

- So which is more—20 and 6 tenths or 20 and 6 hundredths? (Signal.) *20 and 6/10.*

(Change to show:) [111:1E]

20.6 > 20.06

d. I'm going to read the statement that starts with 20 and 6 tenths. 20 and 6 tenths is **more** than 20 and 6 hundredths.

Here's the statement that starts with 20 and 6 hundredths. 20 and 6 hundredths is less than 20 and 6 tenths.

- Start with 20 and 6 hundredths and read the statement. (Signal.) *20 and 6/100 is less than 20 and 6/10.*
- Start with 20 and 6 tenths and read the statement. (Signal.) *20 and 6/10 is more than 20 and 6/100.*

(Repeat step d until firm.)

WORKBOOK PRACTICE

a. Open your workbook to Lesson 111 and find part 1. ✔

(Teacher reference:) **R Test 13: Part A**

a. 14.17	14.1	c. 32.5	32.48
b. 105	105.2	d. 9.3	9.31

You're going to change one of the decimals in each pair so you can compare them. Then you'll write the sign to show which is more or if the decimals are equal.

- Read the numbers for problem A. (Signal.) *14 and 17/100 (and) 14 and 1/10.*
- Do we change both numbers to whole numbers, tenths, or hundredths? (Signal.) *Hundredths.*
- What do you add after 14 and 1 tenth to make it hundredths? (Signal.) *Zero.*

b. Read the numbers for problem B. (Signal.) *105 (and) 105 and 2/10.*

- Do we change both numbers to whole numbers, tenths, or hundredths? (Signal.) *Tenths.*
- What do you add after 105 to make it tens? (Signal.) *Decimal point, zero.*

(Display:) [111:1F]

a.	14.17	14.10
b.	105.0	105.2

Here are the decimals for problems A and B changed. Change one of the numbers in problems A through D so you can compare them. Put your pencil down when you're ready to write the signs.

(Observe students and give feedback.)

from Lesson 111, Exercise 1

Students recognize that comparisons are only valid when decimals refer to the same whole. Students are presented with four pictures of decimals, but only two of the pictures refer to the same whole. Students indicate which two of the decimals refer to the same whole and can be compared.

Here are two examples of pictures of decimals students are presented with in Student Practice Software, Block 2 Activity 6 and Block 4 Activity 4.

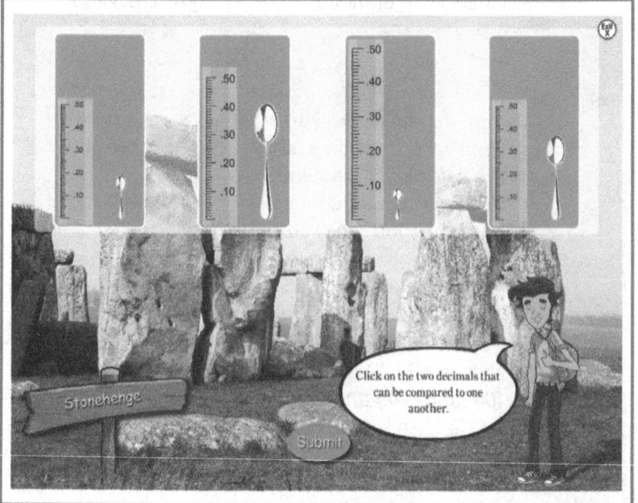

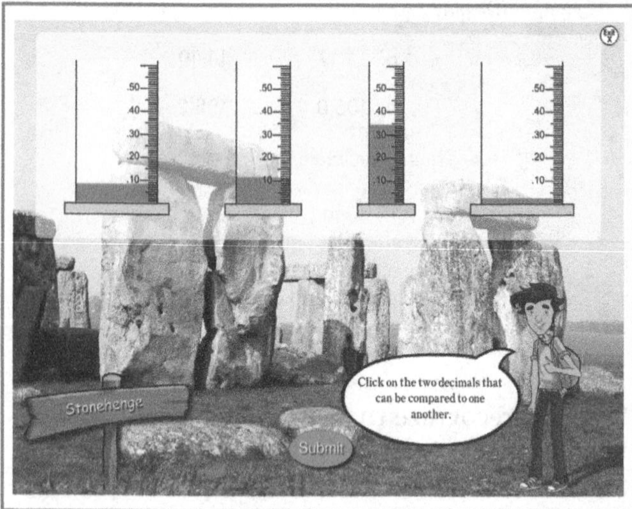

Note that the pictures for the first example are rulers that show hundredths and the pictures for the other example are beakers that show hundredths. After students indicate the pictures that can be compared, they are presented with just those pictures. Students select the correct sign to compare the pictures.

Here is the second part of the first example shown above with the signs shown that students use to compare the pictures:

Measurement and Data (MD)

Common Core State Standards

4.MD 1	Know relative sizes of measurement units within one system of units including km, m, cm; kg, g; lb, oz; l, ml; hr, min, sec. Within a single system of measurement, express measurements in a larger unit in terms of a smaller unit. Record measurement equivalents in a two-column table. *For example, know that 1 ft is 12 times as long as 1 in. Express the length of a 4 ft snake as 48 in. Generate a conversion table for feet and inches listing the number pairs (1, 12), (2, 24), (3, 36), . . .*

Students work extensively learning the relative sizes of measurement within one system of units. Here is the table of measurement facts students work with:

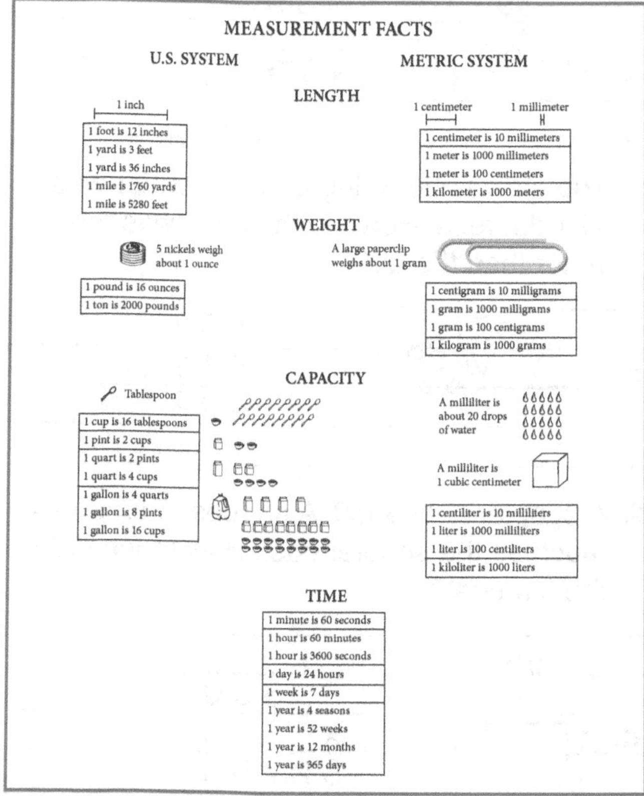

See **Multiplication Number Families,** Word Problems, Unit Conversion on pages 72–74 for a detailed description of how students express the relative sizes of pairs of units within a measurement system and how they use these expressions to solve word problems involving different units.

Later in the program, students learn about functions and complete function tables having both X and Y values missing. See **Coordinate Systems and Functions,** Functions on pages 155–165 for a detailed description of what students learn about functions and how they apply these skills.

Students learn how to apply these skills to recording measurement equivalences in a two-column table.

Here's the introductory exercise:

(Teacher reference:)

	feet	inches
ruler		12
yardstick	3	
rope	24	
chain		120

This table works just like function tables, but the function column isn't shown. You know the function for this table. You're going to figure out the missing values.

- Read the heading for the first column. Get ready. (Signal.) *Feet.*
- Read the heading for the other column. (Signal.) *Inches.*
- If you had X feet, what would you do to X to figure out how many inches you had? (Call on a student. Idea:) *Multiply by 12.*

(Repeat until firm.)

b. Read the heading for the first **row** in table 1. Get ready. (Signal.) *Ruler.*
- Is a number for inches or feet shown for the ruler? (Signal.) *Inches.*
- Start with the number of inches and tell me the problem for figuring out the feet. Get ready. (Signal.) *12 ÷ 12.*

c. Read the next heading. Get ready. (Signal.) *Yardstick.*
- Is a number for inches or feet shown for the yardstick? (Signal.) *Feet.*
- Start with the number of feet and tell me the problem for figuring out the inches. Get ready. (Signal.) *3 × 12.*

d. Read the next heading. Get ready. (Signal.) *Rope.*
- Is a number for inches or feet shown for the rope? (Signal.) *Feet.*
- Start with the number of feet and tell me the problem for figuring out the inches. Get ready. (Signal.) *24 × 12.*

e. Read the next heading. Get ready. (Signal.) *Chain.*
- Is a number for inches or feet shown for the chain? (Signal.) *Inches.*
- Start with the number of inches and tell me the problem for figuring out the feet. Get ready. (Signal.) *120 ÷ 12.*

(Repeat steps b through e that were not firm.)

f. Write the missing values in the table. If you need to work a problem to figure out a missing number, use the lines next to the table. (Observe students and give feedback.)

g. Check your work.
- The ruler is 12 inches long. How many feet long is it? (Signal.) *1.*
- The yardstick is 3 feet long. How many inches long is it? (Signal.) *36.*
- The rope is 24 feet long. How many inches long is it? (Signal.) *288.*
- The chain is 120 inches long. How many feet long is it? (Signal.) *10.*

(Display:) [126:1A]

	feet	inches
ruler	1	12
yardstick	3	36
rope	24	288
chain	10	120

Here's what you should have for the table in part 1.

Lesson 126, Exercise 1

Common Core State Standards

4.MD 2 Use the four operations to solve word problems involving distances, intervals of time, liquid volumes, masses of objects, and money, including problems involving simple fractions or decimals, and problems that require expressing measurements given in a larger unit in terms of a smaller unit. Represent measurement quantities using diagrams such as number line diagrams that feature a measurement scale.

Students learn to work word problems that add and subtract. Many of these problems involve measurement units. See **Addition Number Families,** Comparison Word Problems, on pages 52–53, **Addition Number Families,** Sequence Word Problems, on pages 53–58, and **Addition Number Families,** Classification Word Problems, on pages 59–60 for a detailed description of the initial strategies students learn to solve word problems containing measurement units involving addition and subtraction.

The first strategy students learn for solving word problems involving multiplication and division involves multiplication number families. Many of these problems involve measurement units. See **Multiplication Number Families,** Multiplication Number Family Word Problems, on pages 72–74, for a detailed description of the strategies students learn to solve word problems containing measurement units involving multiplication and division.

Here are some samples of problems that appear in the program that show the student's work:

a. The tank held 128 quarts. How many gallons did the tank hold?

a. 4 | g → 128 q

4 | 128 = 3 2 gallons

b. A tree was 39 yards tall. How many feet tall was the tree?

b. 3 | 39 y → f

39 × 3 = 1 1 7 feet

c. The air was 96.3 degrees. The water was 47.1 degrees. How much cooler was the water than the air?

c. 47.1 w → 96.3 a

96.3 − 47.1 = 49.2°

d. A scooter cost $550. A car cost 5 times as much as the scooter cost. How much did the car cost?

d. 5 | 550 s → c

550 × 5 = $2750

e. Steven rides his bicycle 307.2 minutes a week. Steven rides his bike 45.9 fewer minutes a week than Julia rides her bike. How long does Julia ride her bike a week?

e. 45.9 307.2 S → J

45.9 + 307.2 = 353.1 minutes

f. A dog weighed 32 pounds. How many ounces did the dog weigh?

f. 16 | 32 p → o

32 × 16: 192 + 320 = 512 ounces

- Problems *a* and *b* require students to convert quantities from one measurement unit to another. Problem a expresses a quantity in the smaller unit, quarts, and students convert the quantity into the larger unit, gallons. Both units are liquid volume. Problem *b* expresses a quantity in the larger unit, yards, and students convert the quantity into the smaller unit, feet. Both units are distance. Students solve problem *a* using the division operation and solve problem b using the multiplication operation.
- Problem *c* expresses measurements in units of temperature involving simple decimals. Students solve problem *c* using the subtraction operation.
- Problem *d* expresses measurements in units of money. Students solve problem *d* using the multiplication operation.
- Problem *e* expresses measurements in units of time involving simple decimals. Students solve problem *e* using the addition operation.
- Problem *f* requires students to convert quantities from one measurement unit to another. Problem *f* expresses a quantity in the larger unit, pounds, and students convert the quantity into the smaller unit, ounces. Both units are mass/weight. Students solve problem *f* using the multiplication operation.

After students work problems of the types shown above for many lessons, students learn about equivalent fractions and how to use them to solve ratio word problems. See **Fractions,** Fraction Multiplication, Equivalent Fractions, on pages 92–94, for a detailed description of the skills students learn about equivalent fractions. See **Ratio Word Problems,** on pages 161–165, for a detailed description of how students apply the equivalent fraction skills to solving word problems.

After students master equivalent fraction skills and solving ratio word problems, they are introduced to a problem type that requires them to use all four operations to solve.

Here's a problem from Textbook Lesson 106 that expresses measurements in units of time, and requires students to use all four operations to solve:

b. A pilot does 9 minutes of paperwork for every 6 minutes of flying. The pilot flies 396 minutes.

1) How much time did she have to spend doing paperwork?

2) How much more time did the pilot spend doing paperwork than flying?

3) How much total time did she spend doing paperwork and flying?

Students write a ratio equation for the first part of the problem.

$$\frac{\text{paperwork}}{\text{flies}} \quad \frac{9}{6} \times \frac{\quad}{\quad} = \frac{\quad}{396}$$

Students solve for the missing number in the denominator by working the division problem 396 ÷ 6. The missing value is 66. The starting fraction of this ratio equation and the fraction after the equal sign are equal, so the middle fraction equals 1. The middle fraction is 66/66.

$$\frac{\text{paperwork}}{\text{flies}} \quad \frac{9}{6} \times \frac{66}{66} = \frac{\quad}{396}$$

Students solve for the missing number for paperwork by working the multiplication problem 66 × 9. The answer is 594.

Here's all of the student's work so far:

b. $\frac{\text{paperwork}}{\text{flies}} \quad \frac{9}{6} \times \frac{66}{66} = \frac{594}{396}$ $\quad 6\overline{)396} = 66$, 36 $\quad 66 \times 9 = 594$ (carry 5)

Question 1 is how much time did she have to spend doing paperwork? The number after the equals for paperwork is 594. The units are minutes. So students write the answer 594 minutes for question 1.

Question 2 is how much more time did the pilot spend doing paperwork than flying? Students work the subtraction problem 594 – 396. The answer is 198 minutes.

Question 3 is how much total time did she spend doing paperwork and flying? Students work the addition problem 594 + 396. The answer is 990 minutes.

Here's the rest of the students work:

1) 594 | 2) $\overset{4}{\cancel{5}}\overset{18}{\cancel{9}}{}^{1}4 - 396 = 198$ | 3) $\overset{1}{5}\overset{1}{9}4 + 396 = 990$

Students work many problems like these that express quantities in measurement units and require using all four operations to solve.

Students also work problems with simple fractions and mixed numbers. See **Fractions** on pages 86–91 for a detailed description of the fraction and mixed number computation skills students learn.

Here are problems from Textbook Lesson 127 that students work that contain a mixed number or a fraction:

a. A man was 6 feet tall. A tree was $12\frac{1}{2}$ times as tall as the man. How tall was the tree?

b. A tree was 74 feet tall. A woman was $68\frac{1}{2}$ feet shorter than the tree. How tall was the woman?

Both problems express measurements in distance units. To solve the first problem, students subtract 12 1/2 by 6. To solve the other problem, students subtract 68 1/2 from 74.

Students work area and perimeter problems that involve simple fractions, too. See **Geometry,** Perimeter-Area on pages 115–118 for a detailed description of the area and perimeter skills that are reviewed and taught in this level. After students have mastered area and perimeter problems and fraction and mixed number computation, students work area problems with simple fractions, and make a diagram to represent the measurement quantities.

Here's a problem and the student's work from Textbook Lesson 125:

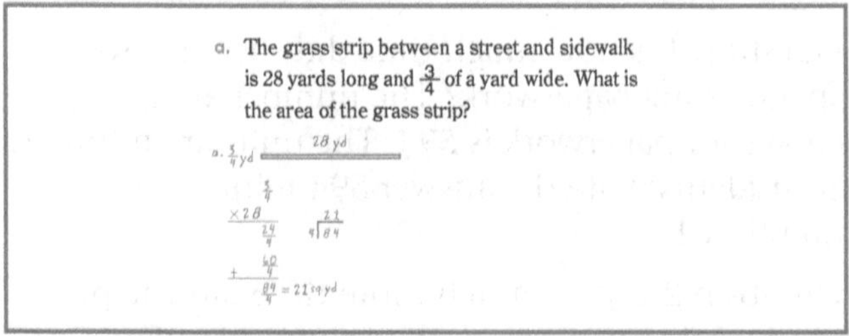
a. The grass strip between a street and sidewalk is 28 yards long and $\frac{3}{4}$ of a yard wide. What is the area of the grass strip?

Students work the multiplication problem 28 × 3/4. The improper fraction they end up with is 84/4. Students divide 84 by 4 to get the answer, 21 square yards. Note again that the students represent the measurement quantities in a diagram.

Common Core State Standards

4.MD 3 Apply the area and perimeter formulas for rectangles in real world and mathematical problems. *For example, find the width of a rectangular room given the area of the flooring and the length, by viewing the area formula as a multiplication equation with an unknown factor.*

Students extensively apply the area and perimeter formulas for rectangles for real world problems. See **Geometry,** Perimeter-Area on pages 115–118 for a detailed description of the area and perimeter skills that are reviewed and taught in this level and for the range of real world and mathematical problems that they are applied to.

Here's a set from Textbook Lesson 126 with a couple of other real-world area and perimeter problems students work:

a. A farmer wants to put a fence around a rectangular garden. The garden is 132 yards wide and 58 yards long. How many yards of fencing does the farmer need to put a fence around that garden?

b. Jim has enough paint to cover 100 square feet of a wall. The wall he wants to paint is 8 feet tall. If he has just enough paint to cover the wall, how long is the wall?

Problem *a* is simple to solve but tough for students who follow procedures without understanding the problem. Students multiply to find the area for most problems that give the lengths for the sides of the rectangle. This problem asks about the perimeter without requiring students to multiply.

Problem *b* requires students to divide the area of the wall by the length that's given to find the missing length.

Common Core State Standards

4.MD 4 Make a line plot to display a data set of measurements in fractions of a unit (1/2, 1/4, 1/8). Solve problems involving addition and subtraction of fractions by using information presented in line plots. *For example, from a line plot find and interpret the difference in length between the longest and shortest specimens in an insect collection.*

Students make line plot displays for sets of data expressed as fractions of a unit. Then students solve problems involving addition and subtraction of fractions by using information presented in their line plot.

Here's the Workbook material and the student work from Workbook Lesson 130 for one of the line plot exercises:

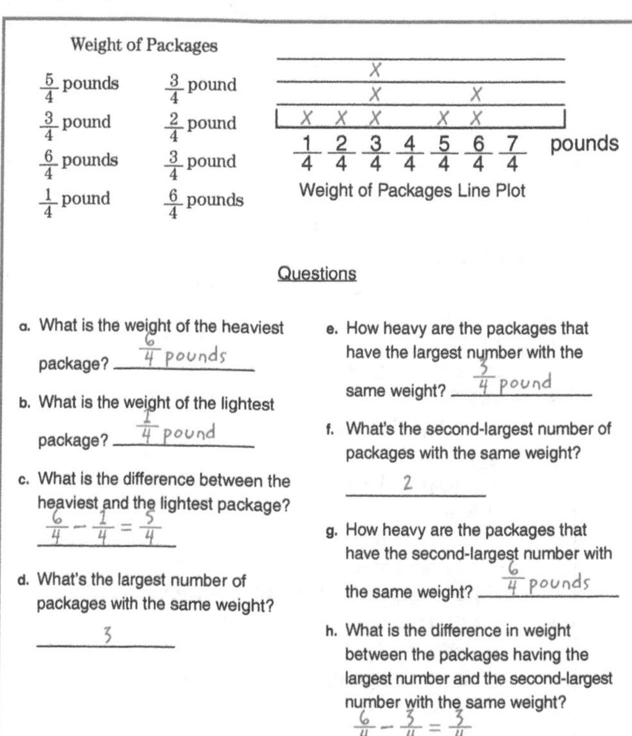

Weight of Packages

$\frac{5}{4}$ pounds	$\frac{3}{4}$ pound
$\frac{3}{4}$ pound	$\frac{2}{4}$ pound
$\frac{6}{4}$ pounds	$\frac{3}{4}$ pound
$\frac{1}{4}$ pound	$\frac{6}{4}$ pounds

$\frac{1}{4}$ $\frac{2}{4}$ $\frac{3}{4}$ $\frac{4}{4}$ $\frac{5}{4}$ $\frac{6}{4}$ $\frac{7}{4}$ pounds

Weight of Packages Line Plot

Questions

a. What is the weight of the heaviest package? $\frac{6}{4}$ pounds

b. What is the weight of the lightest package? $\frac{1}{4}$ pound

c. What is the difference between the heaviest and the lightest package? $\frac{6}{4} - \frac{1}{4} = \frac{5}{4}$

d. What's the largest number of packages with the same weight? 3

e. How heavy are the packages that have the largest number with the same weight? $\frac{3}{4}$ pound

f. What's the second-largest number of packages with the same weight? 2

g. How heavy are the packages that have the second-largest number with the same weight? $\frac{6}{4}$ pounds

h. What is the difference in weight between the packages having the largest number and the second-largest number with the same weight? $\frac{6}{4} - \frac{3}{4} = \frac{3}{4}$

Common Core State Standards

4.MD 5 Recognize angles as geometric shapes that are formed wherever two rays share a common endpoint, and understand concepts of angle measurement:

a. An angle is measured with reference to a circle with its center at the common endpoint of the rays, by considering the fraction of the circular arc between the points where the two rays intersect the circle. An angle that turns through 1/360 of a circle is called a "one-degree angle," and can be used to measure angles.

Students understand that an angle is measured with reference to a circle with its center at the common endpoints of rays, line segments, or at the intersection of lines. See **Geometry,** Angles on pages 119–126 for a detailed description of the instruction students receive on angles. See **Geometry,** Protractors on pages 124–126 for a detailed description of what students learn about measuring angles and how students apply their knowledge of angles and degrees to determine the measure of angles.

Common Core State Standards

4.MD 5 Recognize angles as geometric shapes that are formed wherever two rays share a common endpoint, and understand concepts of angle measurement:

b. An angle that turns through n one-degree angles is said to have an angle measure of n degrees.

Students understand that an angle is measured with reference to a circle with its center at the common endpoints of rays, line segments, or at the intersection of lines. See **Geometry,** Angles on pages 121–123 for a detailed description of the instruction students receive on angles. See **Geometry,** Protractors on pages 124–126 for a detailed description of what students learn about measuring angles and how students apply their knowledge of angles and degrees to determine the measure of angles.

Common Core State Standards

4.MD 6 Measure angles in whole-number degrees using a protractor. Sketch angles of specified measure.

Students measure angles in whole-number degrees using a protractor. Students also draw angles of specified measures using a protractor. See **Geometry,** Protractors on pages 124–126 for a detailed description of how students apply their knowledge of angles and degrees to draw angles and determine the measurement of angles that are shown.

Common Core State Standards

4.MD 7 Recognize angle measure as additive. When an angle is decomposed into non-overlapping parts, the angle measure of the whole is the sum of the angle measures of the parts. Solve addition and subtraction problems to find unknown angles on a diagram in real world and mathematical problems, e.g., by using an equation with a symbol for the unknown angle measure.

Students learn to add and subtract to find the measure of unknown angles. Students work from diagrams to compose or decompose non-overlapping angles. See **Geometry,** Angles on pages 121–123 for a detailed description of the variety of problems students work and the strategies students learn to work them.

Geometry (G)

Common Core State Standards

4.G 1 Draw points, lines, line segments, rays, angles (right, acute, obtuse), and perpendicular and parallel lines. Identify these in two-dimensional figures.

Students learn to measure angles that are given. See **Geometry,** Protractor on pages 124–126 for a detailed description of the procedure students learn about using a protractor to measure angles.

Students learn about points, lines, line segments, rays, and acute, right, and obtuse angles. The vocabulary is introduced one term at a time so students can apply the meaning of the words to different examples before a new vocabulary word is introduced. After students master identifying these angle types, they learn rules for categorizing triangles and other figures based on the presence of these angles. See **Geometry,** Angles on pages 122–123 for a detailed description of what students learn about angles and the tasks students perform to apply what they've learned.

After students learn what a right angle is, teaching them what perpendicular means is easy.

Here's the introduction to the word *perpendicular:*

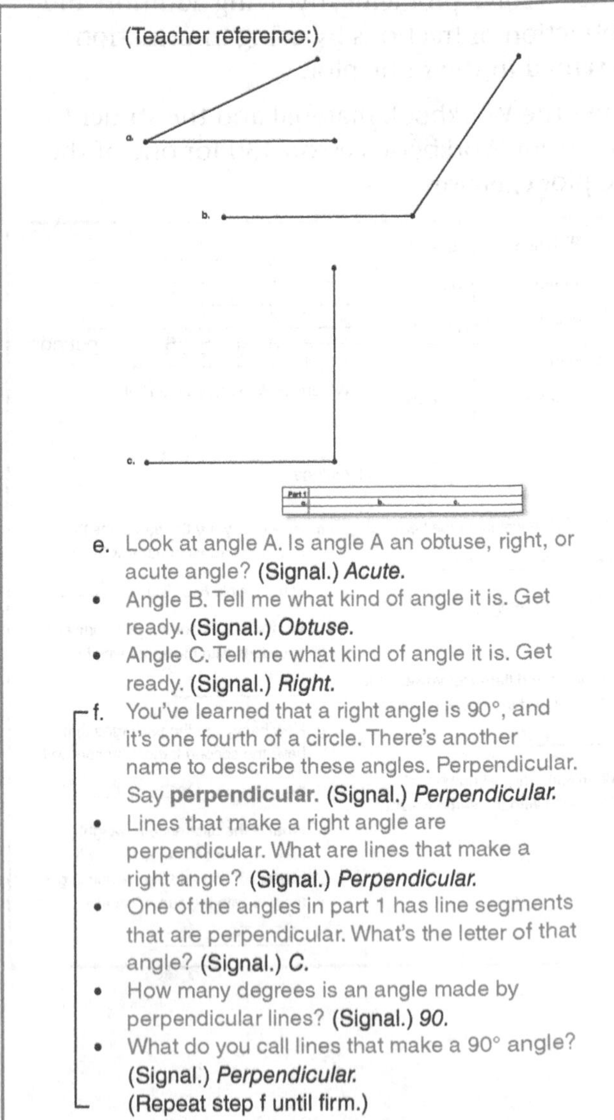

e. Look at angle A. Is angle A an obtuse, right, or acute angle? (Signal.) *Acute.*
- Angle B. Tell me what kind of angle it is. Get ready. (Signal.) *Obtuse.*
- Angle C. Tell me what kind of angle it is. Get ready. (Signal.) *Right.*

f. You've learned that a right angle is 90°, and it's one fourth of a circle. There's another name to describe these angles. Perpendicular. Say **perpendicular.** (Signal.) *Perpendicular.*
- Lines that make a right angle are perpendicular. What are lines that make a right angle? (Signal.) *Perpendicular.*
- One of the angles in part 1 has line segments that are perpendicular. What's the letter of that angle? (Signal.) *C.*
- How many degrees is an angle made by perpendicular lines? (Signal.) *90.*
- What do you call lines that make a 90° angle? (Signal.) *Perpendicular.*

(Repeat step f until firm.)

from Lesson 124, Exercise 2

Initially, students do not have to discriminate between lines and line segments when they work with angles, shapes, and lines. First, students learn the functional difference between a line and a line segment. Later, students learn what a ray is. After students have learned about lines, line segments, and rays and what it means when pairs of them intersect, are parallel or not parallel, students answer the following questions. This is from Textbook Lesson 129:

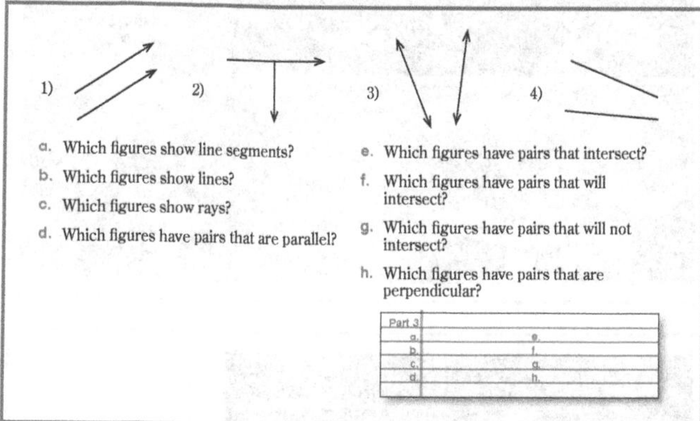
1) 2) 3) 4)

a. Which figures show line segments?
b. Which figures show lines?
c. Which figures show rays?
d. Which figures have pairs that are parallel?
e. Which figures have pairs that intersect?
f. Which figures have pairs that will intersect?
g. Which figures have pairs that will not intersect?
h. Which figures have pairs that are perpendicular?

Part 3		
	a.	e.
	b.	f.
	c.	g.
	d.	h.

As students learn and apply the vocabulary concepts to different geometric examples, students also learn how to draw lines, line segments, and rays that form specific angle-measures. Students use the converse of the rules they learned for measuring angles to draw the angles. After students draw these angles, they identify them as acute, obtuse, or right.

Here is part of an exercise, which directs students to draw angles of specific measures formed by lines:

e. Angle B is 126°. What's angle B? **(Signal.)** *126°.*
- Is 126° an acute, right, or obtuse angle? **(Signal.)** *Obtuse.*
- Write B on your paper. ✔
- Make a point at zero on your protractor. Make a point at the center. Then make a point at 126°. **(Observe students and give feedback.)** **(Display:)** [128:4H]

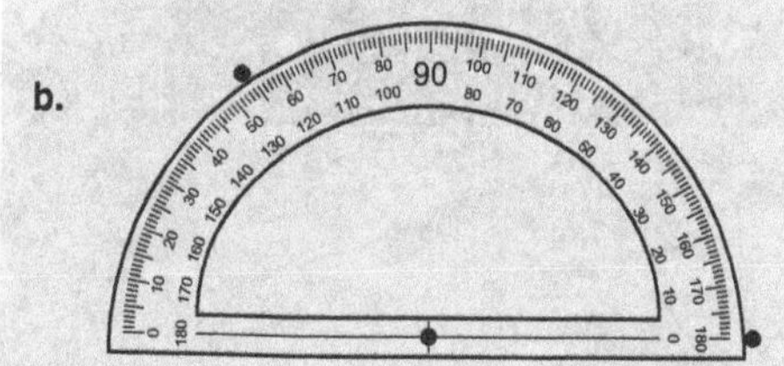

Here's what you should have for angle B so far.

f. Now you'll make lines for the angle. After you make the line segments, what do you draw on the end of each segment? **(Call on a student.)** *Arrowheads.*
- For each line segment, make it through both points. Then make arrowheads on the ends. Then write the degrees for angle B. **(Observe students and give feedback.)** **(Change to show:)** [128:4I]

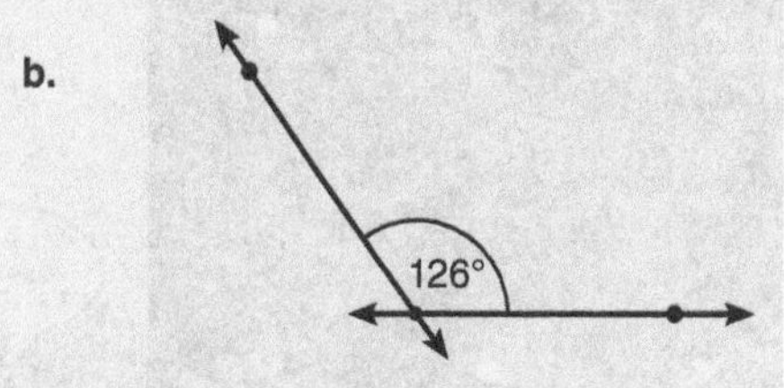

Here are the lines and angle for B.

g. Angle C is 75°. What's angle C? **(Signal.)** *75°.*
- Is 75° an acute, right, or obtuse angle? **(Signal.)** *Acute.*
- Write C on your paper. ✔
- Make a point at zero on your protractor. Make a point at the center. Then make a point at 75°. **(Observe students and give feedback.)** **(Display:)** [128:4J]

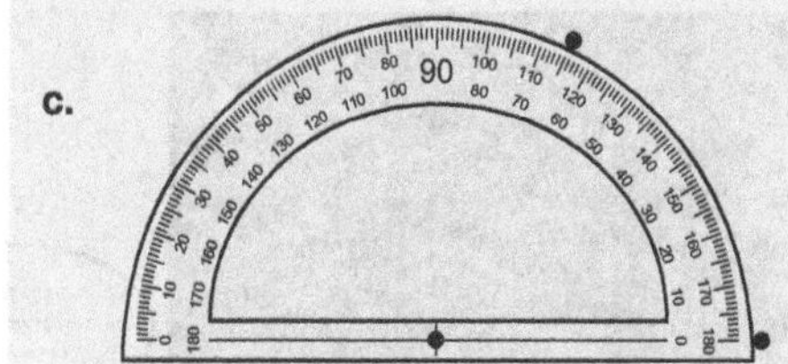

Here's what you should have for angle C so far.

h. Now you'll make rays for the angle. After you make the line segments, what do you draw on one end of each line segment? **(Signal.)** *An arrowhead.*
- Start the rays for angle C at the point they intersect. Make arrowheads to show they go on forever. Then write the degrees for angle C. **(Observe students and give feedback.)** **(Change to show:)** [128:4K]

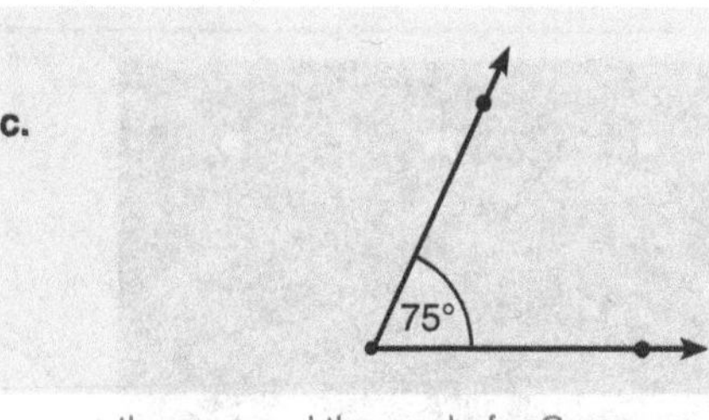

Here are the rays and the angle for C.

from Lesson 128, Exercise 4

Students also draw lines, rays, and line segments in the *Level E* Practice Software activities. The first part of each example for this activity requires students to draw a line, a line segment, or a ray.

Here are the first parts of three different examples of this task. Next to each example is the frame with the correct response.

In the second part of this activity, students make a line, line segment, or ray that is parallel or perpendicular to what they drew in the first part. Here are correct responses for examples of the second part:

Common Core State Standards

4.G 2	Classify two-dimensional figures based on the presence or absence of parallel or perpendicular lines, or the presence or absence of angles of a specified size. Recognize right triangles as a category, and identify right triangles.

Students classify two-dimensional figures based on the presence or absence of parallel or perpendicular lines. Students use the presence of angles of specified sizes to classify many figures. Students learn about acute, right, and obtuse angles. The vocabulary is introduced one term at a time so students can apply the meaning of the words to different examples before a new vocabulary word is introduced. After students master identifying these angle types, they learn rules for categorizing triangles and other figures based on the presence of these angles. See **Geometry,** Angles on pages 121–123 for a detailed description of what students learn about those angles and the tasks students perform to apply what they've learned.

Students also learn to identify lines, line segments, or rays that are parallel or perpendicular. Students then identify parallelograms and rectangles based on the presence of specific angles.

Here's part of an exercise that directs students to classify two-dimensional shapes based on the presence or absence of parallel or perpendicular line segments:

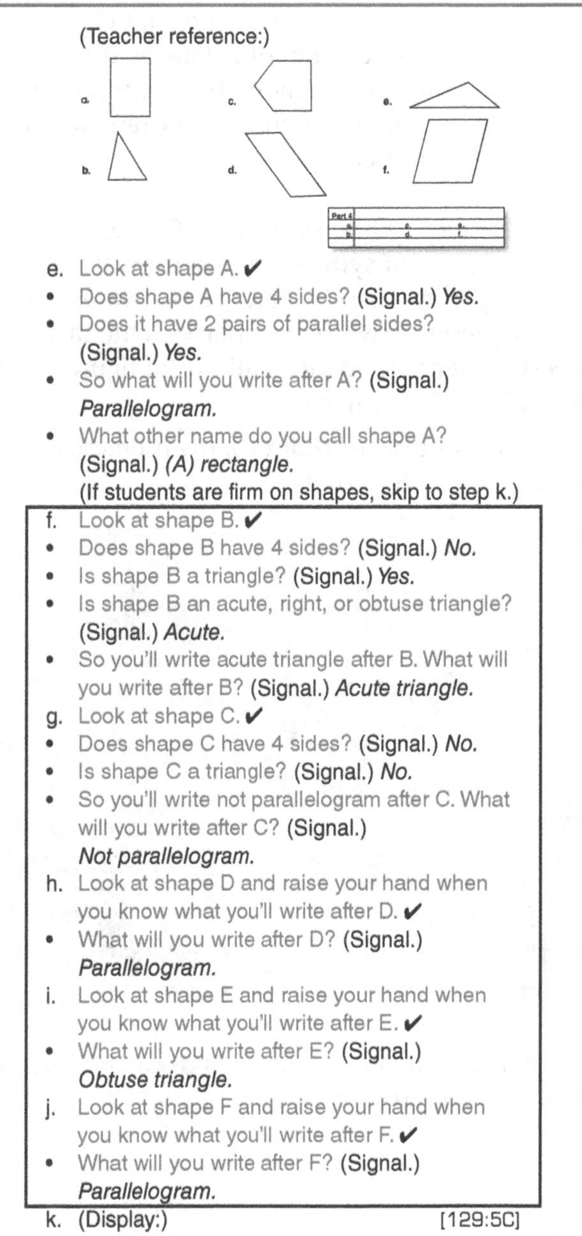

(Teacher reference:)

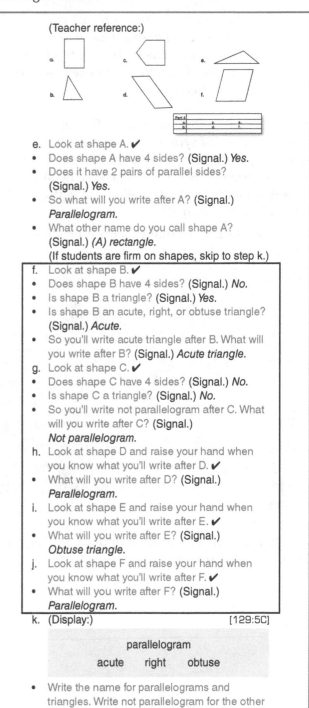

e. Look at shape A. ✔
- Does shape A have 4 sides? (Signal.) *Yes.*
- Does it have 2 pairs of parallel sides? (Signal.) *Yes.*
- So what will you write after A? (Signal.) *Parallelogram.*
- What other name do you call shape A? (Signal.) *(A) rectangle.*

(If students are firm on shapes, skip to step k.)

f. Look at shape B. ✔
- Does shape B have 4 sides? (Signal.) *No.*
- Is shape B a triangle? (Signal.) *Yes.*
- Is shape B an acute, right, or obtuse triangle? (Signal.) *Acute.*
- So you'll write acute triangle after B. What will you write after B? (Signal.) *Acute triangle.*

g. Look at shape C. ✔
- Does shape C have 4 sides? (Signal.) *No.*
- Is shape C a triangle? (Signal.) *No.*
- So you'll write not parallelogram after C. What will you write after C? (Signal.) *Not parallelogram.*

h. Look at shape D and raise your hand when you know what you'll write after D. ✔
- What will you write after D? (Signal.) *Parallelogram.*

i. Look at shape E and raise your hand when you know what you'll write after E. ✔
- What will you write after E? (Signal.) *Obtuse triangle.*

j. Look at shape F and raise your hand when you know what you'll write after F. ✔
- What will you write after F? (Signal.) *Parallelogram.*

k. (Display:) [129:5C]

parallelogram
acute right obtuse

- Write the name for parallelograms and triangles. Write not parallelogram for the other shapes. Put your pencil down when you've completed part 4.

(Observe students and give feedback.)

from Lesson 129, Exercise 5

Common Core State Standards

4.G 3	Recognize a line of symmetry for a two-dimensional figure as a line across the figure such that the figure can be folded along the line into matching parts. Identify line-symmetric figures and draw lines of symmetry.

Students identify line-symmetric figures, recognize lines of symmetry, and draw lines of symmetry in the *Level E* Practice Software. The activity presents two types of tasks. One task directs students to identify all of the figures that have a line of symmetry.

Here's an example of that task from Practice Software Block 5 Activity 3:

Another task directs students to draw the line or lines of symmetry for a figure.

Here are some examples of that task from the same software activity. The correct responses are shown below:

Appendix A

Placement Test

Appendix A: Placement Test

The Placement Test provides for three outcomes:

- The student lacks the necessary skills to place in *CMC Level E.*
- The student places at Lesson 1 of *CMC Level E.*
- The student places at Lesson 31 of *CMC Level E.*

If possible, present the Placement Test on the first day of instruction. Pass out a test to each student. Present the wording in the test Administration Directions script.

Note: What you say is shown in blue type. When observing students, make sure that they are working on the correct part of the test. Do not prompt them in a way that would let them know the answer to an item.

Reproducible copies of the test appear on pages 198–201 of this guide.

CONNECTING MATH CONCEPTS—LEVEL E

PLACEMENT TEST

Administration Directions

> ***Note:*** You will need a stopwatch or a clock with a second hand for Parts 2 and 3.

a. (Hand out Placement Test to students. Direct students to put their names at the top of the test.)

- Everybody, find Part 1. ✔

  (Teacher reference:)

a. ______	c. ______	e. ______	g. ______
b. ______	d. ______	f. ______	h. ______

- I'm going to say numbers. You'll write them on the lines in Part 1.
- Touch line A. ✔
- Write 302 on line A. ✔
- Touch line B. ✔
- Write 217 on line B. ✔
- (Repeat for remaining numbers: C, 409; D, 3,640; E, 1,054.)

b. Now, I'm going to say dollar amounts. You'll write the amounts with a dollar sign and a decimal point.

- Listen: 7 dollars and 45 cents. Say that amount. (Signal.) *Seven dollars and 45 cents.*
- Write 7 dollars and 45 cents on line F. ✔
- Listen: 20 dollars and 16 cents. Say that amount. (Signal.) *20 dollars and 16 cents.*
- Write 20 dollars and 16 cents on line G. ✔
- Listen: 8 cents. Say that amount. (Signal.) *8 cents.*
- Write 8 cents on line H. Write it with a dollar sign and a decimal point. ✔

c. Find Part 2. ✔

(Teacher reference:)

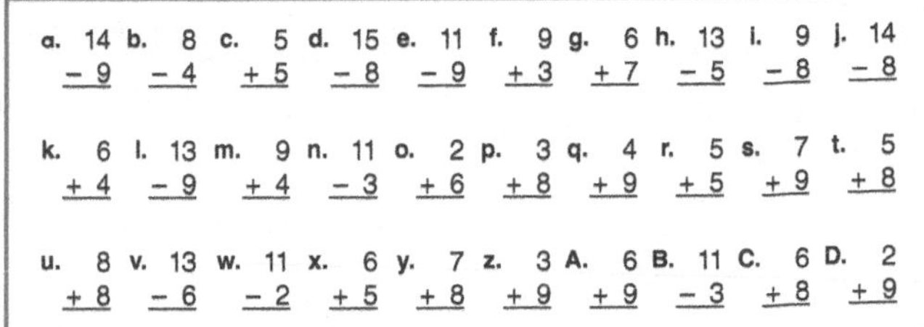

a. 14 − 9	b. 8 − 4	c. 5 + 5	d. 15 − 8	e. 11 − 9	f. 9 + 3	g. 6 + 7	h. 13 − 5	i. 9 − 8	j. 14 − 8
k. 6 + 4	l. 13 − 9	m. 9 + 4	n. 11 − 3	o. 2 + 6	p. 3 + 8	q. 4 + 9	r. 5 + 5	s. 7 + 9	t. 5 + 8
u. 8 + 8	v. 13 − 6	w. 11 − 2	x. 6 + 5	y. 7 + 8	z. 3 + 9	A. 6 + 9	B. 11 − 3	C. 6 + 8	D. 2 + 9

For Part 2, you'll write answers to addition and subtraction problems. I'll time you. You'll have 2 minutes and 30 seconds to write the answers to the problems in Part 2.

- Pencils ready. Go. ✔
- (At the end of 2 minutes and 30 seconds, say:) Everybody, stop and put a circle around the last problem you answered. (Observe students.)

d. Find Part 3. ✔

(Teacher reference:)

a. 5×3 b. 9×2 c. 2×4 d. 7×1 e. 5×6 f. 7×2 g. 5×9 h. 4×3 i. 5×5 j. 2×6

k. 4×4 l. 9×3 m. 2×8 n. 8×5 o. 4×2 p. 9×4 q. 7×5 r. 2×5 s. 3×3 t. 1×8

u. 6×3 v. 6×6 w. 7×3 x. 8×1 y. 4×6 z. 3×8 A. 6×4 B. 2×9 C. 6×5 D. 3×9

For Part 3, you'll write answers to multiplication problems. I'll time you. You'll have 2 minutes and 30 seconds to write the answers to the problems in Part 3.

- Pencils ready. Go. ✔
- **(At the end of 2 minutes and 30 seconds, say:)** Everybody, stop and put a circle around the last problem you answered. **(Observe students.)**

e. Find Part 4. ✔

(Teacher reference:)

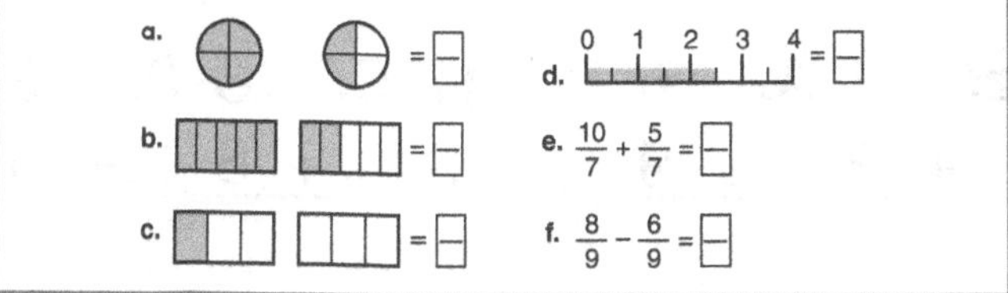

Some of these problems show pictures of fractions. Some of them are problems that add or subtract fractions.

- Write the fraction for each picture. Complete the equation for the fraction problems. Then work the problems in Parts 5 through 12 on your own. Pencils down when you're done.

Passing Criteria for Each Part

Part	Description	#	Pass	Fail
Part 1	Writing 3 and 4 digit numbers and dollar values	8	0–2 errors	3 or more errors
Part 2	Addition and subtraction facts	30	0–5 errors	6 or more errors
Part 3	Multiplication facts	30	0–5 errors	6 or more errors
Part 4	Fraction from pictures and ± fraction problems	6	0–2 errors	3 or more errors
Part 5	Column ± with carrying and borrowing	7	0–3 errors	4 or more errors
Part 6	Division facts	15	0–3 errors	4 or more errors
Part 7	Column multiplication 2 digit × 1 digit and multi-digit × 10	5	0–2 errors	3 or more errors
Part 8	Division: 3 digit ÷ by 1 digit with no remainders	5	0–2 errors	3 or more errors
Part 9	Completing the numerators for whole numbers and equivalent fractions they equal	8	0–2 errors	3 or more errors
Part 10	Comparison and sequence word problems	4	0–1 error	2 or more errors
Part 11	Area and perimeter of rectangles: 1 digit lengths with units	6	0–2 errors	3 or more errors
Part 12	<, >, = for 2 whole numbers or a fraction and 1	6	0–2 errors	3 or more errors

Placement Criteria

Students who pass 10 or 11 parts.	Begin *CMC Level E* at Lesson 31. **Note:** If possible, group students according to the number of parts passed.
Students who pass 6, 7, 8, or 9 parts or who have a total score of 80 points or more.	Begin *CMC Level E* at Lesson 1.
Students who pass 5 or fewer parts.	Administer the *CMC Level D* Placement Test

☐ errors

CMC Level E Placement Test Name ____________________ Date ____________________

Part 1

a. ________ c. ________ e. ________ g. ________

b. ________ d. ________ f. ________ h. ________

Part 2

a. $14 - 9$ b. $8 - 4$ c. $5 + 5$ d. $15 - 8$ e. $11 - 9$ f. $9 + 3$ g. $6 + 7$ h. $13 - 5$ i. $9 - 8$ j. $14 - 8$

k. $6 + 4$ l. $13 - 9$ m. $9 + 4$ n. $11 - 3$ o. $2 + 6$ p. $3 + 8$ q. $4 + 9$ r. $5 + 5$ s. $7 + 9$ t. $5 + 8$

u. $8 + 8$ v. $13 - 6$ w. $11 - 2$ x. $6 + 5$ y. $7 + 8$ z. $3 + 9$ A. $6 + 9$ B. $11 - 3$ C. $6 + 8$ D. $2 + 9$

Part 3

a. 5×3 b. 9×2 c. 2×4 d. 7×1 e. 5×6 f. 7×2 g. 5×9 h. 4×3 i. 5×5 j. 2×6

k. 4×4 l. 9×3 m. 2×8 n. 8×5 o. 4×2 p. 9×4 q. 7×5 r. 2×5 s. 3×3 t. 1×8

u. 6×3 v. 6×6 w. 7×3 x. 8×1 y. 4×6 z. 3×8 A. 6×4 B. 2×9 C. 6×5 D. 3×9

☐ errors

CMC Level E Placement Test Name ________________ Date ________________

Part 4

a. 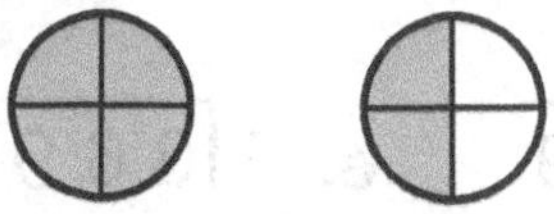 = $\frac{\square}{\square}$

b. 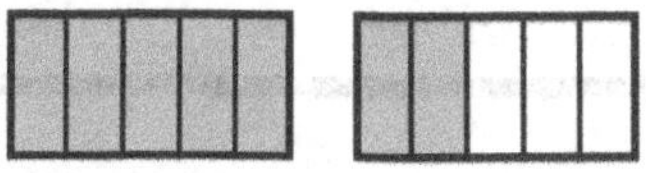 = $\frac{\square}{\square}$

c. 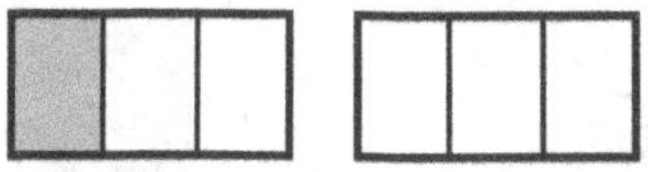 = $\frac{\square}{\square}$

d. 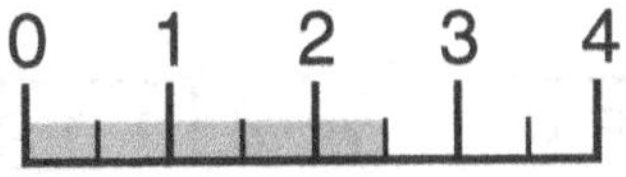

= $\frac{\square}{\square}$

e. $\frac{10}{7} + \frac{5}{7} = \frac{\square}{\square}$

f. $\frac{8}{9} - \frac{6}{9} = \frac{\square}{\square}$

Part 5

a. $\begin{array}{r} 64 \\ +533 \\ \hline \end{array}$

b. $\begin{array}{r} 54 \\ -\ \ 17 \\ \hline \end{array}$

c. $\begin{array}{r} 591 \\ +\ \ 64 \\ \hline \end{array}$

d. $\begin{array}{r} 418 \\ +\ \ 79 \\ \hline \end{array}$

e. $\begin{array}{r} 420 \\ -190 \\ \hline \end{array}$

f. $\begin{array}{r} 752 \\ -\ \ 36 \\ \hline \end{array}$

g. $\begin{array}{r} 149 \\ +353 \\ \hline \end{array}$

Part 6

a. $8\overline{)16}$

b. $4\overline{)36}$

c. $7\overline{)7}$

d. $5\overline{)30}$

e. $2\overline{)14}$

f. $9\overline{)27}$

g. $8\overline{)24}$

h. $5\overline{)40}$

i. $6\overline{)60}$

j. $7\overline{)42}$

k. $7\overline{)35}$

l. $2\overline{)18}$

m. $6\overline{)24}$

n. $5\overline{)20}$

o. $4\overline{)8}$

Part 7

a. $\begin{array}{r} 24 \\ \times\ \ \ 5 \\ \hline \end{array}$

b. $\begin{array}{r} 36 \\ \times\ \ \ 2 \\ \hline \end{array}$

c. $\begin{array}{r} 13 \\ \times\ \ \ 9 \\ \hline \end{array}$

d. $\begin{array}{r} 10 \\ +720 \\ \hline \end{array}$

e. $\begin{array}{r} 38 \\ \times\ \ 10 \\ \hline \end{array}$

errors

CMC Level E Placement Test Name ____________ Date ____________

Part 8

a. $2\overline{)806}$ b. $3\overline{)159}$ c. $4\overline{)208}$ d. $3\overline{)369}$ e. $9\overline{)369}$

Part 9

a. $3 = \frac{}{1} = \frac{}{3} = \frac{}{9} = \frac{}{10}$

b. $4 = \frac{}{2} = \frac{}{8} = \frac{}{1} = \frac{}{3}$

Part 10

a. Carlos is 14 years younger than James. Carlos is 25 years old. How old is James?

c. A bus had some people on it. 34 people got off of the bus. The bus ended up with 43 people on it. How many people were on the bus to start with?

b. Anna read 12 more books than Maria. Anna read 43 books. How many books did Maria read?

d. A train had some people on it. Then 55 people got on the train. The train ended up with 89 people. How many people did the train start with?

errors

CMC Level E Placement Test Name ____________ Date ____________

Part 11 Figure out the area and the perimeter for each rectangle.

a. 6 in. / 2 in.

area: ____________

perimeter: ____________

b. 5 ft / 4 ft

area: ____________

perimeter: ____________

c. 7 yd / 1 yd

area: ____________

perimeter: ____________

Part 12 Write <, >, or = to complete each statement.

a. 12 10

b. 39 47

c. 108 151

d. $\frac{4}{4}$ 1

e. $\frac{3}{2}$ 1

f. $\frac{7}{9}$ 1

CMC Level E Placement Test Answer Key

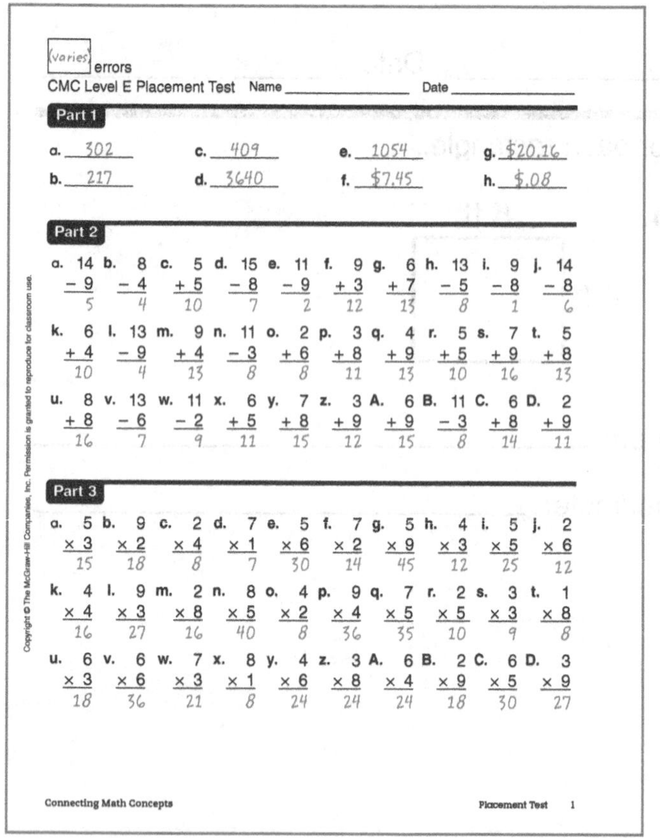

(varies) errors

CMC Level E Placement Test Name ______ Date ______

Part 1

a. 302 c. 409 e. 1054 g. $20.16

b. 217 d. 3640 f. $7.45 h. $.08

Part 2

a. 14 − 9 = 5 b. 8 − 4 = 4 c. 5 + 5 = 10 d. 15 − 8 = 7 e. 11 − 9 = 2 f. 9 + 3 = 12 g. 6 + 7 = 13 h. 13 − 5 = 8 i. 9 − 8 = 1 j. 14 − 8 = 6

k. 6 + 4 = 10 l. 13 − 9 = 4 m. 9 + 4 = 13 n. 11 − 3 = 8 o. 2 + 6 = 8 p. 3 + 8 = 11 q. 4 + 9 = 13 r. 5 + 5 = 10 s. 7 + 9 = 16 t. 5 + 8 = 13

u. 8 + 8 = 16 v. 13 − 6 = 7 w. 11 − 2 = 9 x. 6 + 5 = 11 y. 7 + 8 = 15 z. 3 + 9 = 12 A. 6 + 9 = 15 B. 11 − 3 = 8 C. 6 + 8 = 14 D. 2 + 9 = 11

Part 3

a. 5 × 3 = 15 b. 9 × 2 = 18 c. 2 × 4 = 8 d. 7 × 1 = 7 e. 5 × 6 = 30 f. 7 × 2 = 14 g. 5 × 9 = 45 h. 4 × 3 = 12 i. 5 × 5 = 25 j. 2 × 6 = 12

k. 4 × 4 = 16 l. 9 × 3 = 27 m. 2 × 8 = 16 n. 8 × 5 = 40 o. 4 × 2 = 8 p. 9 × 4 = 36 q. 7 × 5 = 35 r. 2 × 5 = 10 s. 3 × 3 = 9 t. 1 × 8 = 8

u. 6 × 3 = 18 v. 6 × 6 = 36 w. 7 × 3 = 21 x. 8 × 1 = 8 y. 4 × 6 = 24 z. 3 × 8 = 24 A. 6 × 4 = 24 B. 2 × 9 = 18 C. 6 × 5 = 30 D. 3 × 9 = 27

Connecting Math Concepts Placement Test 1

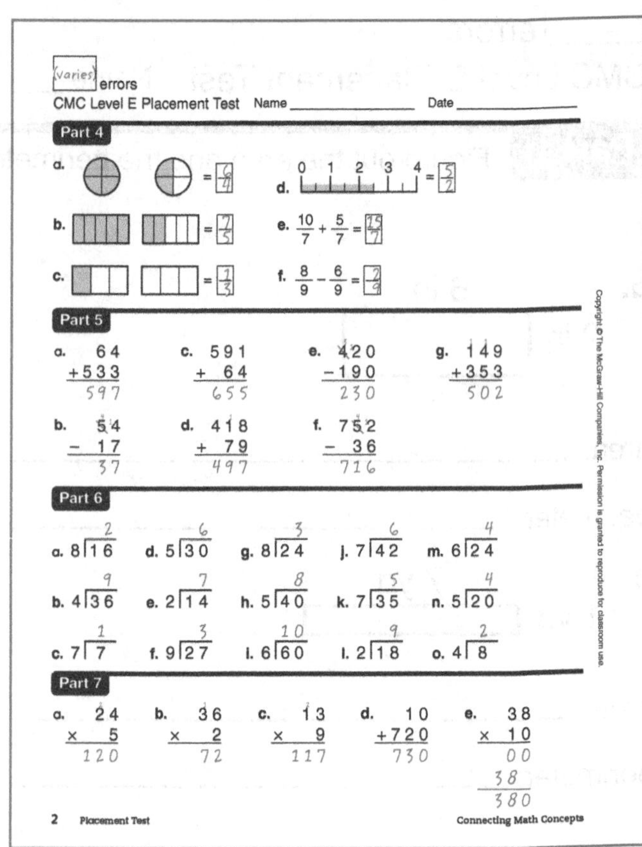

(varies) errors

CMC Level E Placement Test Name ______ Date ______

Part 4

a. = $\frac{6}{4}$ d. 0 1 2 3 4 = $\frac{5}{2}$

b. = $\frac{7}{5}$ e. $\frac{10}{7} + \frac{5}{7} = \frac{15}{7}$

c. = $\frac{1}{3}$ f. $\frac{8}{9} - \frac{6}{9} = \frac{2}{9}$

Part 5

a. 64 + 533 = 597 c. 591 + 64 = 655 e. 420 − 190 = 230 g. 149 + 353 = 502

b. 54 − 17 = 37 d. 418 + 79 = 497 f. 752 − 36 = 716

Part 6

a. 8)16 = 2 d. 5)30 = 6 g. 8)24 = 3 j. 7)42 = 6 m. 6)24 = 4

b. 4)36 = 9 e. 2)14 = 7 h. 5)40 = 8 k. 7)35 = 5 n. 5)20 = 4

c. 7)7 = 1 f. 9)27 = 3 i. 6)60 = 10 l. 2)18 = 9 o. 4)8 = 2

Part 7

a. 24 × 5 = 120 b. 36 × 2 = 72 c. 13 × 9 = 117 d. 10 + 720 = 730 e. 38 × 10 = 00 + 380 = 380

2 Placement Test Connecting Math Concepts

(varies) errors

CMC Level E Placement Test Name ______ Date ______

Part 8

a. 2)806 = 403 b. 3)159 = 53 c. 4)208 = 52 d. 3)369 = 123 e. 9)369 = 41

Part 9

a. $3 = \frac{3}{1} = \frac{9}{3} = \frac{27}{9} = \frac{30}{10}$ b. $4 = \frac{8}{2} = \frac{32}{8} = \frac{4}{1} = \frac{12}{3}$

Part 10

a. Carlos is 14 years younger than James. Carlos is 25 years old. How old is James?

14 C → J (25); 14 + 25 = 39 years

c. A bus had some people on it. 34 people got off of the bus. The bus ended up with 43 people on it. How many people were on the bus to start with?

34 E → S (43); 34 + 43 = 77 people

b. Anna read 12 more books than Maria. Anna read 43 books. How many books did Maria read?

12 M → A (43); 43 − 12 = 31 books

d. A train had some people on it. Then 55 people got on the train. The train ended up with 89 people. How many people did the train start with?

55 S → E (89); 89 − 55 = 34 people

Connecting Math Concepts Placement Test 3

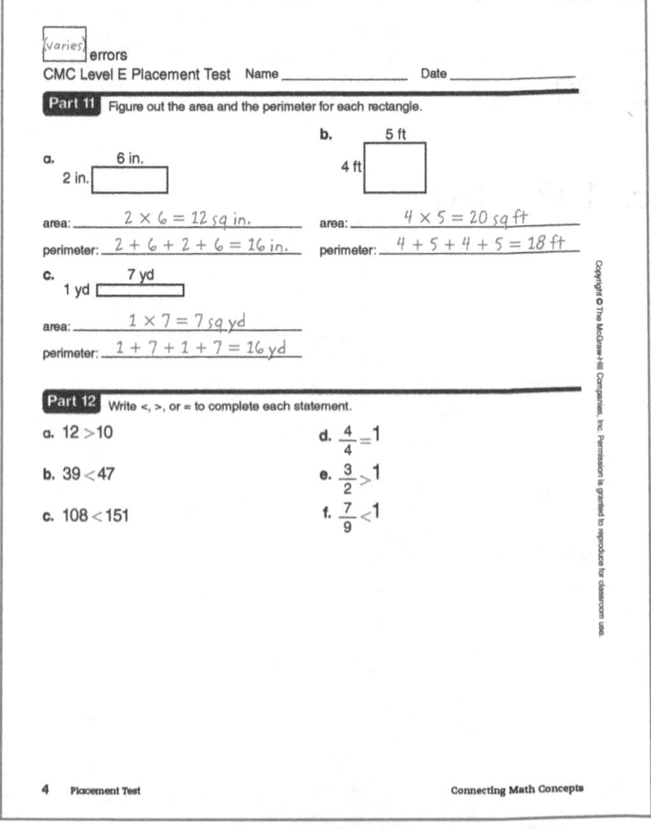

(varies) errors

CMC Level E Placement Test Name ______ Date ______

Part 11 Figure out the area and the perimeter for each rectangle.

a. 6 in., 2 in.

area: 2 × 6 = 12 sq in.

perimeter: 2 + 6 + 2 + 6 = 16 in.

b. 5 ft, 4 ft

area: 4 × 5 = 20 sq ft

perimeter: 4 + 5 + 4 + 5 = 18 ft

c. 7 yd, 1 yd

area: 1 × 7 = 7 sq yd

perimeter: 1 + 7 + 1 + 7 = 16 yd

Part 12 Write <, >, or = to complete each statement.

a. $12 > 10$ d. $\frac{4}{4} = 1$

b. $39 < 47$ e. $\frac{3}{2} > 1$

c. $108 < 151$ f. $\frac{7}{9} < 1$

4 Placement Test Connecting Math Concepts

Appendix B

Reproducible Mastery Test Summary Sheets

Remedy Summary—Group Summary of Test Performance

Note: Test remedies are included in the *Student Assessment Book*. Percent Tables are provided in the *Answer Key*.

	Test 1										Test 2									Test 3										
	Check parts not passed									Total %	Check parts not passed								Total %	Check parts not passed										Total %
Name	1	2	3	4	5	6	7	8	9		1	2	3	4	5	6	7	8		1	2	3	4	5	6	7	8	9	10	
1.																														
2.																														
3.																														
4.																														
5.																														
6.																														
7.																														
8.																														
9.																														
10.																														
11.																														
12.																														
13.																														
14.																														
15.																														
16.																														
17.																														
18.																														
19.																														
20.																														
21.																														
22.																														
23.																														
24.																														
25.																														
26.																														
27.																														
28.																														
29.																														
30.																														
Number of students Not Passed = NP																														
Total number of students = T																														
Remedy needed if NP/T = 25% or more																														

Remedy Summary—Group Summary of Test Performance

Note: Test remedies are included in the *Student Assessment Book*. Percent Tables are provided in the *Answer Key.*

	Test 4													Test 5										
	Check parts not passed												Total %	Check parts not passed										Total %
Name	1	2	3	4	5	6	7	8	9	10	11	12		1	2	3	4	5	6	7	8	9	10	
1.																								
2.																								
3.																								
4.																								
5.																								
6.																								
7.																								
8.																								
9.																								
10.																								
11.																								
12.																								
13.																								
14.																								
15.																								
16.																								
17.																								
18.																								
19.																								
20.																								
21.																								
22.																								
23.																								
24.																								
25.																								
26.																								
27.																								
28.																								
29.																								
30.																								
Number of students Not Passed = NP																								
Total number of students = T																								
Remedy needed if NP/T = 25% or more																								

Remedy Summary—Group Summary of Test Performance

Note: Test remedies are included in the *Student Assessment Book*. Percent Tables are provided in the *Answer Key*.

	Test 6											Test 7												Test 8							
	Check parts not passed										Total %	Check parts not passed											Total %	Check parts not passed							Total %
Name	1	2	3	4	5	6	7	8	9	10		1	2	3	4	5	6	7	8	9	10	11		1	2	3	4	5	6	7	
1.																															
2.																															
3.																															
4.																															
5.																															
6.																															
7.																															
8.																															
9.																															
10.																															
11.																															
12.																															
13.																															
14.																															
15.																															
16.																															
17.																															
18.																															
19.																															
20.																															
21.																															
22.																															
23.																															
24.																															
25.																															
26.																															
27.																															
28.																															
29.																															
30.																															
Number of students Not Passed = NP																															
Total number of students = T																															
Remedy needed if NP/T = 25% or more																															

Remedy Summary—Group Summary of Test Performance

Note: Test remedies are included in the *Student Assessment Book*. Percent Tables are provided in the *Answer Key*.

Name	Test 9										Test 10												Test 11								
	Check parts not passed									Total %	Check parts not passed											Total %	Check parts not passed								Total %
	1	2	3	4	5	6	7	8	9		1	2	3	4	5	6	7	8	9	10	11		1	2	3	4	5	5	7	8	
1.																															
2.																															
3.																															
4.																															
5.																															
6.																															
7.																															
8.																															
9.																															
10.																															
11.																															
12.																															
13.																															
14.																															
15.																															
16.																															
17.																															
18.																															
19.																															
20.																															
21.																															
22.																															
23.																															
24.																															
25.																															
26.																															
27.																															
28.																															
29.																															
30.																															
Number of students Not Passed = NP																															
Total number of students = T																															
Remedy needed if NP/T = 25% or more																															

Remedy Summary—Group Summary of Test Performance

Note: Test remedies are included in the *Student Assessment Book*. Percent Tables are provided in the *Answer Key*.

	Test 12										Test 13									
	Check parts not passed									Total %	Check parts not passed									Total %
Name	1	2	3	4	5	6	7	8	9		1	2	3	4	5	6	7	8	9	
1.																				
2.																				
3.																				
4.																				
5.																				
6.																				
7.																				
8.																				
9.																				
10.																				
11.																				
12.																				
13.																				
14.																				
15.																				
16.																				
17.																				
18.																				
19.																				
20.																				
21.																				
22.																				
23.																				
24.																				
25.																				
26.																				
27.																				
28.																				
29.																				
30.																				
Number of students Not Passed = NP																				
Total number of students = T																				
Remedy needed if NP/T = 25% or more																				

Appendix C

Sample Lessons

- Lesson 42 Presentation Book, Workbook, and Textbook
- Lesson 103 Presentation Book, Workbook, and Textbook

Lesson 42

EXERCISE 1: *NUMBER FAMILIES*
MULTIPLICATION/DIVISION—DIVISION FACTS

a. Open your workbook to Lesson 42 and find part 1. ✔
(Teacher reference:)

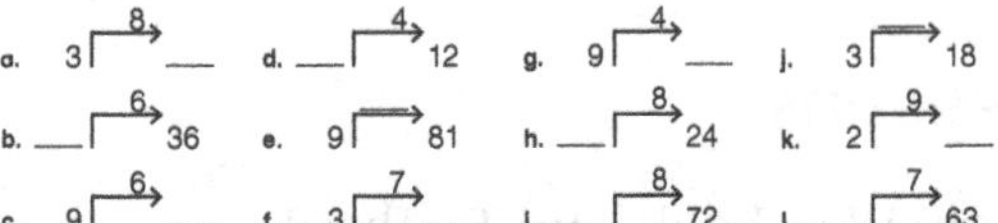

One of the numbers in each family is missing. You'll say the problem and the answer for the missing number in each family. Then you'll complete the families.

- Family A. Say the problem for the missing number. (Signal.) *3 × 8.*
- What's 3 × 8? (Signal.) *24.*

b. (Repeat the following tasks for families B through L:)

Say the problem for family ___.		What's ___?	
B	*36 ÷ 6*	36 ÷ 6	*6*
C	*9 × 6*	9 × 6	*54*
D	*12 ÷ 4*	12 ÷ 4	*3*
E	*81 ÷ 9*	81 ÷ 9	*9*
F	*3 × 7*	3 × 7	*21*
G	*9 × 4*	9 × 4	*36*
H	*24 ÷ 8*	24 ÷ 8	*3*

(If students were 100% on families A through H, skip to step c.)

I	*72 ÷ 8*	72 ÷ 8	*9*
J	*18 ÷ 3*	18 ÷ 3	*6*
K	*2 × 9*	2 × 9	*18*
L	*63 ÷ 7*	63 ÷ 7	*9*

(Repeat families that were not firm.)

c. Write the missing number in each family. Put your pencil down when you've completed the families in part 1.
(Observe students and give feedback.)

d. Check your work. You'll tell me the missing number you wrote for each family.
- Family A. (Signal.) *24.*
- (Repeat for:) B, *6;* C, *54;* D, *3;* E, *9;* F, *21;* G, *36;* H, *3;* I, *9;* J, *6;* K, *18;* L, *9.*

e. Find part 2 in your workbook. ✔
(Teacher reference:)

a. 8 × 3 =	g. 5 × 5 =	m. $4\overline{)16}$	s. $7\overline{)49}$	y. $5\overline{)10}$
b. 9 × 4 =	h. $3\overline{)27}$	n. 9 × 6 =	t. $2\overline{)14}$	z. 7 × 9 =
c. $9\overline{)18}$	i. $8\overline{)64}$	o. $6\overline{)36}$	u. $9\overline{)90}$	A. $3\overline{)27}$
d. $3\overline{)21}$	j. 8 × 9 =	p. $8\overline{)72}$	v. 9 × 9 =	B. 9 × 5 =
e. 4 × 4 =	k. 3 × 7 =	q. 3 × 6 =	w. 5 × 8 =	C. $9\overline{)81}$
f. $3\overline{)9}$	l. $10\overline{)100}$	r. $10\overline{)80}$	x. $9\overline{)54}$	D. $5\overline{)45}$

These multiplication and division problems are from multiplication families you know. You'll read some of the problems and tell me if the answer is the big number or a small number. Then you'll work all of the problems.

f. Read problem A. (Signal.) *8 × 3.*
- Is the answer the big number or a small number? (Signal.) *The big number.*
- What's 8 × 3? (Signal.) *24.*

g. (Repeat the following tasks with problems B through H:)

Read problem ___.		Is the answer the big number or a small number?	What's ___?	
B	*9 × 4*	*The big number.*	9 × 4	*36*
C	*18 ÷ 9*	*A small number.*	18 ÷ 9	*2*
D	*21 ÷ 3*	*A small number.*	21 ÷ 3	*7*
E	*4 × 4*	*The big number.*	4 × 4	*16*
F	*9 ÷ 3*	*A small number.*	9 ÷ 3	*3*
G	*5 × 5*	*The big number.*	5 × 5	*25*
H	*27 ÷ 3*	*A small number.*	27 ÷ 3	*9*

(Repeat problems that were not firm.)

h. Work all the problems in part 2. You have two minutes.
- Get ready. Go.
(Observe students and give feedback.)
- (After 2 minutes say:) Stop.

i. Check your work. You'll read the fact for each problem.

- Problem A. (Signal.) *8 × 3 = 24.*
- (Repeat for:) B, *9 × 4 = 36;* C, *18 ÷ 9 = 2;* D, *21 ÷ 3 = 7;* E, *4 × 4 = 16;* F, *9 ÷ 3 = 3;* G, *5 × 5 = 25;* H, *27 ÷ 3 = 9;* I, *64 ÷ 8 = 8;* J, *8 × 9 = 72;* K, *3 × 7 = 21;* L, *100 ÷ 10 = 10;* M, *16 ÷ 4 = 4;* N, *9 × 6 = 54;* O, *36 ÷ 6 = 6;* P, *72 ÷ 8 = 9;* Q, *3 × 6 = 18;* R, *80 ÷ 10 = 8;* S, *49 ÷ 7 = 7;* T, *14 ÷ 2 = 7;* U, *90 ÷ 9 = 10;* V, *9 × 9 = 81;* W, *5 × 8 = 40;* X, *54 ÷ 9 = 6;* Y, *10 ÷ 5 = 2;* Z, *7 × 9 = 63;* Capital A, *27 ÷ 3 = 9;* B, *9 × 5 = 45;* C, *81 ÷ 9 = 9;* D, *45 ÷ 5 = 9.*

EXERCISE 2: FRACTIONS

ADDING WHOLE NUMBERS AND FRACTIONS REMEDY

a. (Display:) [42:2A]

$$7 + \frac{5}{2}$$

- (Point to $7 + \frac{5}{2}$.) Read this problem. (Signal.) *7 + 5/2.*
- Do 7 and 5/2 have the same bottom number? (Signal.) *No.*
- So you can't work the problem unless you rewrite the whole number as a fraction. What bottom number will you write? (Signal.) *2.*

b. So you rewrite 7 as a fraction with a bottom number of 2.
(Add to show:) [42:2B]

$$\frac{}{2}\not{7} + \frac{5}{2}$$

- Raise your hand when you know the top number of the fraction. ✔
- What's the top number? (Signal.) *14.*
(Add to show:) [42:2C]

$$\frac{14}{2}\not{7} + \frac{5}{2}$$

c. Read the fraction addition problem. (Signal.) *14/2 + 5/2.*

- Can you work that problem? (Signal.) *Yes.*
- What's 14/2 + 5/2? (Signal.) *19/2.*
(Add to show:) [42:2D]

$$\frac{14}{2}\not{7} + \frac{5}{2} = \frac{19}{2}$$

- What does 7 + 5/2 equal? (Signal.) *19/2.*

d. (Display:) [42:2E]

$$3 + \frac{2}{9}$$

- (Point to $3 + \frac{2}{9}$.) Read this problem. (Signal.) *3 + 2/9.*
- Can you add 3 + 2/9? (Signal.) *No.*
- So you have to rewrite 3 as a fraction. What's the bottom number of that fraction? (Signal.) *9.*
(Add to show:) [42:2F]

$$\frac{}{9}\not{3} + \frac{2}{9}$$

- Raise your hand when you know the top number of the fraction. ✔
- What's the top number? (Signal.) *27.*
(Add to show:) [42:2G]

$$\frac{27}{9}\not{3} + \frac{2}{9}$$

e. Read the fraction addition problem. (Signal.) *27/9 + 2/9.*

- What's 27/9 + 2/9? (Signal.) *29/9.*
(Add to show:) [42:2H]

$$\frac{27}{9}\not{3} + \frac{2}{9} = \frac{29}{9}$$

- Read the equation. (Signal.) *27/9 + 2/9 = 29/9.*
- What does 3 + 2/9 equal? (Signal.) *29/9.*
(If students are 100% skip to Exercise 3.)

f. (Display:) [42:2I]

$$4 + \frac{5}{3}$$

- (Point to $4 + \frac{5}{3}$.) Read this problem. (Signal.) *4 + 5/3.*
- Can you add 4 + 5/3? (Signal.) *No.*
- So you have to rewrite 4 as a fraction. What's the bottom number of that fraction? (Signal.) *3.* (Add to show:) [42:2J]

$$\frac{}{3}\cancel{4} + \frac{5}{3}$$

- Raise your hand when you know the top number of the fraction. ✔
- What's the top number? (Signal.) *12.* (Add to show:) [42:2K]

$$\frac{12}{3}\cancel{4} + \frac{5}{3}$$

g. Read the fraction addition problem. (Signal.) *12/3 + 5/3.*

- What's 12/3 + 5/3? (Signal.) *17/3.* (Add to show:) [42:2L]

$$\frac{12}{3}\cancel{4} + \frac{5}{3} = \frac{17}{3}$$

- Read the equation. (Signal.) *12/3 + 5/3 = 17/3.*
- What does 4 + 5/3 equal? (Signal.) *17/3.*

EXERCISE 3: MULTIPLICATION

BY TENS NUMBERS

REMEDY

a. Find part 3 in your workbook. ✔ (Teacher reference:) **R Part I**

a. 7 × 5	b. 2 × 8	c. 7 × 9	d. 5 × 6
e. 7 × 50	f. 2 × 80	g. 7 × 90	h. 5 × 60

The problems in the top row do not multiply by tens numbers. Below are problems that multiply by tens numbers.

b. For each problem in the top row, you'll read the problem and say the answer.

- Problem A. (Signal.) *7 × 5 = 35.*
- Problem B. (Signal.) *2 × 8 = 16.*
- Problem C. (Signal.) *7 × 9 = 63.*
- Problem D. (Signal.) *5 × 6 = 30.*

c. Write answers to the problems in the top row. (Observe students and give feedback.)

d. Touch and read problem E. (Signal.) *7 × 50.*

- Does problem E multiply by a tens number? (Signal.) *Yes.*
- What tens number? (Signal.) *50.* So you'll write zero in the ones column. Then you'll work the problem 7 × 5.
- What will you write in the ones column? (Signal.) *Zero.*
- Then you'll work what problem? (Signal.) *7 × 5.* (Repeat until firm.)

e. Work problem E. (Observe students and give feedback.)

- Problem E: 7 × 5 = 35. So what's 7 × 50? (Signal.) *350.* (Display:) [42:3A]

e. $\begin{array}{r} 7 \\ \times\ 50 \\ \hline 350 \end{array}$

Here's what you should have for problem E.

f. Touch and read problem F. (Signal.) *2 × 80.*

- Does problem F multiply by a tens number? (Signal.) *Yes.*
- What will you write in the ones column? (Signal.) *Zero.*
- Then you'll work what problem? (Signal.) *2 × 8.*

g. Work problem F. (Observe students and give feedback.)

- Problem F: 2 × 8 = 16. So what's 2 × 80? (Signal.) *160.* (If students are 100%, skip to step i.)

h. Touch and read problem G. (Signal.) *7 × 90.*

- Does problem G multiply by a tens number? (Signal.) *Yes.*
- So what will you write in the ones column? (Signal.) *Zero.*
- Then you'll work what problem? (Signal.) *7 × 9.*

i. Work problem G. (Observe students and give feedback.)

- Problem G: 7 × 9 = 63. So what's 7 × 90? (Signal.) *630.* (If students are 100%, skip to step k.)

j. Touch and read problem H. (Signal.) *5 × 60.*

- What will you write in the ones column? (Signal.) *Zero.*
- Then you'll work what problem? (Signal.) *5 × 6.*

k. Work problem H.
(Observe students and give feedback.)
- Problem H: $5 \times 6 = 30$. So what's 5×60? (Signal.) *300.*
(Display:) [42:3B]

f. $2 \times 80 = 160$ g. $7 \times 90 = 630$ h. $5 \times 60 = 300$

Here's what you should have for problems F, G, and H.

EXERCISE 4: DIVISION

WORKING REMAINDER PROBLEMS

a. Find part 4 in your workbook. ✔
(Teacher reference:)

a. 9⟌37 b. 2⟌11 c. 4⟌15 d. 3⟌22

These are division problems that have leftovers.
- Read problem A. (Signal.) *37 ÷ 9.*
- Can you divide 37 by 9? (Signal.) *No.*
- Write the largest part below and write the leftovers. Stop when you've done that much.
(Observe students and give feedback.)
- What's the largest part of 37 you can divide by 9? (Signal.) *36.*
- How many leftovers are there? (Signal.) *1.*
(Display:) [42:4A]

a. 9⟌37, 36 below, leftover 1

Here's the largest part and the leftovers.

b. Now you have to work a division problem and write the answer above.
- Say the division problem you'll work. (Signal.) *36 ÷ 9.*
- Write the answer. ✔
(Add to show:) [42:4B]

a. 9⟌37, 36 below, answer 4, leftover 1

Here's what you should have for problem A. 37 divided by 9 equals 4 and 1 leftover.

c. Read problem B. (Signal.) *11 ÷ 2.*
- Can you divide 11 by 2? (Signal.) *No.*
- Write the largest part below and write the remainder. The remainder is the number for the leftovers. Stop when you've done that much.
(Observe students and give feedback.)
- What's the largest part of 11 you can divide by 2? (Signal.) *10.*
- How many leftovers are there? (Signal.) *1.*
(Display:) [42:4C]

b. 2⟌11, 10 below, leftover 1

d. Now you have to work a division problem and write the answer above.
- Say the division problem you'll work. (Signal.) *10 ÷ 2.*
- Write the answer. ✔
(Add to show:) [42:4D]

b. 2⟌11, 10 below, answer 5, leftover 1

Here's what you should have for problem B.

e. Work problem C. First write the largest part and the remainder. Then write the answer to the division problem you work.
(Observe students and give feedback.)

f. Check your work.
- Problem C is 15 ÷ 4. Say the division problem you worked. (Signal.) *12 ÷ 4.*
- What's the answer? (Signal.) *3.*
- How many leftovers are there? (Signal.) *3.*
(Display:) [42:4E]

c. 4⟌15, 12 below, answer 3, leftover 3

Here's what you should have for problem C.

g. Work problem D. First write the largest part and the remainder. Then write the answer to the division problem you work.
(Observe students and give feedback.)

h. Check your work.
- Problem D is 22 ÷ 3. Say the division problem you worked. (Signal.) *21 ÷ 3.*
- What's the answer? (Signal.) *7.*
- How many leftovers are there? (Signal.) *1.*
(Display:) [42:4F]

$$3\overline{)22}\ \ \text{quotient } 7,\ \text{remainder } 1;\ 21$$

d. 3|22 → 7 r 1, 21

Here's what you should have for problem D.

EXERCISE 5: MULTIPLICATION FACTS

a. Find part 5 in your workbook. ✔
(Teacher reference:)

a. 8 × 9 =	g. 9 × 4 =	m. 3 × 3 =	s. 9 × 10 =	y. 9 × 7 =
b. 3 × 7 =	h. 8 × 0 =	n. 5 × 7 =	t. 3 × 9 =	z. 8 × 5 =
c. 4 × 6 =	i. 6 × 4 =	o. 3 × 8 =	u. 7 × 3 =	A. 4 × 4 =
d. 7 × 9 =	j. 9 × 9 =	p. 4 × 9 =	v. 9 × 2 =	B. 6 × 9 =
e. 8 × 3 =	k. 4 × 3 =	q. 6 × 3 =	w. 10 × 10 =	C. 9 × 8 =
f. 7 × 2 =	l. 6 × 6 =	r. 7 × 10 =	x. 1 × 6 =	D. 3 × 6 =

Some of these multiplication problems are from families we worked with in this lesson. You'll read some of the problems and say the answer.
- Read problem A. (Signal.) *8 × 9.*
- What's the answer? (Signal.) *72.*

b. (Repeat the following tasks for problems B through L:)

Problem __.		What's the answer?
B	3 × 7	21
C	4 × 6	24
D	7 × 9	63
E	8 × 3	24

(If students are 100%, skip to step c.)

F	7 × 2	14
G	9 × 4	36
H	8 × 0	0
I	6 × 4	24
J	9 × 9	81
K	4 × 3	12
L	6 × 6	36

(Repeat problems that were not firm.)

c. Write all of the answers to the problems in part 5. You have two minutes.
- Get ready. Go.
(Observe students and give feedback.)
- (After 2 minutes say:) Stop.

d. Check your work. You'll read the fact for each problem.
- Problem A. (Signal.) *8 × 9 = 72.*
- (Repeat for:) B, *3 × 7 = 21;* C, *4 × 6 = 24;* D, *7 × 9 = 63;* E, *8 × 3 = 24;* F, *7 × 2 = 14;* G, *9 × 4 = 36;* H, *8 × 0 = 0;* I, *6 × 4 = 24;* J, *9 × 9 = 81;* K, *4 × 3 = 12;* L, *6 × 6 = 36;* M, *3 × 3 = 9;* N, *5 × 7 = 35;* O, *3 × 8 = 24;* P, *4 × 9 = 36;* Q, *6 × 3 = 18;* R, *7 × 10 = 70;* S, *9 × 10 = 90;* T, *3 × 9 = 27;* U, *7 × 3 = 21;* V, *9 × 2 = 18;* W, *10 × 10 = 100;* X, *1 × 6 = 6;* Y, *9 × 7 = 63;* Z, *8 × 5 = 40;* Capital A, *4 × 4 = 16;* B, *6 × 9 = 54;* C, *9 × 8 = 72;* D, *3 × 6 = 18.*

EXERCISE 6: FRACTIONS

AS DIVISION

REMEDY

a. (Display:) [42:6A]

$\frac{9}{6}$ $\frac{3}{10}$

$\frac{12}{4}$ $\frac{45}{6}$

- (Point to $\frac{9}{6}$.) Read this fraction. (Signal.) *9/6.*
- (Point to $\frac{12}{4}$.) Read this fraction. (Signal.) *12/4.*
- (Point to $\frac{3}{10}$.) Read this fraction. (Signal.) *3/10.*
- (Point to $\frac{45}{6}$.) Read this fraction. (Signal.) *45/6.*

b. Here's a rule about fractions: You can write any fraction as a division problem.
- (Point to $\frac{9}{6}$.) I'll say the division problem for 9 sixths. 9 divided by 6.
- Your turn: Say the division problem. (Signal.) *9 ÷ 6.*
- (Point to $\frac{12}{4}$.) Say the division problem for 12/4. (Signal.) *12 ÷ 4.*
- (Point to $\frac{3}{10}$.) Say the division problem for 3/10. (Signal.) *3 ÷ 10.*
- (Point to $\frac{45}{6}$.) Say the division problem for 45/6. (Signal.) *45 ÷ 6.*

(Repeat step b until firm.)

c. (Add to show:) [42:6B]

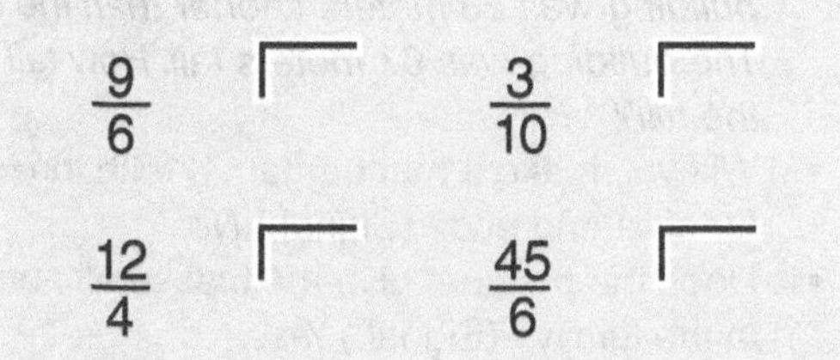

This time, you'll say the division problem, and I'll write it.

- (Point to $\frac{9}{6}$.) Say the division problem for 9/6. (Signal.) *9 ÷ 6.*
  (Add to show:) [42:6C]

$\frac{9}{6}$ $6\overline{)9}$ $\frac{3}{10}$

$\frac{12}{4}$ $\frac{45}{6}$

d. (Point to $\frac{12}{4}$.) Say the division problem for 12/4. (Signal.) *12 ÷ 4.*
(Add to show:) [42:6D]

$\frac{9}{6}$ $6\overline{)9}$ $\frac{3}{10}$

$\frac{12}{4}$ $4\overline{)12}$ $\frac{45}{6}$

e. (Point to $\frac{3}{10}$.) Say the division problem for 3/10. (Signal.) *3 ÷ 10.*
(Add to show:) [42:6E]

$\frac{9}{6}$ $6\overline{)9}$ $\frac{3}{10}$ $10\overline{)3}$

$\frac{12}{4}$ $4\overline{)12}$ $\frac{45}{6}$

f. (Point to $\frac{45}{6}$.) Say the division problem for 45/6. (Signal.) *45 ÷ 6.*
(Add to show:) [42:6F]

$\frac{9}{6}$ $6\overline{)9}$ $\frac{3}{10}$ $10\overline{)3}$

$\frac{12}{4}$ $4\overline{)12}$ $\frac{45}{6}$ $6\overline{)45}$

TEXTBOOK PRACTICE

a. Open your textbook to Lesson 42 and find part 1. ✔
(Teacher reference:)

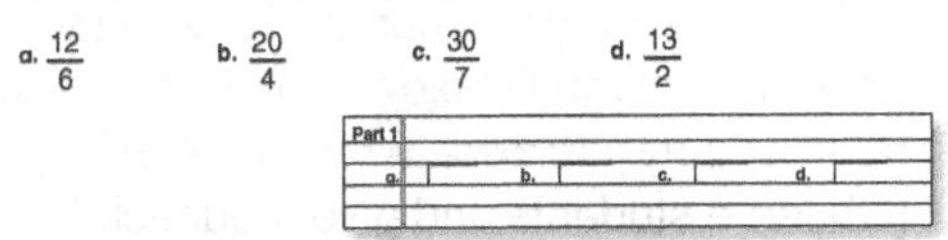

For each fraction, you'll write the division problems on your lined paper.

- Read fraction A. (Signal.) *12/6.*
- Say the division problem for 12/6. (Signal.) *12 ÷ 6.*

b. Read fraction B. (Signal.) *20/4.*

- Say the division problem for 20/4. (Signal.) *20 ÷ 4.*
  (If students have been 100% on saying division problems for fractions for the last two lessons, skip to step e.)

c. Read fraction C. (Signal.) *30/7.*

- Say the division problem for 30/7. (Signal.) *30 ÷ 7.*

d. Read fraction D. (Signal.) *13/2.*

- Say the division problem for 13/2. (Signal.) *13 ÷ 2.*
  (Repeat problems that were not firm.)

e. Write part 1 on your lined paper with the letters A through D below. Then write the division problem for each fraction. Do not work the division problems. Just write them.
(Observe students and give feedback.)

f. Check your work. You'll read the division problem you wrote.

- Problem A. (Signal.) *12 ÷ 6.*
- Problem B. (Signal.) *20 ÷ 4.*
- Problem C. (Signal.) *30 ÷ 7.*
- Problem D. (Signal.) *13 ÷ 2.*

EXERCISE 7: WORD PROBLEMS

ADDITION/SUBTRACTION—MISSING FIRST SMALL NUMBER MIX

a. Find part 2 in your textbook. ✔
(Teacher reference:)

Problems

a. The dog weighed 319 ounces. The cat weighed 374 ounces. How much more did the cat weigh than the dog?

b. Dessi had some money. Dessi spent $113. Dessi ended up with $197. How much money did Dessi have to begin with?

c. There were 543 bottles on a shelf. Some of those bottles were taken off of the shelf. The shelf now has 261 bottles on it. How many bottles were taken off of the shelf?

d. The building was 28 meters shorter than the hill. The building was 67 meters tall. How tall was the hill?

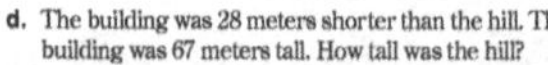

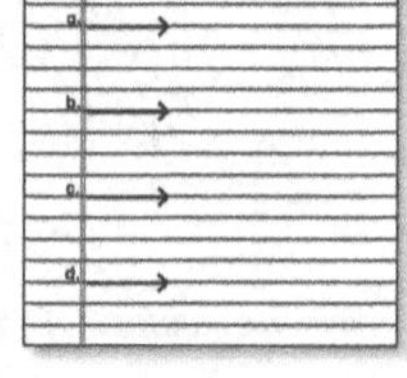

You'll make addition number families to work these problems.

- Write part 2 on your lined paper with the letters A through D below. Make an addition number family arrow after each letter.
(Observe students and give feedback.)

b. Some of the problems in part 2 do not give the first small number. For each problem, you'll tell me if you'll write a family with the letters E and S. Then you'll tell me if you'll write the first small number in the family.

- Read problem A. (Call on a student.) *The dog weighed 319 ounces. The cat weighed 374 ounces. How much more did the cat weigh than the dog?*
- Will you make a number family with the letters for start and end? (Signal.) *No.*
- Does the problem give the first small number in the family? (Signal.) *No.*
(If students have been 100% on word problems for the last two lessons, skip to step f.)

c. Read problem B. (Call on a student.) *Dessi had some money. Dessi spent 113 dollars. Dessi ended up with 197 dollars. How much money did Dessi have to begin with?*

- Will you make a number family with the letters for start and end? (Signal.) *Yes.*
- Does the problem give the first small number in the family? (Signal.) *Yes.*

d. Read problem C. (Call on a student.) *There were 543 bottles on a shelf. Some of those bottles were taken off of the shelf. The shelf now has 261 bottles on it. How many bottles were taken off of the shelf?*

- Will you make a number family with the letters for start and end? (Signal.) *Yes.*
- Does the problem give the first small number in the family? (Signal.) *No.*

e. Read problem D. (Call on a student.) *The building was 28 meters shorter than the hill. The building was 67 meters tall. How tall was the hill?*

- Will you make a number family with the letters for start and end? (Signal.) *No.*
- Does the problem give the first small number in the family? (Signal.) *Yes.*
(Repeat problems that were not firm.)

f. Work all the problems. Put your pencil down when you've completed part 2.
(Observe students and give feedback.)

g. Check your work for problem A.

- What letter did you write for the big number? (Signal.) *C.*
- What letter did you write for a small number? (Signal.) *D.*
- Read the column problem and the answer. (Signal.) *374 – 319 = 55.*
- How much more did the cat weigh than the dog? (Signal.) *55 ounces.*
(Display:) [42:7A]

Part 2

a. 319 d → 374 c   3 6 7 1 4
−3 1 9
5 5 ounces

Here's what you should have for problem A.

h. Check your work for problem B.

- What letter did you write for the big number? (Signal.) *S.*
- What letter did you write for a small number? (Signal.) *E.*
- What number did you write for the first small number? (Signal.) *113.*
- Read the column problem and the answer. (Signal.) *113 + 197 = 310.*
- How much did Dessi have to begin with? (Signal.) *310 dollars.*
(Display:) [42:7B]

b. 113 197 E → S   1 1
$1 1 3
+1 9 7
$3 1 0

Here's what you should have for problem B.

i. Check your work for problem C.

- What letter did you write for the big number? (Signal.) *S.*
- What letter did you write for a small number? (Signal.) *E.*
- Read the column problem and the answer. (Signal.) *543 – 261 = 282.*
- How many bottles were taken off of the shelf? (Signal.) *282.*

(Display:) [42:7C]

c. 261 E → 543 S

543 − 261 = 282 bottles

Here's what you should have for problem C.

j. Check your work for problem D.

- What letter did you write for the big number? (Signal.) *H.*
- What letter did you write for a small number? (Signal.) *B.*
- What number did you write for the first small number? (Signal.) *28.*
- Read the column problem and the answer. (Signal.) *28 + 67 = 95.*
- How tall was the hill? (Signal.) *95 meters.*

(Display:) [42:7D]

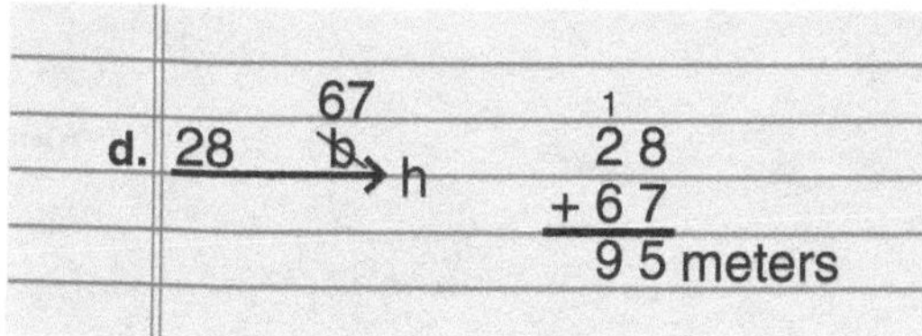

Here's what you should have for problem D.

EXERCISE 8: DIVISION

WRITE ANSWER OR LARGEST PART

a. Find part 3 in your textbook. ✔
(Teacher reference:)

Part 3

a. 4⟌18	c. 9⟌18	e. 8⟌16
b. 3⟌18	d. 5⟌18	f. 10⟌54

- Copy part 3 on your lined paper and work all of the problems. If you can divide, write the answer above. If you can't divide, write the largest part below. Write only the largest part or the answer. Do not write leftovers. (Observe students and give feedback.)

b. Check your work.

- Problem A is 18 ÷ 4. What number did you write? (Signal.) *16.*
- Problem B is 18 ÷ 3. What number did you write? (Signal.) *6.*
- Problem C is 18 ÷ 9. What number did you write? (Signal.) *2.*
- Problem D is 18 ÷ 5. What number did you write? (Signal.) *15.*
- Problem E is 16 ÷ 8. What number did you write? (Signal.) *2.*
- Problem F is 54 ÷ 10. What number did you write? (Signal.) *50.*

EXERCISE 9: DECIMALS

TENTHS AND HUNDREDTHS—WRITING

a. Find part 4 in your textbook. ✔
(Teacher reference:)

Descriptions

a. Decimal number a is thirteen and twelve hundredths.
b. Decimal number b is five hundred and six tenths.
c. Decimal number c is five hundred and sixty hundredths.
d. Decimal number d is eighteen and three hundredths.
e. Decimal number e is twenty and eight tenths.
f. Decimal number f is four hundred seventeen and zero tenths.

Part 4

a.	d.
b.	e.
c.	f.

You're going to write decimal numbers. You'll write some tenths numbers and some hundredths numbers. Each sentence tells the number you'll write.

- Write part 4 on your lined paper with the letters A through F below. (Observe students and give feedback.)

b. Listen: For tenths numbers, how many digits do you write after the decimal point? (Signal.) *1.*

- For hundredths numbers, how many digits do you write after the decimal point? (Signal.) *2.*

(Repeat step b until firm.)

c. Now you'll write decimal numbers. I'll read the sentence for each number.

- Decimal number A is 13 and 12 hundredths. What number? (Signal.) *13 and 12 hundredths.*
- Write the number for A on your lined paper. ✔
- Everybody, touch and read the symbols you wrote for number A. (Signal.) *1, 3, decimal point, 1, 2.* (Repeat until firm.)

d. Decimal number B is 500 and 6 tenths. What number? (Signal.) *500 and 6 tenths.*
- Write the number for B. ✔
- Touch and read the symbols you wrote for decimal number B. (Signal.) *5, 0, 0, decimal point, 6.*

e. Decimal number C is 500 and 60 hundredths. What number? (Signal.) *500 and 60 hundredths.*
- Write the number for C. ✔
- Touch and read the symbols you wrote for decimal number C. (Signal.) *5, zero, zero, decimal point, 6, zero.*

f. Decimal number D is 18 and 3 hundredths. What number? (Signal.) *18 and 3 hundredths.*
- Write the number for D. ✔
- Touch and read the symbols you wrote for decimal number D. (Signal.) *1, 8, decimal point, zero, 3.*

g. Decimal number E is 20 and 8 tenths. What number? (Signal.) *20 and 8 tenths.*
- Write the number for E. ✔
- Touch and read the symbols you wrote for decimal number E. (Signal.) *2, zero, decimal point, 8.*

h. Decimal number F is 417 and zero tenths. What number? (Signal.) *417 and zero tenths.*
- Write the number for F. ✔
- Touch and read the symbols you wrote for decimal number F. (Signal.) *4, 1, 7, decimal point, zero.*

(Display:) [42:9A]

Part 4		
a.	13.12	d. 18.03
b.	500.6	e. 20.8
c.	500.60	f. 417.0

Here's what you should have for part 4.

i. Now you'll read all the decimal numbers you wrote for part 4.
- Read number A. (Signal.) *13 and 12 hundredths.*
- Read number B. (Signal.) *500 and 6 tenths.*
- Read number C. (Signal.) *500 and 60 hundredths.*
- Read number D. (Signal.) *18 and 3 hundredths.*
- Read number E. (Signal.) *20 and 8 tenths.*
- Read number F. (Signal.) *417 and zero tenths.*

EXERCISE 10: INDEPENDENT WORK

MIXED COMPUTATION

a. Find part 5 in your textbook. ✔
(Teacher reference:)

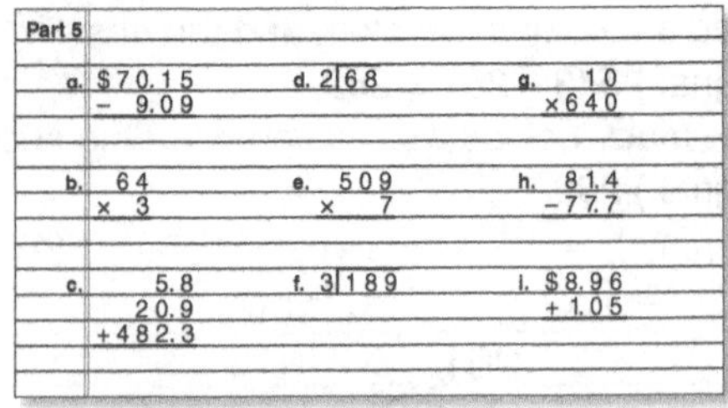

You'll copy part 5 and work the problems. Be careful. Part 5 has addition, subtraction, multiplication, and division problems. Some of the problems are money problems, and some are decimal problems.

Assign Independent Work, Textbook parts 5–9.

Optional extra math-fact practice worksheets for all lessons are available on ConnectED.

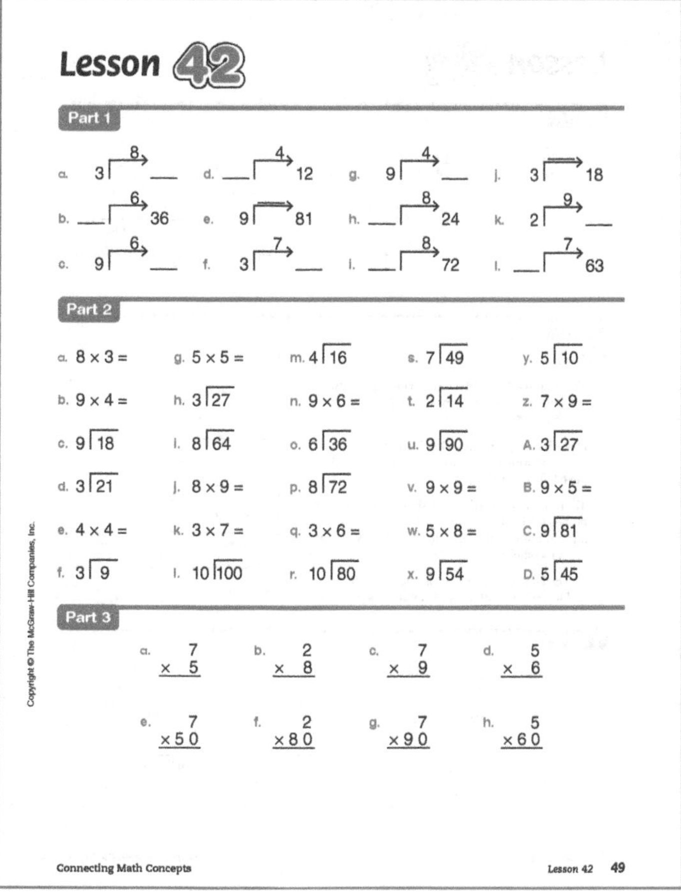

Lesson 42

Part 1

a. 3 ⟶8 ___	d. ___ ⟶4 12	g. 9 ⟶4 ___	j. 3 ⟶ 18
b. ___ ⟶6 36	e. 9 ⟶ 81	h. ___ ⟶8 24	k. 2 ⟶9 ___
c. 9 ⟶6 ___	f. 3 ⟶7 ___	i. ___ ⟶8 72	l. ___ ⟶7 63

Part 2

a. 8 × 3 =	g. 5 × 5 =	m. $4\overline{)16}$	s. $7\overline{)49}$	y. $5\overline{)10}$
b. 9 × 4 =	h. $3\overline{)27}$	n. 9 × 6 =	t. $2\overline{)14}$	z. 7 × 9 =
c. $9\overline{)18}$	i. $8\overline{)64}$	o. $6\overline{)36}$	u. $9\overline{)90}$	A. $3\overline{)27}$
d. $3\overline{)21}$	j. 8 × 9 =	p. $8\overline{)72}$	v. 9 × 9 =	B. 9 × 5 =
e. 4 × 4 =	k. 3 × 7 =	q. 3 × 6 =	w. 5 × 8 =	C. $9\overline{)81}$
f. $3\overline{)9}$	l. $10\overline{)100}$	r. $10\overline{)80}$	x. $9\overline{)54}$	D. $5\overline{)45}$

Part 3

a. 7 × 5	b. 2 × 8	c. 7 × 9	d. 5 × 6
e. 7 × 50	f. 2 × 80	g. 7 × 90	h. 5 × 60

Copyright © The McGraw-Hill Companies, Inc.

Connecting Math Concepts Lesson 42 49

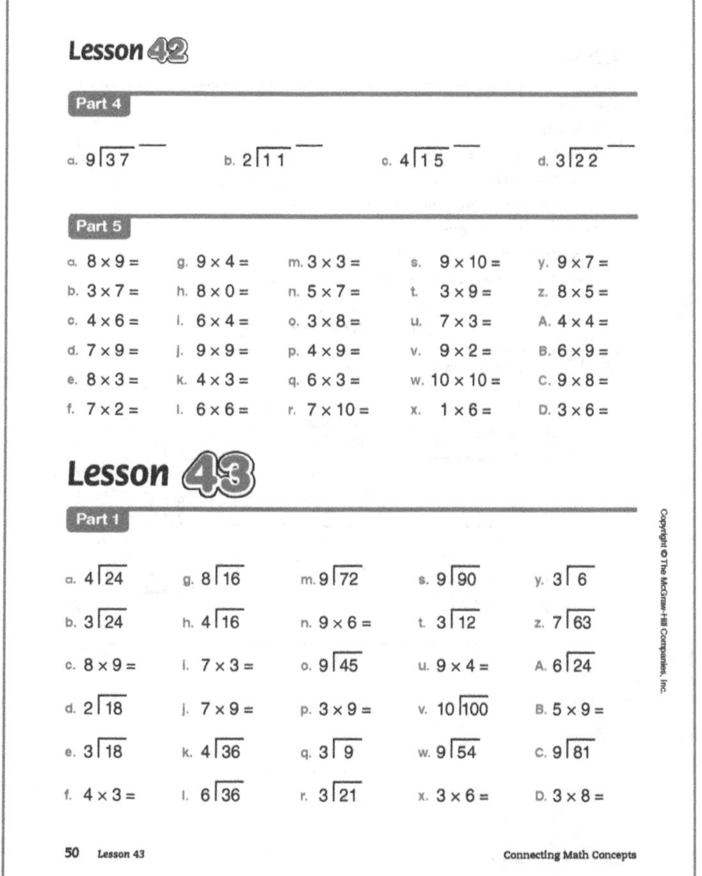

Lesson 42

Part 4

a. $9\overline{)37}$ ___ b. $2\overline{)11}$ ___ c. $4\overline{)15}$ ___ d. $3\overline{)22}$ ___

Part 5

a. 8 × 9 =	g. 9 × 4 =	m. 3 × 3 =	s. 9 × 10 =	y. 9 × 7 =
b. 3 × 7 =	h. 8 × 0 =	n. 5 × 7 =	t. 3 × 9 =	z. 8 × 5 =
c. 4 × 6 =	i. 6 × 4 =	o. 3 × 8 =	u. 7 × 3 =	A. 4 × 4 =
d. 7 × 9 =	j. 9 × 9 =	p. 4 × 9 =	v. 9 × 2 =	B. 6 × 9 =
e. 8 × 3 =	k. 4 × 3 =	q. 6 × 3 =	w. 10 × 10 =	C. 9 × 8 =
f. 7 × 2 =	l. 6 × 6 =	r. 7 × 10 =	x. 1 × 6 =	D. 3 × 6 =

Lesson 43

Part 1

a. $4\overline{)24}$	g. $8\overline{)16}$	m. $9\overline{)72}$	s. $9\overline{)90}$	y. $3\overline{)6}$
b. $3\overline{)24}$	h. $4\overline{)16}$	n. 9 × 6 =	t. $3\overline{)12}$	z. $7\overline{)63}$
c. 8 × 9 =	i. 7 × 3 =	o. $9\overline{)45}$	u. 9 × 4 =	A. $6\overline{)24}$
d. $2\overline{)18}$	j. 7 × 9 =	p. 3 × 9 =	v. $10\overline{)100}$	B. 5 × 9 =
e. $3\overline{)18}$	k. $4\overline{)36}$	q. $3\overline{)9}$	w. $9\overline{)54}$	C. $9\overline{)81}$
f. 4 × 3 =	l. $6\overline{)36}$	r. $3\overline{)21}$	x. 3 × 6 =	D. 3 × 8 =

Copyright © The McGraw-Hill Companies, Inc.

50 Lesson 43 Connecting Math Concepts

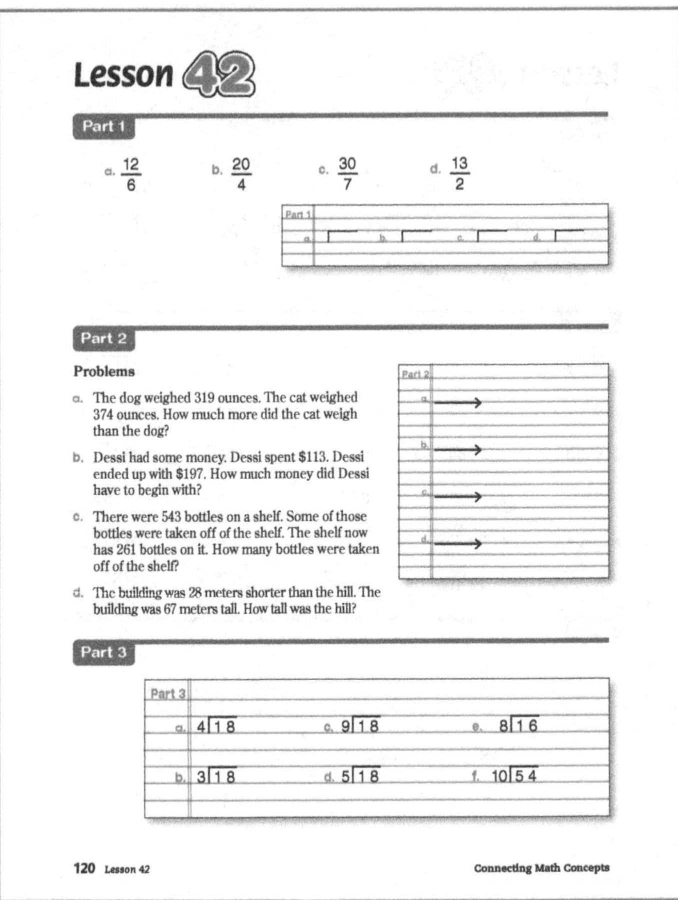

Lesson 42

Part 1

a. $\frac{12}{6}$ b. $\frac{20}{4}$ c. $\frac{30}{7}$ d. $\frac{13}{2}$

Part 2

Problems

a. The dog weighed 319 ounces. The cat weighed 374 ounces. How much more did the cat weigh than the dog?

b. Dessi had some money. Dessi spent $113. Dessi ended up with $197. How much money did Dessi have to begin with?

c. There were 543 bottles on a shelf. Some of those bottles were taken off of the shelf. The shelf now has 261 bottles on it. How many bottles were taken off of the shelf?

d. The building was 28 meters shorter than the hill. The building was 67 meters tall. How tall was the hill?

Part 3

a. 4⟌18 c. 9⟌18 e. 8⟌16

b. 3⟌18 d. 5⟌18 f. 10⟌54

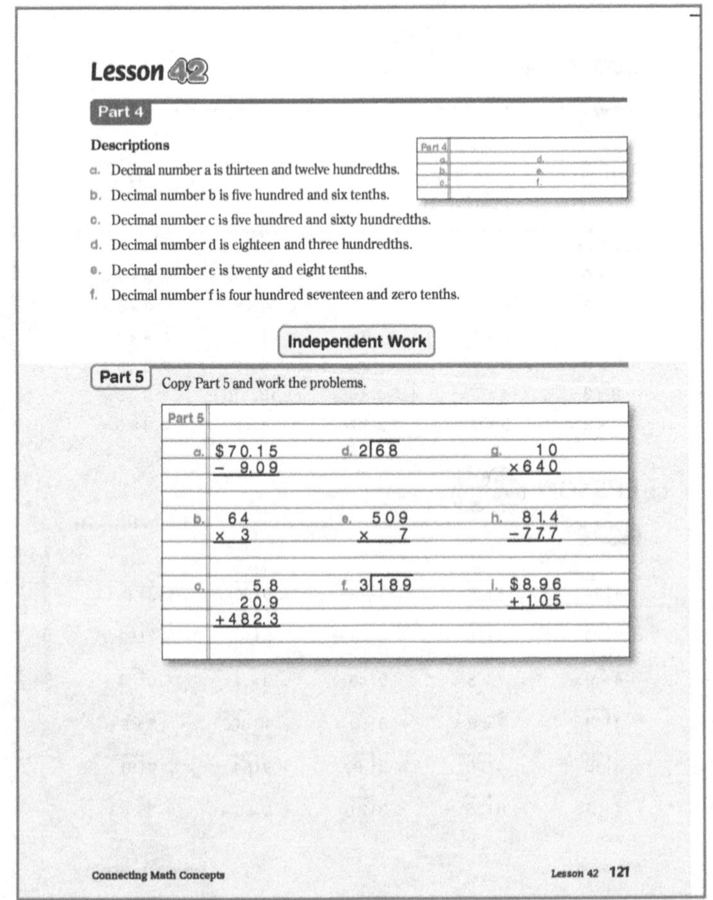

Lesson 42

Part 4

Descriptions

a. Decimal number a is thirteen and twelve hundredths.

b. Decimal number b is five hundred and six tenths.

c. Decimal number c is five hundred and sixty hundredths.

d. Decimal number d is eighteen and three hundredths.

e. Decimal number e is twenty and eight tenths.

f. Decimal number f is four hundred seventeen and zero tenths.

Independent Work

Part 5

Copy Part 5 and work the problems.

a. $70.15 − 9.09

b. 64 × 3

c. 5.8 + 20.9 + 482.3

d. 2⟌68

e. 509 × 7

f. 3⟌189

g. 10 × 640

h. 81.4 − 77.7

i. $8.96 + 1.05

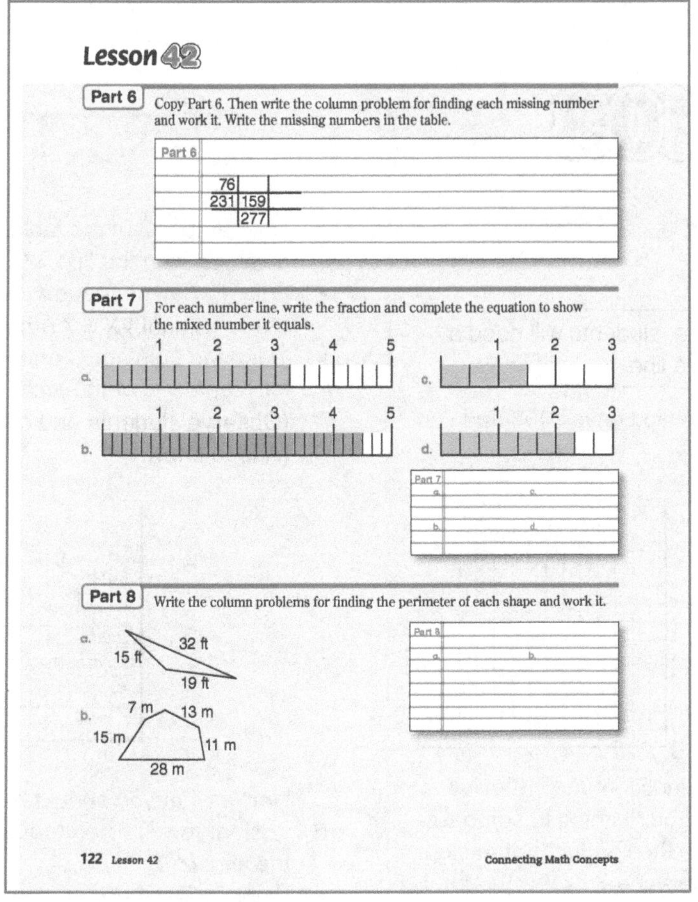

Lesson 42

Part 6 Copy Part 6. Then write the column problem for finding each missing number and work it. Write the missing numbers in the table.

Part 6

76		
231	159	
	277	

Part 7 For each number line, write the fraction and complete the equation to show the mixed number it equals.

a. 1 2 3 4 5

b. 1 2 3 4 5

c. 1 2 3

d. 1 2 3

Part 7 a. c. b. d.

Part 8 Write the column problems for finding the perimeter of each shape and work it.

Part 8 a. b.

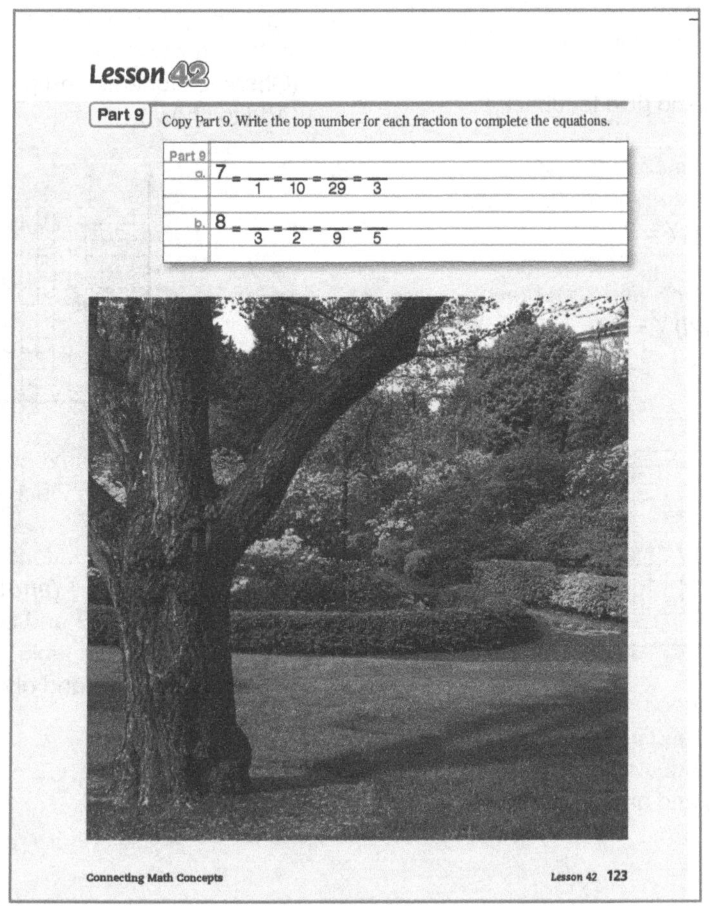

Lesson 42

Part 9 Copy Part 9. Write the top number for each fraction to complete the equations.

Part 9

a. $7 = \frac{\ }{1} = \frac{\ }{10} = \frac{\ }{29} = \frac{\ }{3}$

b. $8 = \frac{\ }{3} = \frac{\ }{2} = \frac{\ }{9} = \frac{\ }{5}$

Lesson 103

EXERCISE 1: FUNCTIONS

POINTS AND LINES

Note: For this exercise, students will need a straight edge to draw a line.

a. Open your workbook to Lesson 103 and find part 1. ✔

(Teacher reference:)

X	Function 2 + X	Y
a. 0		
b. 10		
c.		9
d. 4		
e.		3

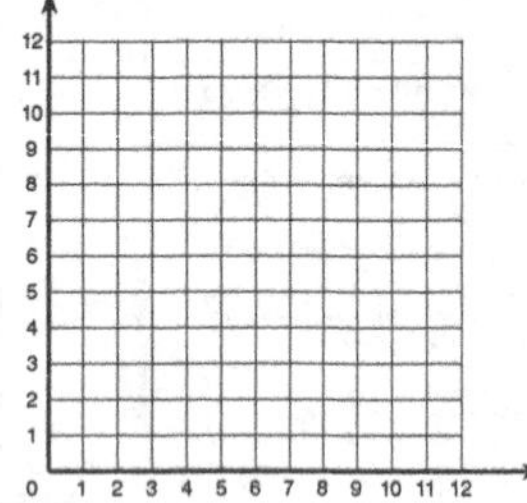

This table shows a function and an X or Y value for each row. You're going to complete rows A and B, make the line for the function, and figure out the missing X or Y value for the rest of the rows.

- Complete rows A and B. Then make the points for A and B on the grid. Put your pencil down when you've done that much.

(Observe students and give feedback.)

b. Check your work.

- Touch the point you made for A. ✔
- Say the equations for X and Y. Get ready. (Signal.) *X = 0 (and) Y = 2.*
- Touch point B. ✔
- Say the equations for X and Y. Get ready. (Signal.) *X = 10 (and) Y = 12.*

(Display:) [103:1A]

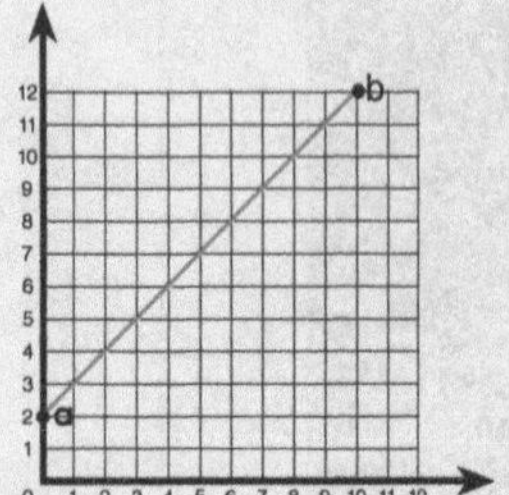

Here's the line that goes through A and B. That's the line for Y equals 2 + X.

- Make the line for Y equals 2 + X.

(Observe students and give feedback.)

c. Look at row C of the table. Then touch the point for C on the line. ✔

- Point C. Say the equations for X and for Y. Get ready. (Signal.) *X = 7 (and) Y = 9.*
- Make the point for C and label it. Then write 7 for X in row C of the table.

(Observe students and give feedback.)

(Add to show:) [103:1B]

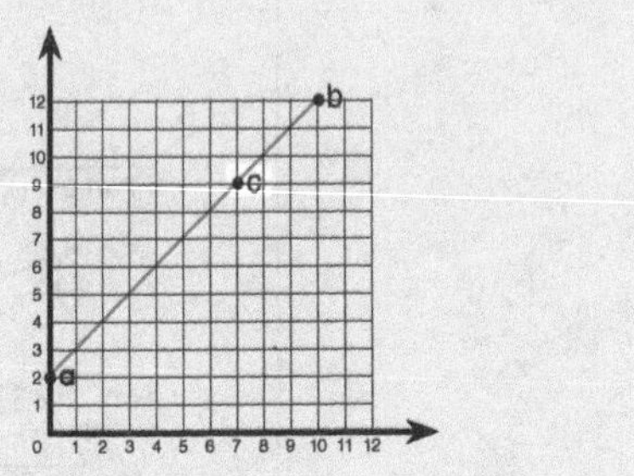

Here's what you should have for point C.

d. Look at row D. Then touch the point for D on the line. ✔

- Point D. Say the equations for X and for Y. Get ready. (Signal.) *X = 4 (and) Y = 6.*
- Make the point for D and label it. Then write 6 for Y in row D of the table.

(Observe students and give feedback.)

(Add to show:) [103:1C]

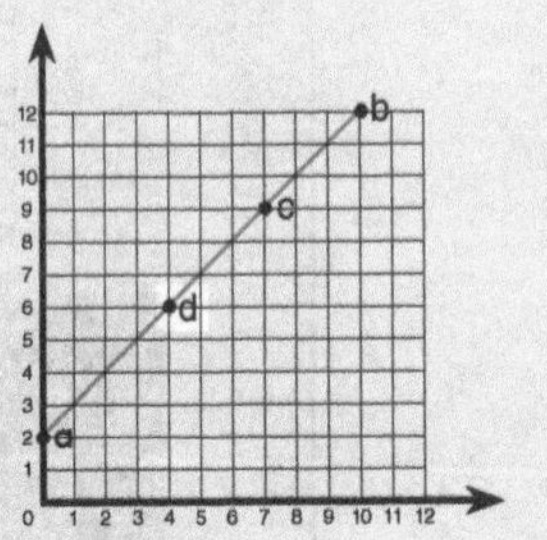

Here's what you should have for point D.

e. Look at row E. Then touch the point for E on the line. ✔

- Point E. Say the equations for X and for Y. Get ready. (Signal.) *X = 1 (and) Y = 3.*
- Make the point for E and label it. Then write 1 for X in row E of the table.

(Observe students and give feedback.)

(Add to show:)

[103:1D]

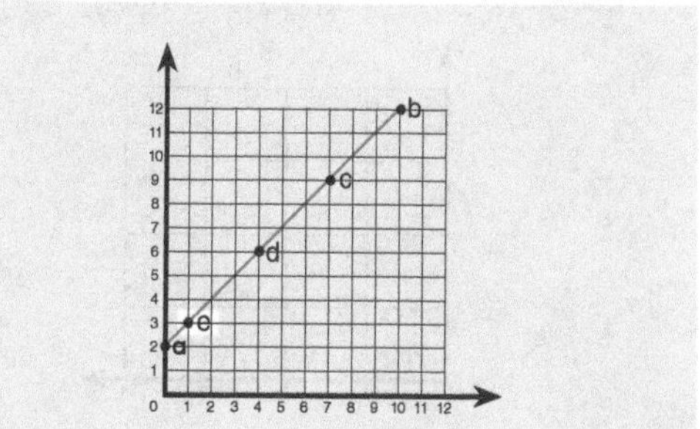

Here's what you should have. Later, you'll write the problem for rows C, D, and E and check to make sure the X and Y values are correct.

EXERCISE 2: MULTIPLES

DIVISIBLE BY 2 AND 5

a. You learned you can divide some numbers by 2 and get a whole-number answer.

- Can you divide odd numbers by 2 and get a whole-number answer? **(Signal.)** *No.*
- What kind of numbers can you divide by 2 and get a whole-number answer? **(Signal.)** *Even (numbers).*
- You can divide some numbers by 5 and get a whole-number answer. Tell me the digits those numbers end in. Get ready. **(Signal.)** *Zero (and) 5.*
- Here's a hard question. You can divide some numbers by 5 and by 2 and get a whole-number answer. What digit do those numbers end in? **(Signal.)** *Zero.*
  Yes, numbers that end in zero are even, and you can divide them by 5.

b. Find part 2 in your workbook. ✔

(Teacher reference:)

a. 390	d. 158	g. 450
b. 71	e. 185	h. 504
c. 645	f. 581	i. 405

2	
5	

You can divide some of these numbers by 2, some of these numbers by 5, and some of these numbers by 2 and 5 to get a whole-number answer. You'll tell me about each number and write it in the rows where it belongs.

- Read number A. Get ready. **(Signal.)** *390.*
- Read number B. **(Signal.)** *71.*
- C. **(Signal.)** *645.*
- **(Repeat for:)** D, *158;* E, *185;* F, *581;* G, *450;* H, *504;* I, *405.*

c. Number A is 390. Raise your hand when you know which rows you'll write 390 in. ✔

- Everybody, tell me which rows you will write 390 in. Get ready. **(Signal.)** *2 (and) 5.*
  Yes, the last digit of 390 is zero, so you can divide it by 5 and by 2 and get a whole-number answer.

d. Number B is 71. Raise your hand when you know which rows you'll write 71 in. ✔

- Everybody, tell me if you'll write 71 in the row for 2, 5, or neither. Get ready. **(Signal.)** *Neither.*
  Yes, 71 is odd and it doesn't end in zero or 5, so you can't divide it by 2 or by 5 and get a whole-number answer.
  (If students are firm, skip to step g.)

e. Number C is 645. Raise your hand when you know which rows you'll write 645 in. ✔

- Everybody, tell me which rows you will write 645 in. Get ready. **(Signal.)** *5.*
  Yes, the last digit of 645 is 5, so you can divide it by 5 and get a whole-number answer.

f. Number D is 158. Raise your hand when you know which rows you'll write 158 in. ✔

- Everybody, tell me which rows you will write 158 in. Get ready. **(Signal.)** *2.*
  Yes, 158 is an even number, so you can divide it by 2 and get a whole-number answer.

g. Write all of the numbers in the table where they belong. Put your pencil down when you've completed the table for part 2.
(Observe students and give feedback.)

h. Check your work.

- Read the numbers in your table that you can divide by 2 and get a whole-number answer. Get ready. **(Signal.)** *390, 158, 450, 504.*
- Read the numbers in your table that you can divide by 5 and get a whole-number answer. Get ready. **(Signal.)** *390, 645, 185, 450, 405.*
- Tell me the numbers you can divide by 2 and 5 and get a whole-number answer. Get ready. **(Signal.)** *390 (and) 450.*

(Display:) [103:2A]

2	390	158	450	504	
5	390	645	185	450	405

Here's what you should have for part 2.

EXERCISE 3: COORDINATE SYSTEM

POINTS FOR MIXED NUMBERS

a. Find part 3 in your workbook. ✔
(Teacher reference:)

X	Y
a. 3	$1\frac{3}{4}$
b. $2\frac{2}{4}$	4
c. $1\frac{1}{4}$	$\frac{3}{4}$
d. $3\frac{2}{4}$	$2\frac{1}{4}$
e. $\frac{1}{4}$	$3\frac{2}{4}$

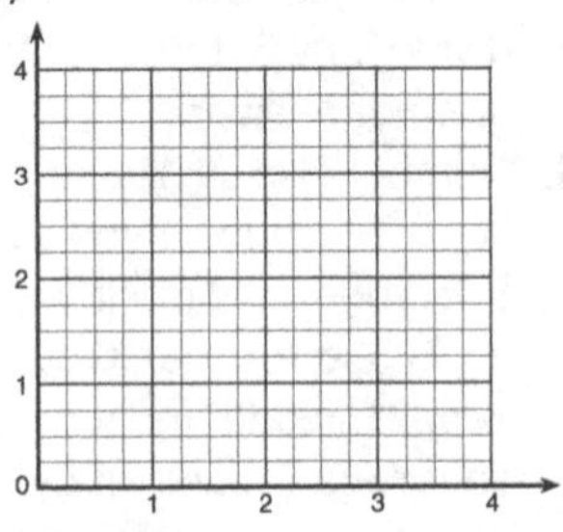

The grid shows fourths. The rows in the table show the X and Y values for points A through E. For each point, the X value or the Y value is a mixed number. You'll make the point for each row.

- Touch the row for point A. ✔
- Say the equations for X and for Y. **(Signal.)** *X = 3 (and) Y = 1 and 3/4.*
- Touch the X distance of 3. ✔
- Now go up to Y = 1 and 3/4. Keep your finger on the point X = 3, Y = 1 and 3/4 when you find it.
(Observe students and give feedback.)
(Display:) [103:3A]

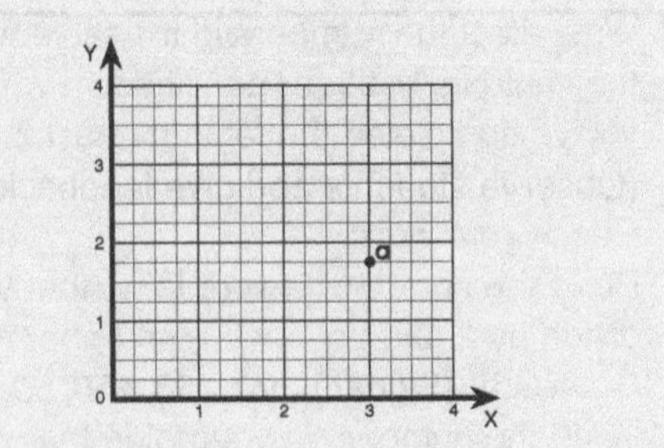

- Here's the point for A. Make the point and write the letter.
(Observe students and give feedback.)

b. Touch the row for point B. ✔

- Say the equations for X and for Y. **(Signal.)** *X = 2 and 2/4 (and) Y = 4.*
- Touch the X distance of 2 and 2/4.
- Now go up to Y = 4. Keep your finger on the point X = 2 and 2/4, Y = 4 when you find it.
(Observe students and give feedback.)

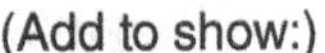
(Add to show:)

[103:3B]

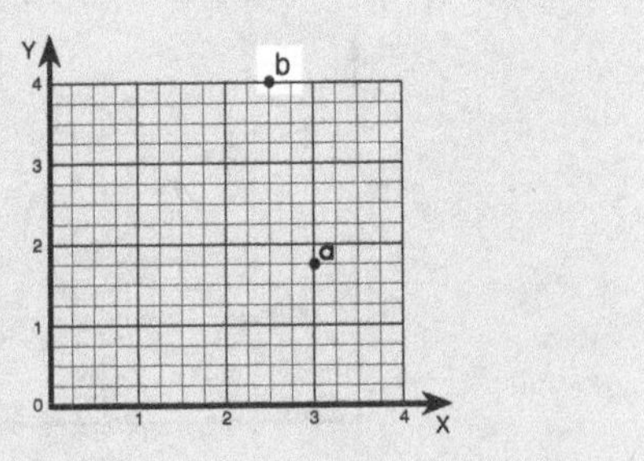

- Here's the point for B. Make the point and write the letter.
(Observe students and give feedback.)

c. Look at the row for point C. Then find the point for C on the grid. Keep your finger on the point for C when you find it.
(Observe students and give feedback.)
(Add to show:) [103:3C]

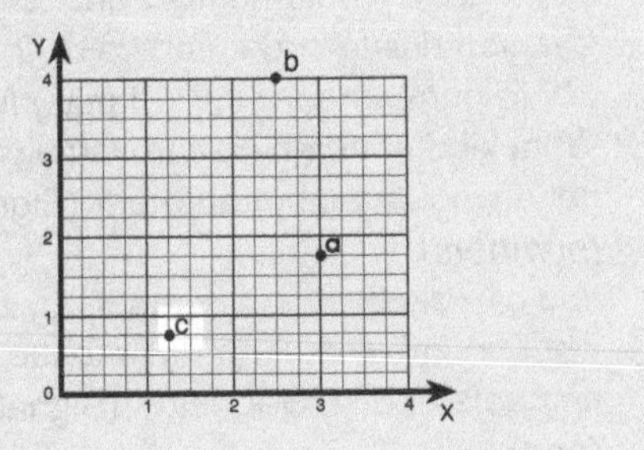

- Here's the point for C. Make the point and write the letter.
(Observe students and give feedback.)

d. Look at the row for point D. Then find the point for D on the grid. Keep your finger on the point for D when you find it.
(Observe students and give feedback.)
(Add to show:) [103:3D]

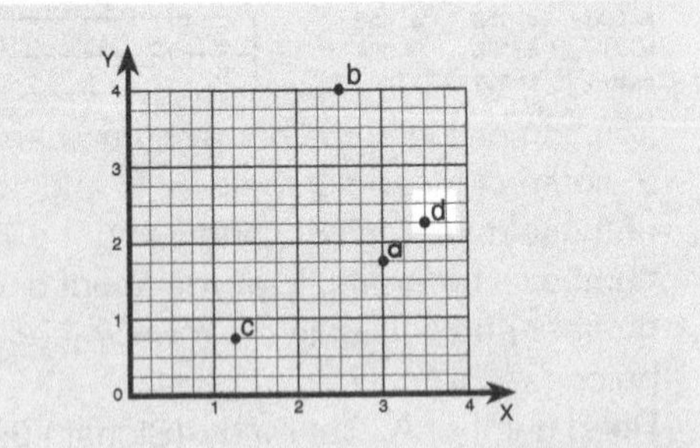

- Here's the point for D. Make the point and write the letter.
(Observe students and give feedback.)

(If students are 100%, skip to Exercise 4.)

e. Look at the row for point E. Then find the point for E on the grid. Keep your finger on the point for E when you find it.
(Observe students and give feedback.)
(Add to show:) [103:3E]

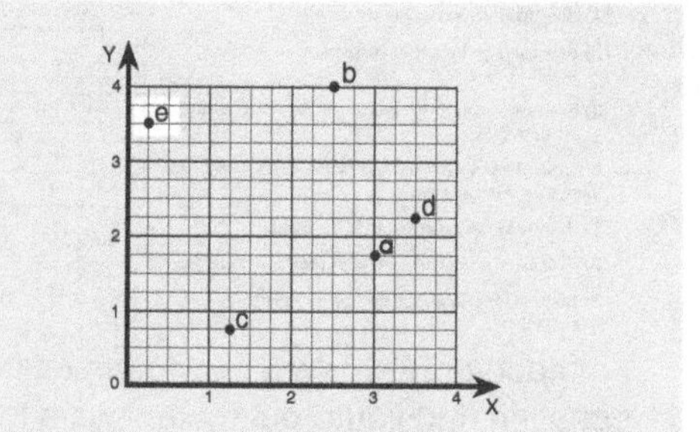

- Here's the point for E. Make the point and write the letter.
(Observe students and give feedback.)

EXERCISE 4: FRACTIONS

COMPARE BY DIVISION **REMEDY**

a. Open your textbook to Lesson 103 and find part 1. ✔
(Teacher reference:)

Part 1

a. $\frac{17}{2}$ $\frac{39}{5}$ c. $\frac{20}{3}$ $\frac{50}{7}$

b. $\frac{24}{4}$ $\frac{54}{9}$ d. $\frac{65}{8}$ $\frac{36}{4}$

You can't compare each pair of fractions in part 1 by looking at the fractions.

- Listen: When you can do something just by looking at it, you can do it by inspection. Say **inspection.** (Signal.) *Inspection.*
- If you can compare fractions by looking at them, you can compare fractions by inspection. What's another way of saying you can compare fractions by looking at them? (Signal.) *You can compare fractions by inspection.*

b. What's another way of saying you can solve a problem by looking at it? (Signal.) *You can solve a problem by inspection.*

- What's another way of saying you can figure out an angle by looking at it? (Signal.) *You can figure out an angle by inspection.*
- What's another way of saying you can compare fractions by looking at them? (Signal.) *You can compare fractions by inspection.*
- What's another way of saying you **cannot** compare fractions by looking at them? (Signal.) *You cannot compare fractions by inspection.*
(Repeat step b until firm.)

c. You cannot compare each pair of fractions in part 1 by inspection. You **can** compare each pair by figuring out the mixed number or whole number each fraction equals.

- Copy part 1 on your lined paper.
(Observe students and give feedback.)

d. Read the fractions for problem A on your lined paper. Get ready. (Signal.) *17/2 (and) 39/5.*

- Say the division problem for 17/2. Get ready. (Signal.) *17 ÷ 2.*
- Say the division problem for 39/5. Get ready. (Signal.) *39 ÷ 5.*
- Are either of those problems for division facts? (Signal.) *No.*
(Display:) [103:4A]

Part 1

a. $\frac{17}{2}$ $\frac{39}{5}$ $2\overline{)17}$ $5\overline{)39}$

Here's problem A with the division problems you'll work.

- Write the division problems and work them. Then make the sign to show which fraction is more or if they're equal.
(Observe students and give feedback.)

e. Check your work for problem A.

- What does 17 ÷ 2 equal? (Signal.) *8 and 1/2.*
- What does 39 ÷ 5 equal? (Signal.) *7 and 4/5.*
- Are 17/2 and 39/5 equal? (Signal.) *No.*
- Which fraction is more? (Signal) *17/2.*

(Add to show:) [103:4B]

Part 1

a. $\frac{17}{2} > \frac{39}{5}$ $\quad 2\overline{)17}\,$ = $8\frac{1}{2}$ (16) $\quad 5\overline{)39}$ = $7\frac{4}{5}$ (35)

Here's what you should have for problem A.

f. Read the fractions for problem B. Get ready. (Signal.) *24/4 (and) 54/9.*

- Figure out the whole number or mixed number each fraction equals. Then make the sign to show which fraction is more or if they're equal. (Observe students and give feedback.)

g. Check your work for problem B.

- What does 24/4 equal? (Signal.) *6.*
- What does 54/9 equal? (Signal.) *6.*
- Read the statement you wrote for problem B. Get ready. (Signal.) *24/4 = 54/9.*

h. Read the fractions for problem C. Get ready. (Signal.) *20/3 (and) 50/7.*

- Figure out the whole number or mixed number each fraction equals. Then make the sign to show which fraction is more or if they're equal. (Observe students and give feedback.)

i. Check your work for problem C.

- What does 20/3 equal? (Signal.) *6 and 2/3.*
- What does 50/7 equal? (Signal.) *7 and 1/7.*
- Which fraction is more? (Signal.) *50/7.*

(If students are firm, skip to Exercise 5.)

j. Read the fractions for problem D. Get ready. (Signal.) *65/8 (and) 36/4.*

- Figure out the whole number or mixed number each fraction equals. Then make the sign to show which fraction is more or if they're equal. (Observe students and give feedback.)

k. Check your work for problem D.

- What does 65/8 equal? (Signal.) *8 and 1/8.*
- What does 36/4 equal? (Signal.) *9.*
- Which fraction is more? (Signal.) *36/4.*

EXERCISE 5: RATIO WORD PROBLEMS

MULTIPLE OPERATIONS

REMEDY

a. Find part 2 in your textbook. ✔

(Teacher reference:)

a. In glycine, the ratio of hydrogen to carbon atoms is 5 to 2. There are 30 hydrogen atoms.

1) How many carbon atoms are there?
2) How many hydrogen and carbon atoms are there in all?
3) How many more hydrogen atoms than carbon atoms are there?

b. In a mall, there were 500 cashiers. For every 4 cashiers, there were 9 customers.

1) How many customers were in the mall?
2) How many fewer cashiers than customers were there?
3) How many cashiers and customers were there altogether?

Part 2
a.
1) 2) 3)
b.
1) 2) 3)

For each problem, you'll write the ratio equation and figure out the answers to questions 1, 2, and 3.

- Write part 2 on your lined paper with the letters A and B below. Write the numbers 1, 2, and 3 below each letter. (Observe students and give feedback.) (If students are firm, skip to step c.)

b. Read problem A through question 1. (Call on a student.) *In glycine, the ratio of hydrogen to carbon atoms is 5 to 2. There are 30 hydrogen atoms. Question 1: How many carbon atoms are there?*

c. Set up the ratio equation for problem A and complete it. (If students are firm, skip to step g.)

Put your pencil down when you've made the ratio equation for problem A. (Observe students and give feedback.)

d. Check your work for problem A.

- Tell me the names. Get ready. (Signal.) *Hydrogen (and) carbon.*
- Read the fraction equation. (Signal.) *5/2 × 6/6 = 30/12.*

(Display:) [103:5A]

Part 2

a. $\frac{\text{hydrogen}}{\text{carbon}}\ \frac{5}{2} \times \frac{6}{6} = \frac{30}{12}$

1) 2) 3)

Here's what you should have for problem A.

e. Read problem A, question 2. (Call on a student.) *Question 2: How many hydrogen and carbon atoms are there in all?*
- Do you add or subtract to figure out the answer? (Signal.) *Add.*

f. Read problem A, question 3. (Call on a student.) *Question 3: How many more hydrogen atoms than carbon atoms are there?*
- Do you add or subtract to figure out the answer? (Signal.) *Subtract.*

(Repeat steps e and f until firm.)

g. Write the answer to question 1. Then work the problems for questions 2 and 3. Put your pencil down when you've finished problem A.
(Observe students and give feedback.)

h. Check your work for problem A.
- Question 1: How many carbon atoms are there? (Signal.) *12.*
- Read the problem and the answer you wrote for question 2. Get ready. (Signal.) *30 + 12 = 42.*
- How many hydrogen and carbon atoms are there in all? (Signal.) *42.*
- Read the problem and the answer you wrote for question 3. Get ready. (Signal.) *30 – 12 = 18.*
- How many more hydrogen atoms than carbon atoms are there? (Signal.) *18.*

(Add to show:) [103:5B]

Part 2

a. hydrogen / carbon $\frac{5}{2} \times \frac{6}{6} = \frac{30}{12}$

1) 12 2) $30 + 12 = 42$ 3) $30 - 12 = 18$

Here's what you should have for problem A.
(If students are firm, skip to step j.)

i. Read problem B through question 1. (Call on a student.) *In a mall, there were 500 cashiers. For every 4 cashiers, there were 9 customers. Question 1: How many customers were in the mall?*

j. Set up the ratio equation for problem B and complete it.
(If students are firm, skip to step n.)
Put your pencil down when you've made the ratio equation for problem B.
(Observe students and give feedback.)

k. Check your work for problem B.
- Tell me the names. Get ready. (Signal.) *Cashiers (and) customers.*
- Read the fraction equation. (Signal.) *4/9 × 125/125 = 500/1125.*

(Display:) [103:5C]

b. cashiers / customers $\frac{4}{9} \times \frac{125}{125} = \frac{500}{1125}$ $500 \div 4 = 125$ $125 \times 9 = 1125$

1) 2) 3)

Here's what you should have for problem B.

l. Read problem B, question 2. (Call on a student.) *Question 2: How many fewer cashiers than customers were there?*
- Do you add or subtract to figure out the answer? (Signal.) *Subtract.*

m. Read problem B, question 3. (Call on a student.) *Question 3: How many cashiers and customers were there altogether?*
- Do you add or subtract to figure out the answer? (Signal.) *Add.*

(Repeat steps l and m until firm.)

n. Write the answer to question 1. Then work the problems for questions 2 and 3. Put your pencil down when you've finished problem B.
(Observe students and give feedback.)

o. Check your work for problem B.
- Question 1: How many customers were in the mall? (Signal.) *1125.*
- Read the problem and the answer you wrote for question 2. Get ready. (Signal.) *1125 – 500 = 625.*
- How many fewer cashiers than customers were there? (Signal.) *625.*
- Read the problem and the answer you wrote for question 3. Get ready. (Signal.) *500 + 1125 = 1625.*
- How many cashiers and customers were there altogether? (Signal.) *1625.*

(Add to show:) [103:5D]

b. cashiers / customers $\frac{4}{9} \times \frac{125}{125} = \frac{500}{1125}$ $500 \div 4 = 125$ $125 \times 9 = 1125$

1) 1125 2) $1125 - 500 = 625$ 3) $500 + 1125 = 1625$

Here's what you should have for problem B.

EXERCISE 6: ESTIMATION PROBLEMS

NEAREST WHOLE NUMBER FROM DECIMALS

a. Find part 3 in your textbook. ✔
(Teacher reference:)

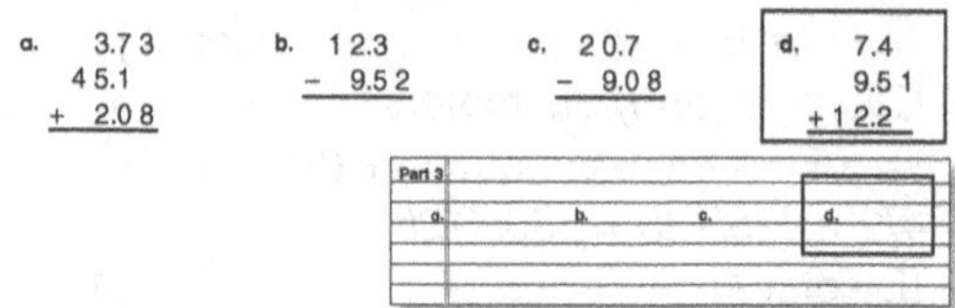

Each of these problems has decimal numbers. You're going to write the estimation problem rounded to the nearest whole number.

- Write part 3 on your lined paper with the letters A through D below.
(Observe students and give feedback.)

b. Problem A is 3 and 73/100 + 45 and 1/10 + 2 and 8/100.
- The first number is 3 and 73/100. What's the first digit after the decimal point? **(Signal.)** *7.*
- Is that digit more than or equal to 5? **(Signal.)** *Yes.*
- So what whole number does 3 and 73/100 round to? **(Signal.)** *4.*

c. What's the first digit after the decimal point of 45 and 1/10? **(Signal.)** *1.*
- Is that digit more than or equal to 5? **(Signal.)** *No.*
- So what whole number does 45 and 1/10 round to? **(Signal.)** *45.*

d. What's the first digit after the decimal point of 2 and 8/100? **(Signal.)** *Zero.*
- Is that digit more than or equal to 5? **(Signal.)** *No.*
- So what whole number does 2 and 8/100 round to? **(Signal.)** *2.*

e. Write the estimation problem for A and work it.
(Observe students and give feedback.)

f. Check your work for problem A.
- Read the estimation problem and the answer. **(Signal.)** *4 + 45 + 2 = 51.*

(If students are firm, skip to step i.)

g. Problem B is 12 and 3/10 − 9 and 52/100.
- The first number is 12 and 3/10. What's the first digit after the decimal point? **(Signal.)** *3.*
- Is that digit more than or equal to 5? **(Signal.)** *No.*
- So what whole number does 12 and 3/10 round to? **(Signal)** *12.*

h. What's the first digit after the decimal point for 9 and 52/100? **(Signal)** *5.*
- Is that digit more than or equal to 5? **(Signal)** *Yes.*
- So what whole number does 9 and 52/100 round to? **(Signal.)** *10.*

(Repeat steps g and h until firm.)

i. Write the estimation problem for B and work it.
(Observe students and give feedback.)

j. Check your work for problem B.
- Read the estimation problem and the answer. **(Signal.)** *12 − 10 = 2.*

k. Write and work the estimation problem to the nearest whole number for C.
(Observe students and give feedback.)

l. Check your work for problem C.
- Read the estimation problem and the answer. **(Signal.)** *21 − 9 = 12.*

(Display:) [103:6A]

Part 3

a.	b.	c.	d.
4 45 + 2 51	12 − 10 2	21 − 9 12	

Here's what you should have for problems A, B, and C.

(If students are firm, skip to Exercise 7.)

m. Write and work the estimation problem to the nearest whole number for D.
(Observe students and give feedback.)

n. Check your work for problem D.
- Read the estimation problem and the answer. **(Signal.)** *7 + 10 + 12 = 29.*

(Display:) [103:6B]

Part 3

a.	b.	c.	d.
4 45 + 2 51	12 − 10 2	21 − 9 12	7 10 + 12 29

Here's what you should have for part 3.

EXERCISE 7: MIXED NUMBERS

TIMES WHOLE NUMBER

a. Find part 4 in your textbook. ✔
(Teacher reference:)

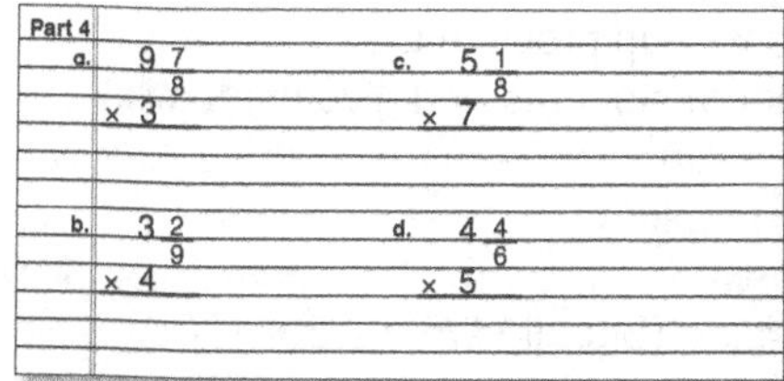

These problems multiply a mixed number by a whole number. For some of the problems, the answer you'll write below the problem will not be a mixed number. For those problems, you'll have to figure out the mixed-number answer. For the other problems, the answer you'll write below the problem will be a mixed number.

- Copy part 4 and work the multiplication for the fraction and the whole numbers. Stop when you've written a whole number and a fraction below each problem.
(Observe students and give feedback.)

b. Check your work.

- Read problem A and the answer you wrote below it. (Signal.) *9 and 7/8 × 3 = 27 and 21/8.*
- Is 27 and 21/8 a mixed number? (Signal.) *No.*
- So are you finished working problem A? (Signal.) *No.*

c. Read problem B and the answer you wrote below it. (Signal.) *3 and 2/9 × 4 = 12 and 8/9.*

- Is 12 and 8/9 a mixed number? (Signal.) *Yes.*
- So are you finished working problem B? (Signal.) *Yes.*
- What does 3 and 2/9 × 4 equal? (Signal.) *12 and 8/9.*

d. Read problem C and the answer you wrote below it. (Signal.) *5 and 1/8 × 7 = 35 and 7/8.*

- Is 35 and 7/8 a mixed number? (Signal.) *Yes.*
- So are you finished working problem C? (Signal.) *Yes.*
- What does 5 and 1/8 × 7 equal? (Signal.) *35 and 7/8.*

e. Read problem D and the answer you wrote below it. (Signal.) *4 and 4/6 × 5 = 20 and 20/6.*

- Is 20 and 20/6 a mixed number? (Signal.) *No.*
- So are you finished working problem D? (Signal.) *No.*

f. Go back and figure out the mixed-number answer to problem A.
(Observe students and give feedback.)

g. Check your work for problem A.

- Tell me the mixed number 21/8 equals. Get ready. (Signal.) *2 and 5/8.*
- Tell me the mixed number 9 and 7/8 × 3 equals. Get ready. (Signal.) *29 and 5/8.*

(Display:) [103:7A]

Part 4

a. $9\frac{7}{8}$

× 3

$27\frac{21}{8} = 27 + 2\frac{5}{8} = 29\frac{5}{8}$

Here's what you should have for problem A.

h. Go back and figure out the mixed-number answer to problem D.
(Observe students and give feedback.)

i. Check your work for problem D.

- Tell me the mixed number 20/6 equals. Get ready. (Signal.) *3 and 2/6.*
- Tell me the mixed number 4 and 4/6 × 5 equals. Get ready. (Signal.) *23 and 2/6.*

(Display:) [103:7B]

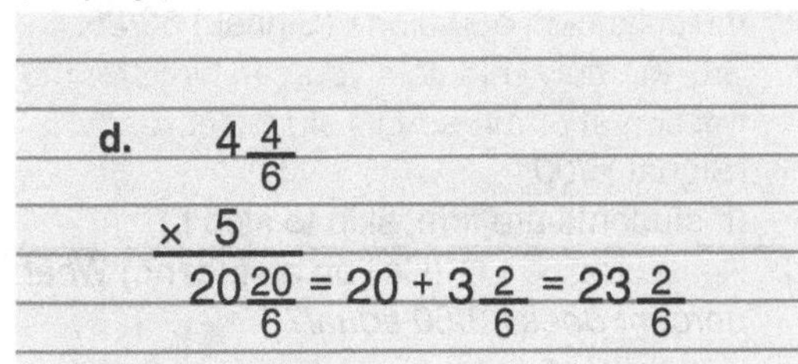

Here's what you should have for problem D.

EXERCISE 8: PERCENT

FROM FRACTION

a. Find part 5 in your textbook. ✔
(Teacher reference:)

Questions

a. What percent does $\frac{3}{5}$ equal?

b. What percent does $\frac{52}{25}$ equal?

c. What percent does $\frac{43}{50}$ equal?

d. What percent does $\frac{1}{2}$ equal?

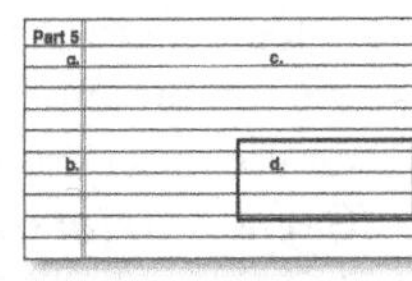

You've completed fraction-multiplication equations and written statements to show the percent fractions equal. For each of the questions in part 5, you'll set up a fraction-multiplication equation to figure out the percent each fraction equals.

- For some of the equations you worked in the last lesson, you had two choices as the denominator of the fraction after the equals sign. Tell me the choices. Get ready. **(Signal.)** *10 (and) 100.*
- Write part 5 on your lined paper with the letters A through D below.
(Observe students and give feedback.)

b. For each question, you'll tell me the first fraction you'll write in the multiplication equation. Then you'll tell me if you can write 10 or 100 as the denominator or just 100.
- Read question A. **(Call on a student.)** *What percent does 9/5 equal?*
- What's the first fraction in the fraction-multiplication equation? **(Signal.)** *9/5.*
- Say the denominators you can write for the fraction after the equals sign. Get ready. **(Signal.)** *10 (and) 100.*
- Raise your hand if you'll write a denominator of 10. ✔
- Raise your hand if you'll write a denominator of 100. ✔

c. Read question B. **(Call on a student.)** *What percent does 52/25 equal?*
- What's the first fraction in the fraction-multiplication equation? **(Signal.)** *52/25.*
- Say the denominators you can write for the fraction after the equals sign. Get ready. **(Signal.)** *100.*
(If students are firm, skip to step f.)

d. Read question C. **(Call on a student.)** *What percent does 43/50 equal?*
- What's the first fraction in the fraction-multiplication equation? **(Signal.)** *43/50.*
- Say the denominators you can write for the fraction after the equals sign. Get ready. **(Signal.)** *100.*

e. Read question D. **(Call on a student.)** *What percent does 1/2 equal?*
- What's the first fraction in the fraction-multiplication equation? **(Signal.)** *1/2.*
- Say the denominators you can write for the fraction after the equals sign. Get ready. **(Signal.)** *10 (and) 100.*
- Raise your hand if you'll write a denominator of 10. ✔
- Raise your hand if you'll write a denominator of 100. ✔

f. Set up the fraction-multiplication equations for part 5 and complete them. Put your pencil down when you're finished.
(If students have been firm on fraction-multiplication equations for the last two lessons, skip problem D.)
(Observe students and give feedback.)

g. Check your work. You'll read the equation for each problem.
- Problem A. If you wrote a denominator of 10, get ready. **(Signal.)** *9/5 × 2/2 = 18/10.*
- If you wrote a denominator of 100, get ready. **(Signal.)** *9/5 × 20/20 = 180/100.*

h. Problem B. Get ready. **(Signal.)** *52/25 × 4/4 = 208/100.*
- Problem C. Get ready. **(Signal.)** *43/50 × 2/2 = 86/100.*
(If students are firm, skip to step j.)

i. Problem D. If you wrote a denominator of 10, get ready. **(Signal.)** *1/2 × 5/5 = 5/10.*
- If you wrote a denominator of 100, get ready. **(Signal.)** *1/2 × 50/50 = 50/100.*

j. Write the statements below each equation to show the percent each fraction equals.
(Observe students and give feedback.)

k. Check your work. You'll read the statement below each equation.
- Read the statement below equation A. **(Signal.)** *9/5 = 180%.*
- Equation B. **(Signal.)** *52/25 = 208%.*
- Equation C. **(Signal.)** *43/50 = 86%.*
(If students are firm, skip to Independent Work.)
- Equation D. **(Signal.)** *1/2 = 50%.*

Assign Independent Work, the rest of Workbook part 1 and Textbook parts 6–10.

Optional extra math-fact practice worksheets for all lessons are available on ConnectED.

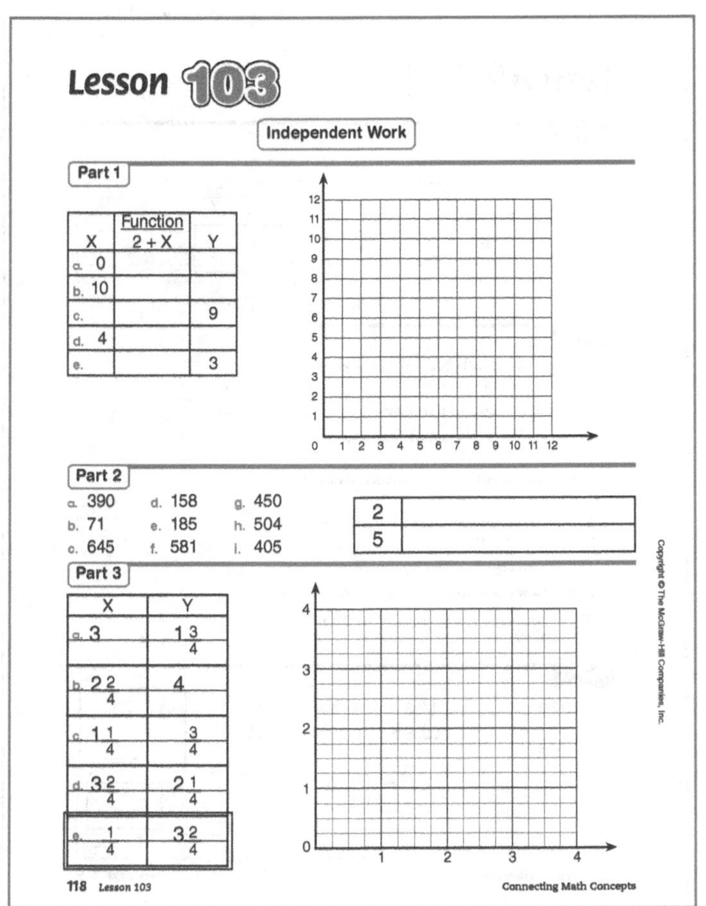

Lesson 103

Independent Work

Part 1

X	Function 2 + X	Y
a. 0		
b. 10		
c.		9
d. 4		
e.		3

Part 2

a. 390 d. 158 g. 450
b. 71 e. 185 h. 504
c. 645 f. 581 i. 405

2	
5	

Part 3

X	Y
a. 3	$1\frac{3}{4}$
b. $2\frac{2}{4}$	4
c. $1\frac{1}{4}$	$\frac{3}{4}$
d. $3\frac{2}{4}$	$2\frac{1}{4}$
e. $\frac{1}{4}$	$3\frac{2}{4}$

118 Lesson 103 Connecting Math Concepts

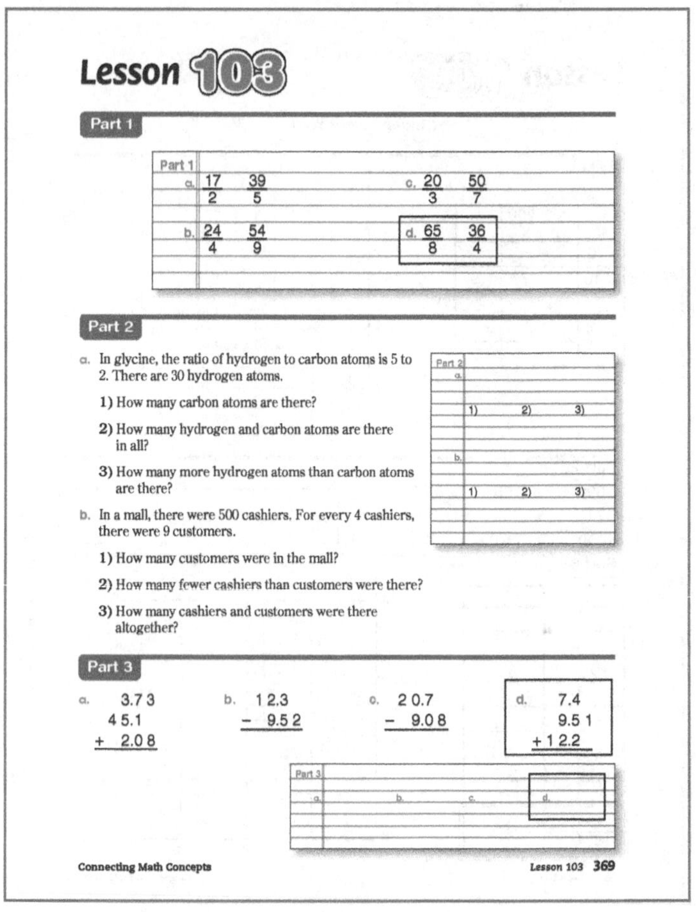

Lesson 103

Part 1

Part 1

a. $\frac{17}{2}$ $\frac{39}{5}$

b. $\frac{24}{4}$ $\frac{54}{9}$

c. $\frac{20}{3}$ $\frac{50}{7}$

d. $\frac{65}{8}$ $\frac{36}{4}$

Part 2

a. In glycine, the ratio of hydrogen to carbon atoms is 5 to 2. There are 30 hydrogen atoms.

1) How many carbon atoms are there?

2) How many hydrogen and carbon atoms are there in all?

3) How many more hydrogen atoms than carbon atoms are there?

b. In a mall, there were 500 cashiers. For every 4 cashiers, there were 9 customers.

1) How many customers were in the mall?

2) How many fewer cashiers than customers were there?

3) How many cashiers and customers were there altogether?

Part 2
a. 1) 2) 3)
b. 1) 2) 3)

Part 3

a.
```
   3.73
  45.1
+  2.08
```

b.
```
  12.3
-  9.52
```

c.
```
  20.7
-  9.08
```

d.
```
   7.4
   9.51
+ 12.2
```

Part 3 a. b. c. d.

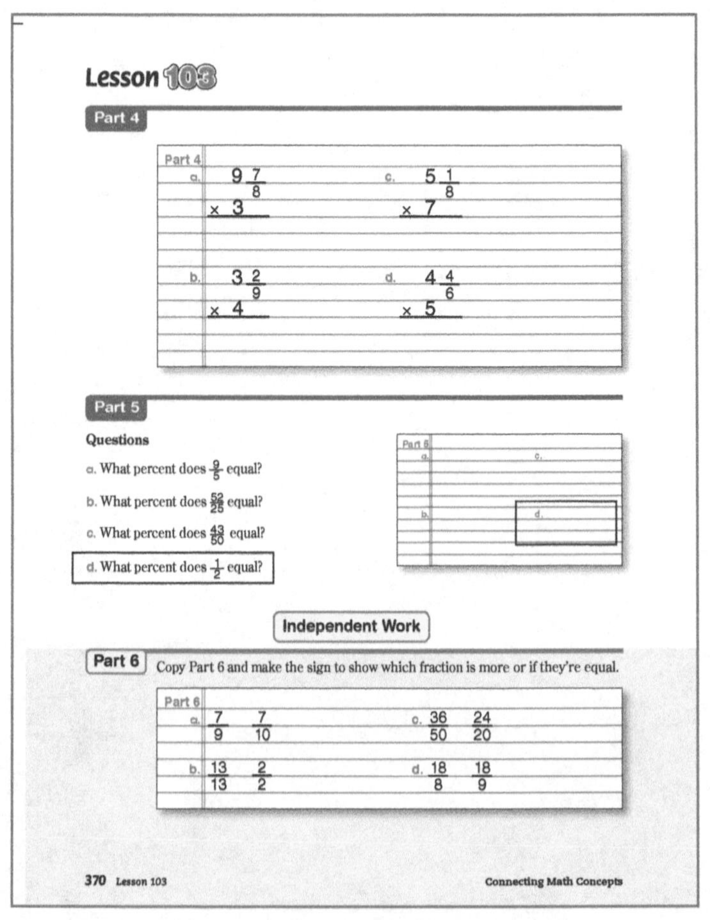

Lesson 103

Part 4

Part 4

a. $9\frac{7}{8} \times 3$

b. $3\frac{2}{9} \times 4$

c. $5\frac{1}{8} \times 7$

d. $4\frac{4}{6} \times 5$

Part 5

Questions

a. What percent does $\frac{9}{5}$ equal?

b. What percent does $\frac{52}{25}$ equal?

c. What percent does $\frac{43}{50}$ equal?

d. What percent does $\frac{1}{2}$ equal?

Part 5 a. b. c. d.

Independent Work

Part 6

Copy Part 6 and make the sign to show which fraction is more or if they're equal.

Part 6

a. $\frac{7}{9}$ $\frac{7}{10}$

b. $\frac{13}{13}$ $\frac{2}{2}$

c. $\frac{36}{50}$ $\frac{24}{20}$

d. $\frac{18}{8}$ $\frac{18}{9}$

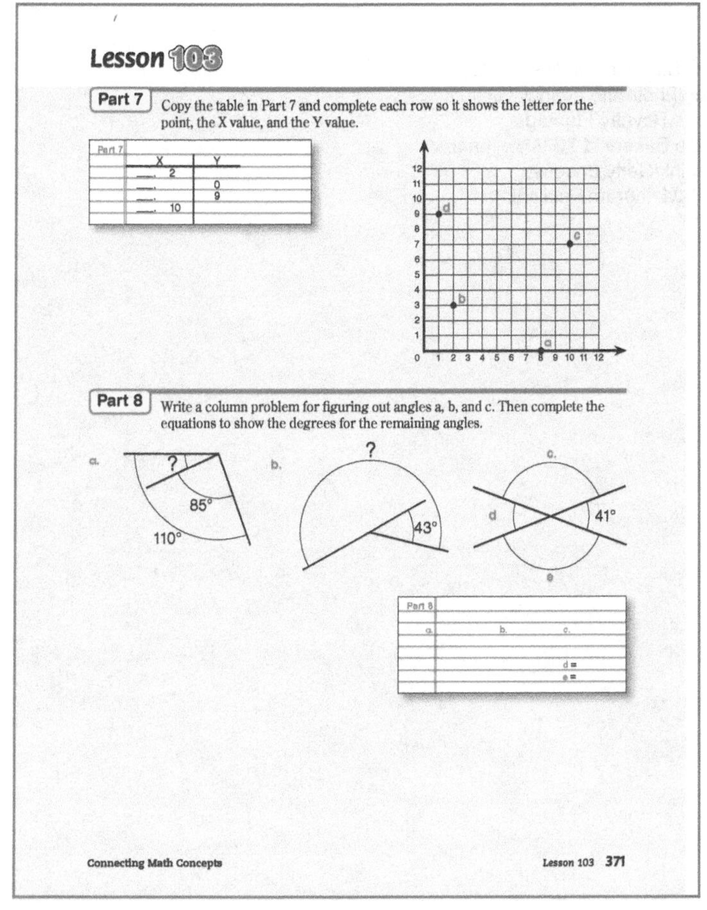

Lesson 103

Part 7 Copy the table in Part 7 and complete each row so it shows the letter for the point, the X value, and the Y value.

Part 7	X	Y
	___, 2	
	___,	0
	___,	9
	___, 10	

Part 8 Write a column problem for figuring out angles a, b, and c. Then complete the equations to show the degrees for the remaining angles.

Part 8			
	a.	b.	c.
			d =
			e =

Connecting Math Concepts Lesson 103 371

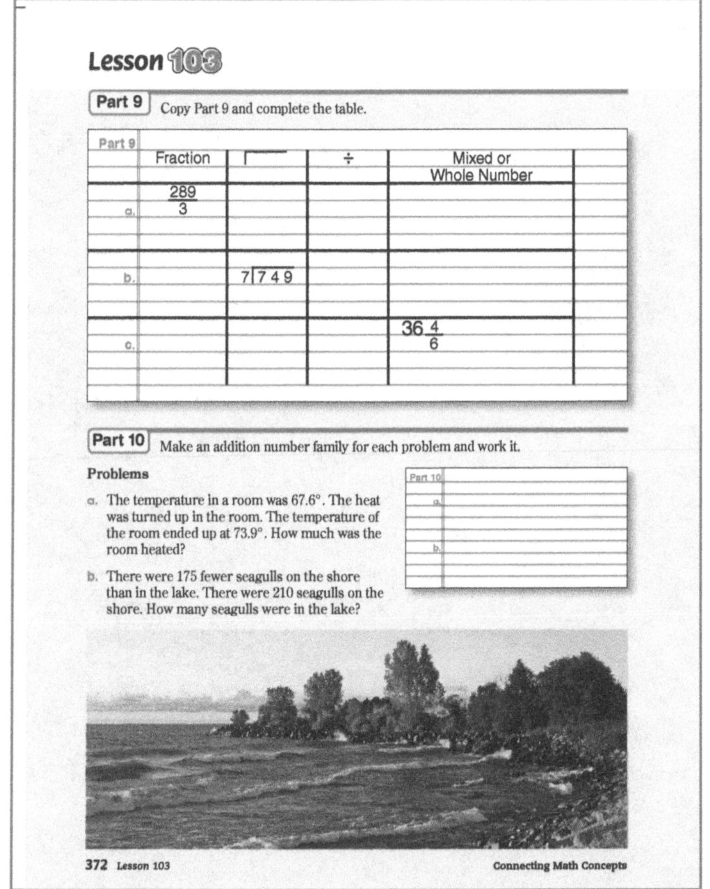

Lesson 103

Part 9 Copy Part 9 and complete the table.

Part 9	Fraction	$\overline{)}$	÷	Mixed or Whole Number
a.	$\frac{289}{3}$			
b.		$7\overline{)749}$		
c.				$36\frac{4}{6}$

Part 10 Make an addition number family for each problem and work it.

Problems

a. The temperature in a room was 67.6°. The heat was turned up in the room. The temperature of the room ended up at 73.9°. How much was the room heated?

b. There were 175 fewer seagulls on the shore than in the lake. There were 210 seagulls on the shore. How many seagulls were in the lake?

Part 10	
a.	
b.	

372 Lesson 103 Connecting Math Concepts

Photo Credits

114 Comstock Images/Alamy; **117** C Squared Studios/Photodisc/Getty Images; **171 172** James Hardy/PhotoAlto Agency RF Collections/Getty Images; **177** (l) Wave Royalty Free/age footstock, (r) Media Bakery; **178** Media Bakery; **179** Glowimages/Getty Images; **184** R. Strange/PhotoLink/Getty Images; **192** Lissa Harrison; **194** Stockdisc; **221** Ingram Publishing; **233** legge/Alamy.